MATERIALS SCIENCE AND TECHNOLOGIES

SCHEELITE AND ZIRCON: BRIGHTNESS, COLOR AND NIR REFLECTANCE IN CERAMICS

MATERIALS SCIENCE AND TECHNOLOGIES

Additional books and e-books in this series can be found on Nova's website under the Series tab.

MATERIALS SCIENCE AND TECHNOLOGIES

SCHEELITE AND ZIRCON: BRIGHTNESS, COLOR AND NIR REFLECTANCE IN CERAMICS

GUILLERMO MONRÓS

nova science publishers
New York

Library of Congress Cataloging-in-Publication Data

ISBN: 978-1-53619-332-9

Published by Nova Science Publishers, Inc. † New York

CONTENTS

PREFACE

The subject of the book is the analysis of the behavior of scheelite ($CaWO_4$) and zircon ($ZrSiO_4$) in ceramics. Currently, the zircon is the main opacifier of ceramic glazes and the basis of a family of pigments widely used for ceramic coloring; these pigments are analyzed as cool pigments and photocatalytic agents when applied on buildings and urban environments by reflecting NIR radiation and removing air pollutants, as well as their photocatalytic capacity on organic substrates, CO and NOx. Scheelite used as an opacifier by addition to glazes is the basis of another family of pigments that can compete with the classic zircon family as cool pigments with photocatalytic capacity.

The aesthetics of glazed ceramics and paints depend to a large extent on the so-called opacifiers that give it brightness; likewise, suitably modified by chemical doping, they are capable of imparting an intense color to the glazes, acting as ceramic pigments. At present, zircon $ZrSiO_4$ is the main opacifier of industrial ceramic glazes (replacing the classic cassiterite SnO_2). Zircon lattice is the basis of a widely used pigment palette for ceramic coloring: cyan of vanadium in zircon, magenta of hematite in zircon and yellow of praseodymium in zircon. These pigments are analyzed from the perspective of their cool effect on buildings and urban environments, associated with their high reflectance of solar NIR radiation (700-2500 nm), as well as their photocatalytic capacity on organic substrates, CO and NOx;

therefore the pigmented surfaces with these pigments purify the air, producing cleaner and cooler environments, allowing savings in the use of air conditioning and a moderation of the so-called urban heat island effect. Scheelite $CaWO_4$ has been used as an opacifier by addition to glazes and is the basis of another family of inorganic pigments that can compete with the classic zircon family as cool pigments, when doped with suitable chromophores, showing photocatalytic capacity. In summary, brightness, color, photocatalytic activity and NIR reflectance of cool surfaces in paints and ceramics based on the chemistry of Scheelite and Zircon are discussed.

Chapter 1

SCHEELITE AND ZIRCON: STRUCTURE, PROPERTIES AND APPLICATIONS

1.1. THE CRYSTAL STRUCTURE OF SCHEELITE AND ZIRCON

A number of structure types, especially those of ABX_4 compounds with a large difference between the sizes of the A and B atoms, consist of AX_8 bisdisphenoids and BX_4 tetrahedra. They include some important mineral structures such as zircon ($ZrSiO_4$), anhydrite ($CaSO_4$) and scheelite ($CaWO_4$). It is shown that these structures are closely and simply related via crystallographic twin operations. A bisdisphenoid sharing an edge with a tetrahedron is easily deformed to a pair of edge-sharing octahedra or to a row of four edge-sharing tetrahedra. The former reveals a simple relation between ABX_n structures and MX_2 (= MMX_4) octahedral structures [1].

The eight-coordination polyhedron known as a bisdisphenoid or dodecadeltahedron is still rather unfamiliar; this is strange because it is a not uncommon coordination figure and, as we shall see, is observed in some important mineral structure types. This polyhedron, shown in Figure 1.a, has 12 triangular faces, 18 edges and, of course, 8 vertices [1]. A mathematical bisdisphenoid is a deltahedron with symmetry D_{2d} in Schönflies notation and vertex symbol $(3^4)_4(3^5)_4$ [2]. Wells [3] has discussed its simple topological relations to the cube, square antiprism and bicapped trigonal prism. Like the cube it may be described as consisting of two interpenetrating, concentric tetrahedra (disphenoids). But, unlike the cube, these are irregular; one being elongated and the other compressed along the common -4 axis. It is possible to have all vertices equidistant from the center of the bisdisphenoid but, in most crystal structures in which it occurs, there are two sets each of four equivalent bond lengths (i.e., center to vertex) corresponding to the two tetrahedra which interpenetrate. The two vertex types are differentiated in Figure 1.1; the filled circles being somewhat closer to the center than the open ones.

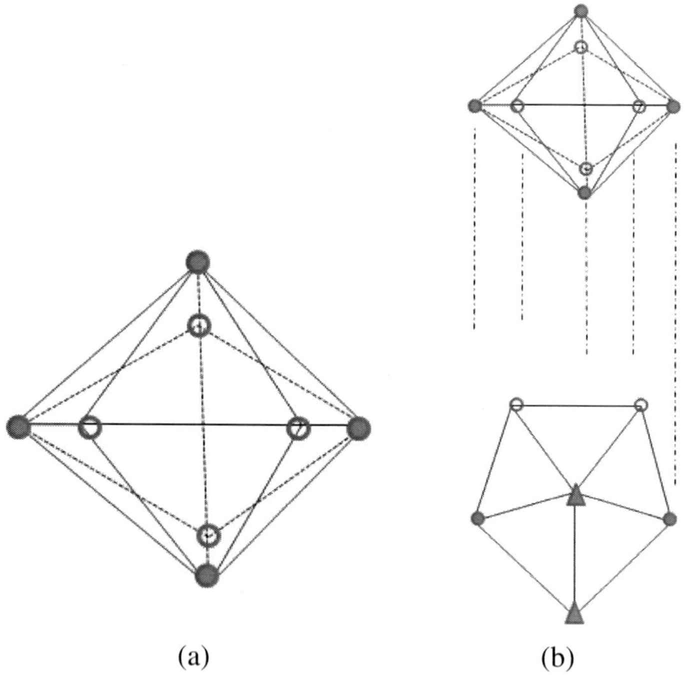

Figure 1.1. (a) Bisdisphenoid or dodecadeltahedron indicating the two kinds of atoms/vertex (the filled circles being somewhat closer to the centre than the open ones), (b) Bisdisphenoid and its projection on (100) (Triangles denote superimposed atoms/vertex) [1].

The structure of actinide tetrahalides such as β-$ThCl_4$ show a straightforward array of $ThCl_8$ bisdisphenoids sharing edges in the two equivalent directions of its tetragonal *(I4t/amd)* unit cell *(c/a=0.881),* with two different Th-C1 bond lengths of 2.72 and 2.90 Å respectively (Figure 1.2). Examination the projection of the array of the bisdisphenoids on (100) in Figure 1.2 reveals the presence of empty (rather regular, see dashed lines) tetrahedra between the rows of bisdisphenoids: they are edge connected to bisdisphenoids in adjacent rows. If these tetrahedra are occupied by B atoms the AX_4 structure of the $ThC1_4$ type becomes the ABX_4 structure of the $ZrSiO_4$ (zircon) type: also tetragonal (*$I4_1$/amd*, *c/a* =0.905) with Zr atoms in the O_8 bisdisphenoids (bond lengths 2.43 and 2.75 Å) and Si atoms in the O_4 tetrahedra. In the case of scheelite Ca occupies the O_8 bisdisphenoids and W atom the O_4 tetrahedra sites but in a different distribution that in zircon: in zircon bisdisphenoids of ZrO_8 and tetrahedrons of SiO_4 alternate in a column and in the case of scheelite in a column alternate two bisdisphenoids of CaO_8 and two tetrahedrons of WO_4 (Figure 1.3) [4]. Scheelite is tetragonal and belongs to the *I41/a* space group (network parameters: a = 5.25 Å b = 11.40 Å).

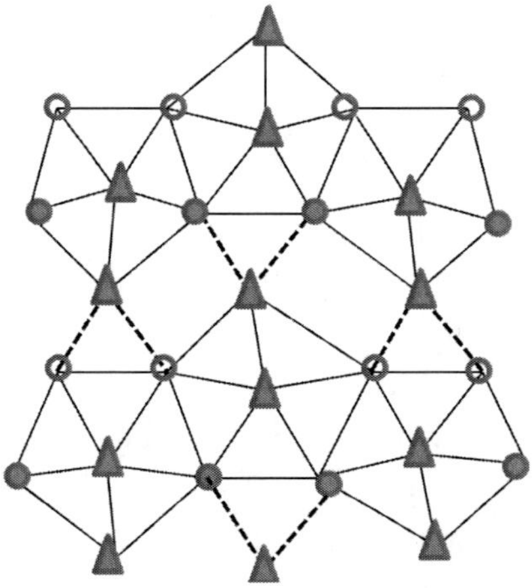

Figure 1.2. Array of bisdisphenoids sharing edges in the two equivalent directions of its tetragonal unit cell ad its projection on (100). In zircon, Zr^{4+} occupies the disphenoids center and Ca^{2+} the tetrahedron center.

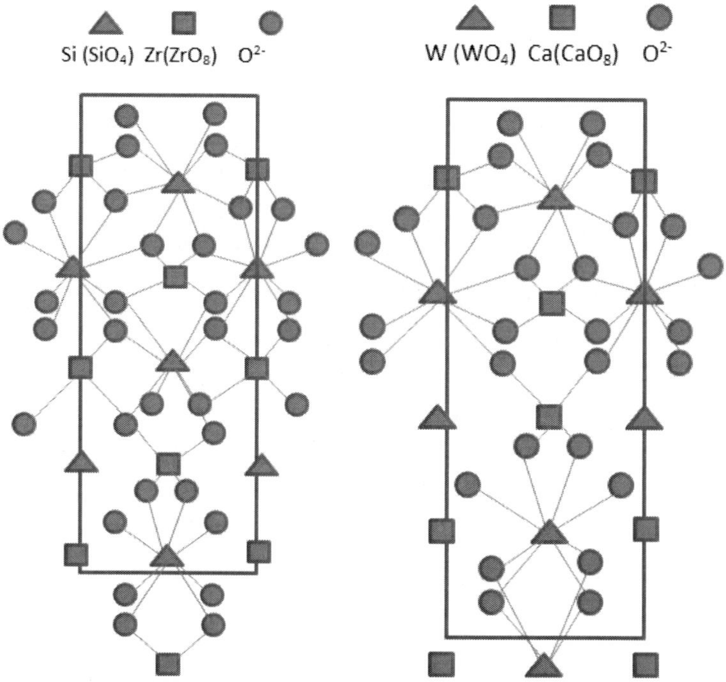

Figure 1.3. Structure of zircon and scheelite, showing the respective bisdisphenoids and tetrahedrons [4].

1.2. Sources, Properties and Applications of Zircon

1.2.1. Sources of Zircon

Zircon ($ZrSiO_4$) is one of the two main product streams from "mineral sands," the other being titanium minerals (ilmenite, leucoxene and rutile). The name Zircon derives from the Persian *zargun*, meaning "gold-hued." The English word "zircon" is derived from *Zirkon*, which is the German adaptation of this word. It is a nesosilicate or orthosilicate (constituted by isolated $[SiO_4]^{4-}$ tetrahedra that are connected only by interstitial cations), whose chemical formula is $ZrSiO_4$ although usually shows the presence of rare earth and other elements in solid solution substituting Zr^{4+}, e.g., hafnium and zirconium are the two most chemically similar in the Periodic Table, therefore Hf is almost always present in quantities ranging from 1 to 4%. The presence of trace amounts of uranium and thorium (from 10 ppm up to 1 wt.%) has played an important role during the evolution of radiometric dating due to it high stability, making them the oldest minerals so far dated on Earth [5].

Deposits of mineral sands are formed along ancient coastlines where the heavier minerals have been concentrated by wave and wind action. Most mineral sand deposits are found in unconsolidated fossil shorelines several hundred meters to tens of kilometers (or occasionally hundreds of kilometers) inland from the present coastline. Major deposits are found in Australia and Southern Africa, but also in South and Southeast Asia, China, East and West Africa, Ukraine and in North and South America. Zircon forms economic concentrations within heavy mineral sands ore deposits, within certain pegmatites, and within some rare alkaline volcanic rocks, for example the Toongi Trachyte, Dubbo, New South Wales Australia in association with the zirconium-hafnium minerals eudialyte and armstrongite. Australia leads the world in zircon mining, producing 37% of the world total and accounting for 40% of world EDR (economic demonstrated resources) for the mineral. South Africa is Africa's main producer, with 30% of world production, second after Australia [6]. The production of zircon has increased regularly from 1940 when the ceramics industry substitutes SnO_2 by zircon as opacifier; likewise to obtain the zircon pigments it is necessary to start from the zirconia, zirconium dioxide (ZrO_2), which is found in its most natural form in the mineral baddeleyite as will be discussed later.

The reactivity of ZrO_2 with SiO_2 in presence of V_2O_5 and using non-conventional methods was studied by Monrós et al. [7]. Figure 1.4 shows the XRD patterns of several zircon powders prepared by the ceramic method and Sol-Gel methods:

a) Pure $ZrSiO_4$ samples (CE, CG or PG samples).
b) Vanadium doped samples $(Zr_{0.9}V_{0.1})SiO_4$ (VCE, VCG or VPG samples),

c) Fluorine mineralized samples $0.2NaF.(Zr_{0.9}V_{0.1})SiO_4$ (VFCE, VFCG or VFPG samples).

The solid state method or ceramic route (CE) starting from oxides or carbonates, colloidal route (CG) starting from soluble salts or colloids of precursors, and polymeric gels (PG) obtained by hydrolysis-condensation of alkoxides, were used for the synthesis of samples.

Zirconia ZrO_2 (baddeleyite) and quartz SiO_2 of a commercial technical grade supplied by SEPR, NH_4VO_3 and NaF analytical grade (from Aldrich) and Pr_6O_{11} (from Rhone Poulenc) were used as Zr, V and Pr precursors on ceramic method. Commercial colloidal silica (75% SiO_2) supplied by Quimidroga S.A., $ZrOCl_2.8H_2O$, $VOSO_4.5H_2O$ and $Pr(CH_3COO)_3.3H_2O$, supplied by Merck, Aldrich and Johnson Matthey respectively, were employed instead of commercial technical grades on sol-gel procedures. In the case of polymeric gels, zirconium (IV) n-propoxide ($Zr(OCH_2CH_2CH_3)_4$ from Fluka) and tetraethyl orthosilicate TEOS ($Si(OCH_2CH_3)_4$ from Fluka) were used as Zr and Si precursors with vanadyl sulfate pentahydrate ($VOSO_4.5H_2O$ from Aldrich) and praseodymium (III) acetate trihydrate ($Pr(CH_3COO)_3.3H_2O$ from Aldrich).

In order to describe the methodology of ceramic and non conventional methods usually used in the pigments synthesis, the methods used in this case are described:

i. Ceramic method (CE); the appropriate amounts of commercial oxides, were mixed and ball-milled in acetone for 20 minutes. Residual acetone was then removed by evaporation, and the resulting dried powder was finally homogenized in an agate mortar.
ii. Colloidal gel method (CG): in this method a colloidal dispersion was obtained by dissolving zirconium oxychloride in water (50 mL) and colloidal silica maintained at 70ºC with continuous stirring followed by addition of vanadium and praseodymium precursors thirty minutes later. The resulting mixture was left for 1 hour under the same stirring and heating conditions to provide the basic CG solution. The vanadium-doped solutions (b) were prepared by the

appropriate addition of vanadyl sulfate pentahydrate. Half hour later NaF powder was added to dispersion in the case of chromophore-fluorine doped samples. When a homogeneous dispersion was reached, a concentrated ammonia solution was added dropwise until gelation occurred at pH= 5-6. The resulting homogenous bluish gel was aged and dried at room temperature in the open air. The dried gels were then crushed and ground in an agate mortar.

iii. Polymeric method (PG): in this method ethanol (ethanol: zirconium (IV) n-propoxide=1:26 molar ratio) was used as solvent. Acid catalyst (3M HNO_3 keeping the molar ratio acid: zirconium (IV) n-propoxide=1:0.18), acetylacetone and water (molar ratios: acetylacetone: zirconium (IV) n-propoxide= 1:1 and water: zirconium (IV) n-propoxide= 4:1) were added and refluxed at 70°C in the ethanol with continuous stirring. Then vanadyl sulfate pentahydrate was added and refluxed. Half-hour later TEOS were added and refluxed for prehydrolysis during 24 hours. Then zirconium (IV) n-propoxide was added continuously stirring to the solution and hydrolyzed at 70°C for 6 hours. The dried gels obtained were dark red colored.

In agreement with the literature, the following behavior on the reactivity of the proposed systems can be pointed out:

i. Using ceramic method (CE) the ZrO_2 (baddeleyite) and SiO_2 (quartz) begin to react at 1350°C using a soaking time of 8 hours (at 1200°C the presence of zircon is not detected by XRD).

ii. The addition of vanadium as NH_4VO_3 (starting formula $Zr_{1-x}V_xSiO_4$ x=0.05, 0.1 and 0.2) decreases the temperature of synthesis of zircon and a narrowing of the complete development of the zircon phase was observed. The sample x=0.05 crystallize zircon at 800°C which is detected as single phase at 1350°C, x=0.1 at 800°C and 1200°C respectively and for x=0.2 at 800°C the reaction is practically completed. This behavior is in agreement with the known performance of vanadium oxides as mineralizing agents or fluxes.

iii. The use of colloidal method (CG) starting from colloidal silica (Aerosil 200, Degussa), $ZrOCl_2.6H_2O$ and NH_4VO_3 which are coprecipitated in ammonia media also decreases the temperature of synthesis of zircon and a narrowing its complete development, stabilizing tetragonal zirconia previously. The x=0 sample shows a similar reactivity than the ceramic method and the zircon is detected at 1350°C. At lower temperatures, stabilized-ZrO_2 (tetragonal, T) and ZrO_2 (monoclinic, M) in equilibrium are detected (Figure 1.4). For x=0.05 the zircon appears as single crystalline phase at 1000°C (at lower temperatures T+M in equilibrium are detected), for x=0.1 zircon crystallizes at 750°C and the reaction is completed at 900°C, and for x=0.2 the reaction is completed at 800°C.

iv. By the use of a polymeric gel (PG) method starting from TEOS, $ZrOCl_2.6H_2O$ and NH_4VO_3 dissolved in ethanol media and hydrolyzed by water addition at 60°C in refluxing and continuous stirring conditions, the gelling is obtained after 17 h for x=0.1 sample and crystallize zircon al 750°C (T+M in equilibrium are detected at lower temperatures) and the reaction is completed at 900°C.

Figure 1.4 shows the XRD patterns of the powders fired at 1400°C/6h. To directly compare the behavior of the prepared powders no washing and disgregation treatments were carried out in any of the methods. The powders were fired successively at 700, 750, 800, 900, 1000, 1100, 1200 and 1400°C for 6 hours. As described above, the samples without vanadium addition (CE and CG) show unreacted monoclinic zirconia like the ceramic sample with vanadium addition (VCE). The colloidal gel with vanadium addition (VCG) shows zircon as single crystalline phase. In the case of polymeric gel with vanadium (VPG) and vanadium and mineralizer addition (VFPG), cristobalite and monoclinic zirconia appear as unreacted phases respectively.

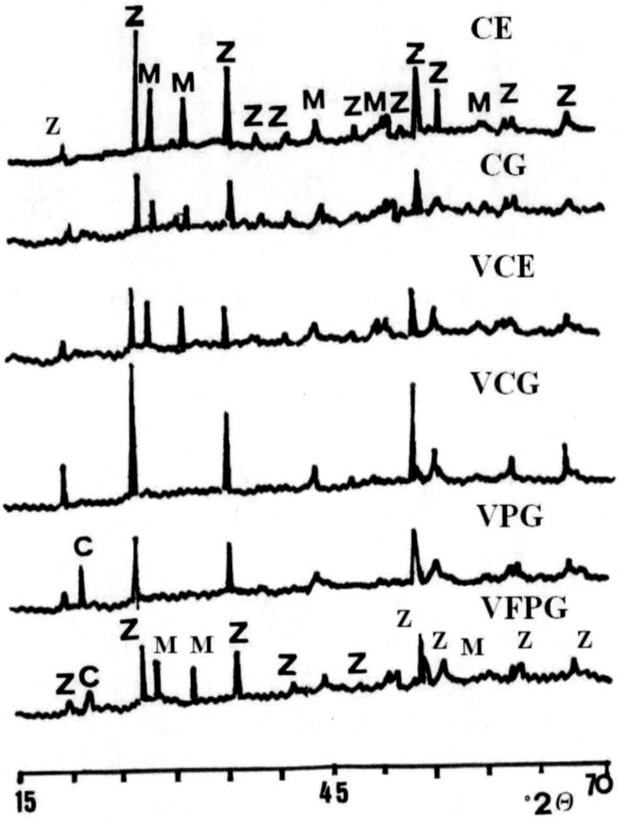

Figure 1.4. XRD evolution of indicated samples fired at 1400°C/6h. CRYSTALLINE PHASES: Z($ZrSiO_4$), M(monoclinic ZrO_2), C(cristobalite).

Figure 1.5 shows the representative 5wt.% powders glazed samples fired at 1200°C/6h in a conventional single firing glaze (1080°C) showing the blue obtained with vanadium addition in polymeric gel (VPG) without fluorine addition (b*=-11), the green color (a*=-2 and b*=-4) obtained by vanadium addition in ceramic sample (VCE) and the commercial blue turquoise observed in vanadium and fluorine addition in polymeric gel (b*=-19) (see the discussion around vanadium turquoise blue in Chapter 6).

Figure 1.5. Representative glazed samples of 5wt.% powders fired at 1200°C/6h.

1.2.2. Properties and Applications of Zircon

More than 50% of all zircon is used in the ceramics industry due to zircon's ability to impart opacity, whiteness, brightness, and to create glazes on products such as tiles, sanitary-ware (baths, sinks and toilets) and crockery. Zircon is a hard and tough material as it maintains its physical and chemical composition when subjected to high temperatures and corrosive environments. It is an extremely good foundry and refractory material, used in refractory bricks, especially in furnaces for steel and glass making.

Apart from its use in ceramics, zircon was used to block X-ray emissions in cathode ray tubes (CRT) found in computers and television monitors, to produce biomedical devices such as hip and knee replacements, rechargeable light weight batteries, television screens, fuses and toothpaste and as a foundry sand for specialist investment casting or high-precision casting. As historical curiosity, zirconium alloy is used as the housing for fuel rods in nuclear power stations, due to its low thermal neutron absorption; however hafnium readily absorbs neutrons and therefore must be completely separated and removed from zirconium before either element can be used in nuclear reactors. A major task of the Manhattan Project was the separation of hafnium from zirconium. The recovered hafnium metal is used to make the control rods of nuclear reactors, as the metal readily absorbs neutrons [8].

On the other hand due the high chemical and thermal stability of zircon, it would be used as immobilization lattice for radioactive actinides in nuclear wastes based on studies of zircons formed by crystallization from the reactor core melt generated in the course of the accident at the Chernobyl Nuclear Power Plant. These zircons have compositions in the range $(Zr_{0.94},U_{0.06})SiO_4$ to $(Zr_{0.9}, U_{0.1})SiO_4$. Hot-pressing of oxides was studied to make zircon-based waste forms and waste streams high in zirconium and actinides could be converted into zircon-based waste forms. The adaptation and modification of the mixed-oxide reactor fuel (MOX) production process is proposed as a process for the production of $(Zr,Pu)SiO_4$, a durable waste form for excess weapons plutonium [9].

The discovery of zircon pigments can be considered an example of the application of scientific knowledge (Research) to technology (Innovation). Indeed, the introduction of zirconium in ceramics was caused by the high increase in the price of cassiterite (SnO_2) around 1914. As discussed, tin oxide was, par excellence; the opacifier of enamel from Rome, only the antimony oxide and titanium oxide was sometimes used as will be discussed in Chapter 2, but volatilization of antimony oxides at relatively high temperatures and yellowing of the enamel with titanium oxide when it contains traces of iron or chromium, discarded them. The high refractive index and the high melting point of zirconia (ZrO_2) and $ZrSiO_4$ zircon, together with its low solubility in enamels, were emphasized by US and European researchers [10-12]. In 1930-1940, the new opacifier was introduced by several manufacturers, such as Titanium Alloy Manufacturing Company [12], which patented the brilliant enamels obtained with zirconia (ZrO_2) and zircon. Zirconia (ZrO_2) and zircon meet the three main characteristics of the ceramic pigments (e.g., high thermal stability, high stability in glazes and porcelain stoneware and obviously high coloring performance) and produce ceramic pigments by doping with chromophores ions. Zirconia yellow doped with vanadium was developed, but is destabilized by reaction with the enamel silica, precipitating the zircon. On the other hand, direct doped zircon is not successful. In 1948, Clarence A. Seabright thought that it was probably necessary to change the synthesis strategy and, following the process of obtaining with the corundum pigments, in which aluminum hydroxides are used instead of corundum, using for the case of zircon a mixture of silica and zirconia with the addition of the corresponding chromophores. In effect, the zircon vanadium turquoise with NaF as mineralizer was very successful [13]. Later, in 1960, Seabright develops praseodymium yellow in zircon (although it had already been handled and cited in Japan) [14], and finally in 1965 the hematite iron coral in zircon was patented by the same author [15].

Ozel et al. [18] synthesized zircon ceramic pigments from intermediate products obtained by decomposing zircon sand with NaOH: at around 850ºC, decomposed products (Na_2ZrSiO_5 and Na_2ZrO_3) were used as raw materials to produce zircon pigments and Pr, Fe and V oxides were used as coloring agents. The crystalline phases formed after decomposition and pigment production steps were determined by X-ray diffraction (XRD) technique. XRD analysis showed that Na_2ZrSiO_5 and Na_2ZrO_3 phases were formed after decomposition and when these products used as a raw material, zircon containing coloring ion can be obtained. Prepared pigments were used in transparent glazes and porcelain tiles and their color properties were determined.

Therefore to obtain the zircon pigments it is necessary to start from the zirconia, zirconium dioxide (ZrO_2), or related compounds of zirconium (such as $ZrOCl_2.6H_2O$ [7] or Na_2ZrSiO_5 [18]. ZrO_2 is found in its most natural form in the mineral baddeleyite.

Baddeleyite is named to honor Joseph Baddeley who was a superintendent of a railroad project in Rakwana (Sri Lanka); in 1982 send several samples to the Museum of Practical Geology in London for its study being identified the $MgTiO_3$ named as geikielite in honor to the Director General of the Geological Survey, Sir Archibald Geikie, and ZrO_2 which was named baddeleyite. Monoclinic baddeleyite has the space group symmetry $P2_1/c$ with four ZrO_2 in the unit cell (cell dimensions: a = 5.169 b = 5.232 c = 5.341 Å and $\beta = 99°15'$). The coordination number for ZrO_2 has been found to be 7 with a Zr-O distance ranging from 2.04 to 2.26 Å, the second Zr-O separation is 3.77 Å. The structure of baddeleyite structure is a combination of tetrahedrally coordinated oxide ions parallel to (100) with triangular coordinated oxide ions. This explains baddeleyite tendency to twin along the (100) planes [19]. It has been observed that baddeleyite without twinning is extremely rare. Studies have shown that zircon and baddeleyite can be recovered from some anorthositic rocks in Proterozoic anorthosite complexes (a common intrusive igneous rock characterized by its composition mostly plagioclase and other with a minimal mafic components [20]. Places where these Proterozoic anorthosite complexes can be found are: the Laramie Anorthosite Complex in Wyoming, the Nain and Grenville provinces of Canada, the Vico Volcanic Complex in Italy, and Minas Gerais and Jacupiranga, São Paulo, Brazil. Baddeleyite forms in igneous rocks low in silica, it can be found in rocks containing potassium feldspar and plagioclase. Associated minerals include ilmenite, apatite, magnetite, perovskite, fluorite, nepheline, pyrochlore and allanite. As mentioned above, because of their refractory nature and stability under diverse conditions, baddeleyite grains, along with zircon, are used for uranium-lead radiometric age determinations.

Baddeleyite can also be chemically derived from zircon. The methods used to obtain zirconia from zircon, although distinct from each other, have three common features:

a) They all involve the decomposition of zircon by chemical, thermal or mechanical means. By thermal treatment zircon is mixed with coke and heated to its fusion point (in excess of 2800ºC) in an electric arc furnace where it dissociates to zirconium oxide and fumed silica. Zircon decomposes at lower temperature by fusion with flux agents such as NaOH, $CaCO_3$, CaO, $MgCO_3$, K_2SiF6, chlorination or carbiding [21].
b) All products obtained from zircon decomposition are subsequently treated by solubility differentiation.
c) Finally in all cases the isolation of zirconium compounds from residual impurities is involved.

Zirconia products are characterized by good mechanical properties and stability at high temperatures, strong thermal and corrosion resistance, chemical inertness and consistent quality. This makes them ideal for use in a wide range of refractory products, ceramic pigments and electronic applications. Other applications include friction materials, welding rods and zirconium metal and alloys. Advanced zirconia ceramics have excellent biocompatibility ensuring zirconia has replaced alumina as the material of choice for prosthesis devices such as hip joints or femoral ball heads. It has superior strength and hardness, wear resistance, stability, resistance to scratching and biocompatibility with the human body. One of zirconia's other most common uses is within dental implants. Cubic zirconia is one of the most well-known forms of zirconia. Commonly known as a synthetic diamond, it has become popular as a gemstone due to its optically clear single crystals and high refractive index, as well as its ability to maintain its color and brilliance [22].

1.3. Sources, Properties and Applications of Scheelite

Scheelite, a calcium tungstate, $CaWO_4$, is an important ore of tungsten, was first described in 1781 for an occurrence in *Mount Bispbergs klack*, Säter, Dalarna, Sweden, and named for Carl Wilhelm Scheele (1742–1786). Owing to its unusual heaviness, it had been given the name *tungsten* by the Swedes, meaning "*heavy stone*." The name was later used to describe the metal originally named wolfram which was predicted in 1779 for the Irish chemist Peter Woulfe from his analysis of the mineral wolframite (an iron manganese tungstate mineral). Tungsten was isolated as tungstic oxide (WO_3) in 1781 [23], in Sweden, by Carl W. Scheele from scheelite (calcium tungstate). However he did not have a suitable furnace to reduce the oxide to the metal.

Tungsten was finally isolated by brothers Fausto and Juan Jose de Elhuyar in 1783, in Spain, by reduction of acidified wolframite with charcoal [24]. The element name is tungsten but the chemical symbol, W, comes from the original name of the element, wolfram.

Scheelite ($CaWO_4$) shows Mohs hardness 5 and refractive index 1.9, it is together with wolframite ($Fe_xMn_{1-x}WO_4$) and stolzite ($PbWO_4$) one of the main ores of tungsten given its high reticular stability. According to a report by the USGS in 2016, the total global mine production of tungsten in 2015 was 87,000 t, of which 71,000 t was from China. Currently, four EU countries including Austria, Portugal, Spain and UK are producing tungsten concentrate and a total of 2,830 t of tungsten was produced in 2015 [24]. The average abundance of tungsten in the earth's crust is estimated to be 1.25–1.5 ppm, about the same as that of tin and molybdenum. It is more abundant in granite (about 2 ppm) than basaltic (1 ppm) and ultra-mafic rocks (0.5 ppm). There are numerous tungsten minerals, but only scheelite ($CaWO_4$) and wolframite ($(Fe,Mn)WO_4$) are of economic importance. However, wolframite is not a mineral species but a series between ferberite ($FeWO_4$) and hübnerite ($MnWO_4$). The domination of either iron or manganese would result in forming one of two minerals. The iron dominated one will result in forming ferberite while the manganese dominated one will result in forming hübnerite [26]. The world reserves of tungsten are estimated at 4.000.000 t, of which China holds about 40%.

The beneficiation process of scheelite and wolframite ores generally consists of pre-concentration after crushing and grinding, followed by roughing, cleaning and final purification stages to produce a concentrate with 65–75% WO_3, to meet the requirements of international trading [27]. Only scheelite is readily amenable to flotation. Wolframite, in contrast to scheelite, is paramagnetic. Thus beneficiation techniques of gravity concentration and flotation are applied to scheelite ore, and gravity and/or magnetic separation applied to wolframite ore.

Scheelite is widely used in the field of ceramic glazes in order to develop the so-called nacre effects by producing variations in brightness depending on the angle of incidence of light. In general, for ceramic purposes, $CaWO_4$ is obtained by adding WO_3 to glazes with certain CaO content, so that the scheelite phase precipitates in firing. In the pearling process, the size of the devitrified scheelite crystals is very important, as well as the tendency to form aggregates thereof, so that long firing and/or high firing temperatures decrease this effect because either the aggregates and of particle size increase. Given the high refractive index of the crystalline phase as well as its reticular stability, scheelite can be used as a bleaching and opacifying additive for ceramic glazes. The bleaching, opacifying and pearlescent capacity of the scheelite depends largely on the particle size and the tendency to form aggregates thereof.

1.3.1. Synthesis of Scheelite

With the aim of knowing the trend towards the formation of aggregates and possibilities of developing scheelite of submicron sizes that allow sintering at low temperatures, the synthesis and sintering of scheelite powders obtained by two methods is analyzed:

a) Ceramic or solid state reaction starting from WO_3 and $CaCO_3$ (CE sample).
b) Coprecipitation method; adding ammonia to a solution of $CaCl_2.4H_2O$ and WO_3 (previously solved in HNO_3 30 wt.%). There are obtained two kinds of coprecipitated samples: the coprecipitate directly dried at 110°C (sample CO) and the sample obtained by washing the precipitate twice by decantation in water, maintaining the mass ratio powder: water = 1: 5 and vigorously stirring every time for 15 minutes the suspension (sample COLav).

The maintenance of the counterions in the gel without washing can be positive in favoring the thermal and hourly performance of the crystalline phase for three basic reasons: (a) the heterogeneous nucleating effect that the presence of salt microcrystals can generate, with respect to the formation of the phase sought, (b) a certain mineralizing effect of these phases, which generally melt at temperatures prior to decomposition (around 500-600°C for most of the developed chlorinated salts such as NH_4Cl that remain in the processes), (c) the modifying effect of the emission of decomposition gases of these salts in the firing process, which can modify the microstructure, generating crystallization nuclei themselves or altering the specific surface of the particles when separated from them. The powders were successively calcined at 600, 900 and 1100°C with soaking times of 6 hours and characterized by XRD and SEM/EXD electron microscopy. Powders fired at 1100°C were screened at 40 μm and pressed uniaxially at 250 kg/cm^2. The tablets thus formed were sintered at 1300°C/6 hours. In the sintered tablets the density was measured by the method of Archimedes. Vickers microhardness is measured in order to evaluate the densification of the material, for this purpose the tablets stuffed in epoxy resin were roughing and carefully polished and studied by SEM/EDX and indented.

Table 1.1 shows the evolution with firing temperature of the crystalline phases detected by XRD by the randomly oriented powder method in the samples. The XRD patterns evolution in the CO sample is shown in Figure 1.6. Results indicate the precipitation of scheelite in the raw coprecipitate sample with wide peaks indicating a low crystallite size, in fired sample a weak crystallization of WO_3 is observed at 900°C that disappears at 1100°C, the peaks of scheelite become acute and grow in intensity with the temperature. The ceramic material develops scheelite from 900°C without apparent presence of free WO_3.

Table 1.1. Evolution with firing temperature of the crystalline phases detected by DRX in the samples

T (°C)	CE	CO	COLav
Raw	WH(591) Cal(188) W(169)	S(554) N(127)	S(623)
600°C	S(504) W(250) Cal(92)	S(1023) W(39)	S(1132) W(33)
900°C	S(1909)	S(1730) W(37)	S(1690) W(36)
1100°C	S(836)	S(1350)	S(570)

CRYSTALLINE PHASES: W: WO_3, Cal: $CaCO_3$, WH: $WO_3{\cdot}H_2O$, S: $CaWO_4$, N: NH_4Cl (Intensity (counts/s) of the most intense peak in parentheses).

Figure 1.7 shows the SEM micrographs of CO powders at 900°C (Figure 1.7.a) and 1100°C (Figure 1.7.b) and CE at 1100°C (Figure 1.7.c) respectively [28]. In CO sample at 900°C, spherical homogeneous particles are observed around 700 nm of particle size. These powders CO sintered in particles of 2 μm particle size at 1100°C (Figure 1.7.b), while the ceramic sample shows particles around 500-800 nm which form aggregates of 5-7 μm. The washed coprecipitated COLav sample shows a monolithic appearance with irregular particles greater than 5 μm of particle size at 1100°C which was not considered suitable for densification.

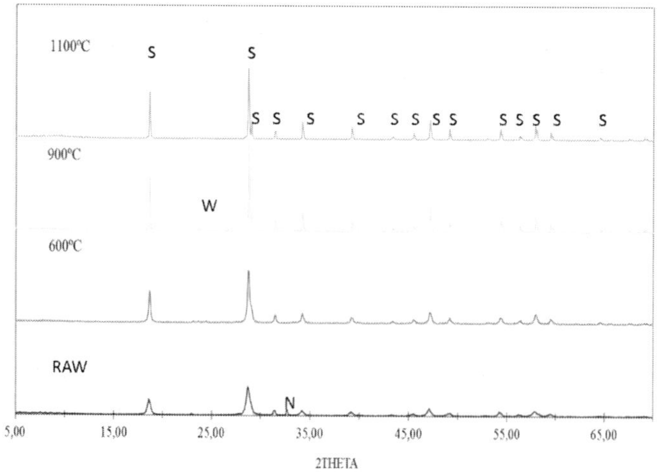

Figure 1.6. XRD evolution with temperature of CO sample. CRYSTALLINE PHASES: W (WO_3), S ($CaWO_4$), N (NH_4Cl).

Figure 1.7. SEM micrographs of powders: (a) CO at 900ºC, (b) CO at 1100ºC, (c) CE at 1100ºC.

The sintered and polished tablets obtained from powders at 1300°C/6h are showed in Figure 1.8 and Table 1.2 shows the measured density and Vickers microhardness.

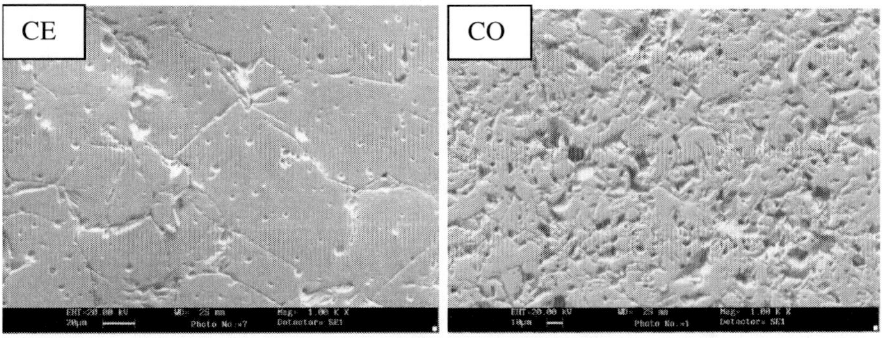

Figure 1.8. SEM micrographs of sintered and polished tablets at 1300ºC.

Table 1.2. Density and Vickers microhardness of sintered and polished samples

Sample	Densification (%)	HV (Kg/mm^2)
CE	88.9	192
CO	88.4	179

The results show a better densification of the CE-1100ºC powders with development of cracks and intergranular microporosity and with grain sizes around 30 μm. This relatively more compact microstructure also shows higher HV values than the CO-1100°C powders, which exhibit open porosity but finer grain size. The results are consistent with the microstructures of the pressed powders in which the ceramic powders maintain the nanometric characteristics [28].

1.3.2. Properties and Applications of Scheelite: Using Scheelite as Opacifier by Devitrification of Glazes

The classic process of opacification and hardening of ceramic glazes is based on the introduction into the vitreous matrix of a reinforcing crystalline particulate of high refractive index [30]. If the glass-ceramic composite material presents a low mean free path (by 0.5-1μm) between the short particles occluded in the vitreous matrix, the interfacial phenomena are of singular importance in order to avoid the disintegration of particulate by the abrasive load and it is very important to reduce the particle-matrix interfacial tension, this generally means working with low surface tension matrices [31].

On the other hand, WO_3 is a luster agent that has always been used in ceramic technology in order to produce these effects by devitrification of different tungstates. It is also a potent surfactant in fused glassy media, so it is expected to reduce the interfacial tension of the vitreous matrix-crystalline particle in the composite material for the development of a small size of devitrified ceramic particles [32]. As above mentioned scheelite ($CaWO_4$) show Mohs hardness 5 and refractive index of 1.9 and can be obtained by devitrification in modern calcium-rich ceramic floor frits by adding WO_3 to the glaze. It can also be obtained previously by adding directly as $CaWO_4$.

Table 1.3. Seger's formula of the glazes

Glass Base		Seger formula	
A	Na_2O 0.576		
	K_2O 0.094	Al_2O_3 0.633	SiO_2 6.916
	CaO 0.193		
	PbO 0.137		
B	K_2O 0.106		
	CaO 0.565	Al_2O_3 0.323	SiO_2 1.972
	ZnO 0.329		
C	Na_2O 0.01		
	K_2O 0.05		SiO_2 0.88
	MgO 0.05	Al_2O_3 0.15	ZrO_2 0.07
	CaO 0.67		
	ZnO 0.22		

Table 1.4. Results obtained by additions to glazes

Sample	WI	Brightness	Mohs	XRD	HV(GPa)
Glaze A					
1% $ZrSiO_4$	38.9	76.3			
Base	42.0	77.2	5	vitreous	5.3
1% WO_3	35.0	78.0	5	S(m)Q(vw)	5.3
3% WO_3	24,1	122.0	5	S(vs)Q(vw)	5.2
6% WO_3	19.6	157.0	5	S(vs)Q(vw)	5.2
1% CE	41.0	77.6	5-6		7.3
1% CO	42.1	77.3	6		7.1
3% CE	33.9	75.9	5		7.1
3% CO	35.7	76.1	6		6.5
Glaze B					
1% $ZrSiO_4$	17.3	76.9			
Base	17.7	82.4	6	vitreous	5.9
1% WO_3	17,7	83.9	6	vitreous	5.7
3% WO_3	14.9	103.1	6	S (vs)	5.8
6% WO_3	11,9	125.0	6	S (vs)	5.9
1% CE	17.2	73.6	6		5.6
1% CO	17.4	80.2	6-7		5.3
3% CE	17.1	78.9	6		5.6
3% CO	17.6	83.7	6-7		5.8
Glaze C					
1% $ZrSiO_4$	11.0	69.9	6		
Base	12.5	95.8	6	Zr(m)	5.9
3% WO_3	10.7	100.0	6	S(vs) Zr(m)	5.9
6% WO_3	12.2	107.0	6	S(vs) Zr(m)	6.0
1% CE	11.6	55.2	6		5.8
1% CO	11.6	59.6	5-6		5.6
3% CE	11.8	62.6	6		5.3
3% CO	12.2	68.7	6		5.6

CRYSTALLINE PHASES: Zr($ZrSiO_4$), S($CaWO_4$), PEAK INTENSITY: vw (very weak), w(weak), m(medium), vs(very strong).

The glass-scheelite composite developed by reaction between WO_3 and $CaCO_3$ (scheelite CE sample) as well as by coprecipitation method (scheelite CO) is compared with the glass-scheelite composite material obtained by adding WO_3 to different calcium ceramic glazes that devitrifies scheelite. As above discussed the powders have 2 μm particles in CO/1100°C sample, while the ceramic sample has nanoparticles around 500-800 nm that form aggregates of 5-7 μm.

Table 1.3 shows the Seger's formula of the three used glazes: A) leaded transparent double-firing glaze, B) transparent single-firing glaze for wall cladding rich in Ca-Zn, and C) matt glaze single-firing for ceramic floor with Ca -Zn-Zr content.

The glazed samples were characterized using the following techniques: X-ray diffraction (XRD), Scanning Electron Microscopy (SEM), Brightness using a Minolta Multi-Gloss Reflectometer 268, Whiteness Index (WI), to characterize the whiteness produced by the devitrification and Vickers microhardness (HV). Table 1.4 summarizes the results obtained with the additions to the different glazes compared with the addition of 1wt.% zircon of 3-5 μm of average size.

With the addition of scheelite, the glaze A moderately opacifies (WI and brightness decrease and increase respectively, similar to the addition of 1% zircon) and the Vickers microhardness increase significantly. However, when the effect is compared with the addition of WO_3, the opacification and brightness effects are higher in the devitrification process by WO_3 addition, and it is effective with only 1 wt.% addition of WO_3.

In the case of additions to glaze B, the scheelite does not affect the opacity or the whiteness of the glass, the Vickers microhardness decreases slightly and also the brightness. In contrast, the addition of WO_3 develops a significant pacifying effect.

In the opaque glaze C, the behavior of additions showed in Table 1.4 indicates that although WO_3 addition shows a certain bleaching effect and increases the brightness of the material, the addition of scheelite does not exert a significant bleaching effect, instead, has a tendency to decrease the brightness and also the HV values.

Data in Tab. 1.4 also indicates that in all glazes, CE scheelite powders show higher bleaching (WI) and slightly higher reinforcing (HV) effect than the sample prepared by chemical precipitation CO, although a lower brightness effect.

Figure 1.9 shows the characteristic microstructures of the devitrified scheelite obtained from WO_3 addition (Figure 1.9.a) and the developed by adding the scheelite powders fired at 1100ºC (Figure 1.9.b,c) to glaze A. While in the first case scheelite clusters develop with free spaces between the clusters of the order of 1 μm, in the second case the scheelite particles are homogeneously dispersed in the glaze; these dispersion of CO powders show minor number of particles and particle size than CE sample. The devitrified clusters of scheelite significantly increase the whiteness (WI) and brightness of the material but without reinforcing activity. Instead, the dispersions of scheelite have less whitening and shine effect but develop a reinforcing effect, likewise CE scheelite shows higher HV values than the more dispersed CO scheelite particle but the bleaching WI is higher for CO scheelite addition (Table 1.4).

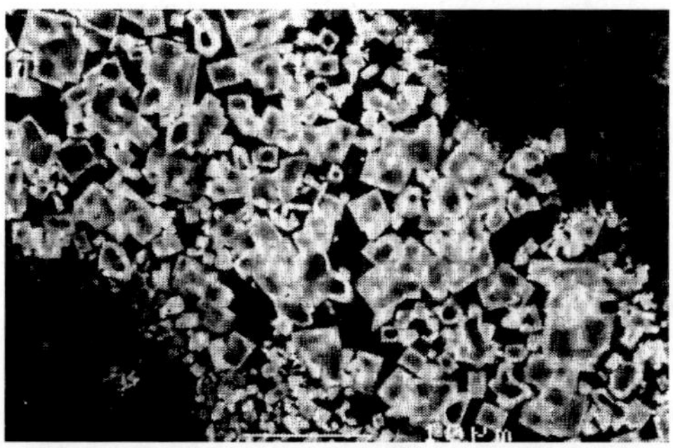

a

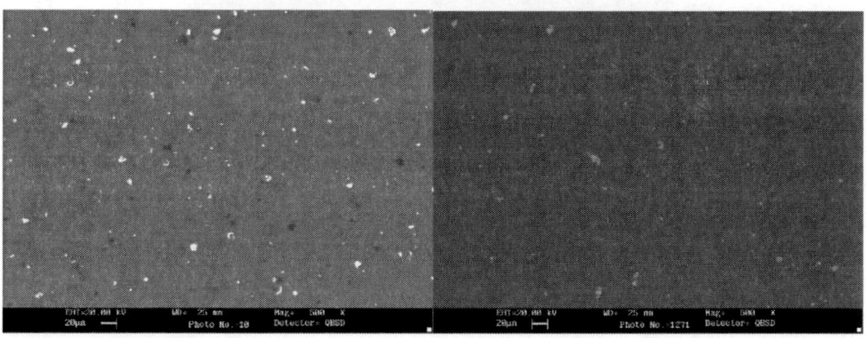

b c

Figure 1.9. Microstructures of glazed samples in glaze A: (a) devitrified scheelite from 6 wt.% WO_3 addition, (b) addition of 3 wt.% of scheelite CE, (c) addition of 3 wt.% of scheelite CO.

In conclusion, the addition of previously synthesized scheelite develops homogeneous dispersions in the glazes that show a reinforcing effect in the leaded double firing glaze, does not shows a significant effect in the single firing glaze rich in Zn-Ca and appears to weaken the silicate-zirconia matte glaze. The relatively large particle size (CO) scheelite shows a slightly effect on brightness that is higher than the observed with the lower and agglomerated particle size of CE scheelite. The addition of WO_3 to the glazes induces the devitrification of clusters of scheelite microcrystals and follows the same trend effect on the whiteness and brightness of the glazes, but is more significant.

REFERENCES

[1] Nyman H., Hyde B. G. "Zircon, Anhydrite, Scheelite and Some Related Structures Containing Bisdisphenoids," *Acta Cryst.*, B40 (1984): 441-447.

[2] Kloß Simon D., Weis Andreas, Wandelt Sophia, Schnick Wolfgang. "High-Pressure Metathesis of the $M_{1-x}PO_{3+4x}N_{1-4x}$ (x ≈0.05) and $M_{0,75}PO_4$ (M = Zr, Hf) Orthophosphates," *Inorg. Chem.* 57 (2018): 4164−4170.

[3] Wells, A. F., *Models in Structural Inorganic Chemistry* (Oxford: Clarendon Press, 1970) 31.

[4] Zalkin, A., Templeton, D. H. (1964). "X-ray diffraction refinement of the calcium tungstate structure," *Journal of Chemical Physics*, 40(1964): 501–504. doi: 10.1063/1.1725143.

[5] Hanchar, John M., Hoskin Paul, W. O. (eds.), *Zircon* (Reviews in Mineralogy and Geochemistry, 53. ISBN 0-939950-65-0, Mineralogical Society of America monograph, 2003) 12.

[6] Dunant, C. F., *Technical handbook on zirconium and zirconium compounds 2019* (published by Zircon Industry Association, fourth edition, www.zircon-association.org/download-handbook.html, 2109) 20.

[7] Monrós, G., Carda, P., Escribano, P., Alarcón, J., "Synthesis of V-$ZrSiO_4$ solid solutions," *J. of Mat. Sci. Lett.*, 9 (1990): 184-186.

[8] Greenwood, N. N., and Earnshaw, A., *Chemistry of the Elements* (New York: Pergamon Press, 1985).

[9] Burakov, B. E., Anderso, E. B., Rovsha, V. S., Ushakov, S. V., "Synthesis of Zircon for Immobilization of Actinides," (*Symposium V – Scientific Basis for Nuclear Waste Management XIX*, Volume 412, 1995) 33, doi: https://doi.org/10.1557/PROC-412-33.

[10] Minton, R. H. (1920) "Use of substitutes for tin oxide in glazes," *J. Amer. Ceram. Soc*, 3, 1(1920): 6-12.

[11] Budnikov, P. P., "Zircon for white glazes," *Tonind. Ztg.*, 47(1923): 1734-1738.

[12] Budnikov, P. P., Ginstling, A. M., Shaw, K., *Principles of Solid State Chemistry*, (Elsevier Science, 1968).

[13] Seabright, C. A., Blue turquoise, *Ceramic Pigments*, U.S. patent, 2,441,407, May 1948.

[14] Seabright C. A., *Yellow of Praseodymium*, US Pat. 2,992,123, July 11, 1961.

[15] Seabright C. A., Pink Koral, *Iron Ceramic Pigment*, U.S. Pat., 3166430, 1965.

[16] Li, C. H., Eppler, D. R., Eppler, R. A., "Iron zircon pigments," *Ceram. Eng. Sci. Proc.,* 16(1992): 109-118.

[17] Llusar, M, Badenes, J., Calbo J, Tena, M.A., Vicente, M., Monrós, G. (2000) "Environmental and colour optimisation of mineraliser addition in synthesis of iron zircon ceramic pigment," *British Ceramic Transaction*, 99, 1(2000): 14-22.

[18] Ozel, E., Turan, S., "Production of coloured zircon pigments from zircon," *Journal of the European Ceramic Society.* 27 (2007): 1751–1757.

[19] Bayanova, T. B., "Baddeleyite: A promising geochronometer for alkaline and basic magmatism," *Petrology*. 14 (2)(2006): 187. doi: 10.1134/S0869591106020032.

[20] Scoates, James, Kevin Chamberlain, "Baddeleyite (ZrOr) and zircon (ZrSiO) from anorthositic rocks of the Laramie anorthosite complex, Wyoming: Petrologic consequences and U-Pb ages." *American Mineralogist.* 80(1995): 1317–1327. doi: 10.2138/am-1995-11-1222.

[21] Ayala, J. M., Verdeja, L. F., Garcia, M. P., Llavona, M. A. and Sancho, J. P., "Production of zirconia powders from the basic disintegration of zircon, and their characterization." *J. Mater. Sci.*, 27(1992): 458–463.

[22] *Zircon Industry Association*, Grenville Court, Britwell Road, Burnham, Bucks SL1 8DF, UK, https://www.zircon-association.org/zirconia.html.

[23] Scheele, C. W., "Tungstens bestånds-delar (Tungsten's constituent components)," *Kongliga Vetenskaps Academiens Nya Handlingar* (New Proceedings of the Royal Academy of Sciences [of Sweden]), 2(1781): 89-95.

[24] de Elhuyar, F., de Elhuyar J. J., "Análisis químico del volfram, y examen de un nuevo metal, que entra en su composición" (Chemical analysis of wolframite, and examination of a new metal, which enters into its composition)," *Extractos de las Juntas Generales celebradas por la Real Sociedad Bascongada de los Amigos del País en la ciudad de Vitoria por setiembre de 1783*, 1783: 46–88.

[25] Yang, X., "Beneficiation studies of tungsten ores – A review," *Minerals Engineering,* 125(2018): 111–119.

[26] Errandonea, D., Segura, A., High-pressure phase transition and compressibility of wolframite-type tungstates. *J. Appl. Phys.,* 107(8) (2010): 127–142.

[27] Lassner, E., Schubert, W. D., *Tungsten: Properties, Chemistry, Technology of the, Element, Alloys, and Chemical Compounds* (Kluwer Academic/Plenum Publishers, 1998).

[28] Chen, P. L., Chen, I. W., "Sintering of Fine Powders I, Microstructural evolution," *J. Am. Ceram. Soc.*, 75(1996):3129-41.

[29] Chen, P. L., Chen, I. W., "Sintering of Fine Powders I, Sintering Mechanism," *J. Am. Ceram. Soc.* 80(1997)637-45.

[30] Breval, E., Dodds, G. C., Mcmillan, N. H., "The Hardness, Stiffness and Thoughness of Diphasic Abrasive Materials prepared by Sol-Gel Techniques," *Mat. Res. Bull.*, 20(1985): 413-429.

[31] McMillan, P. W., *Glass-Ceramics* (Academic Press, 2a. ed., New York 1979).

[32] Xia, K., Langdon, T. G., "The thoughenining and strenthening of ceramic materials though discontinous reinforcement," *J. of Mat. Sci.*, 29(1994):5219-5231.

Chapter 2

Scheelite and Zircon as Ceramic Opacifiers

2.1. Concept of Opacifier

When light passes through a solid, it slows down because solids are denser than air and this changes the path of the light. Refraction is the bending of light as it passes in or out of a media (Figure 2.1): if light reaches a medium with low speed of light when reaches the interface between the materials at an angle one side of the wave will reach the second material first, and therefore slow down earlier, therefore the whole wave will drift to this side and the wave will bend away from the surface or toward the normal when going into a slower material. Conversely, when a wave reaches a medium with higher speed of light, the wave will pivot away the normal and towards the surface.

Refractive index n is the relationship between the speed of light in a vacuum and the speed of light in the considered media. The refractive index n of a transparent medium, introduced by the British physicist Thomas Young in 1807, is defined as:

$$n = \frac{c}{V_{medium}} \quad (2.1)$$

Where c is the speed of light in vacuum. The refraction follows the Snell's law:

$$n_1 \sin i = n_2 \sin r \tag{2.2}$$

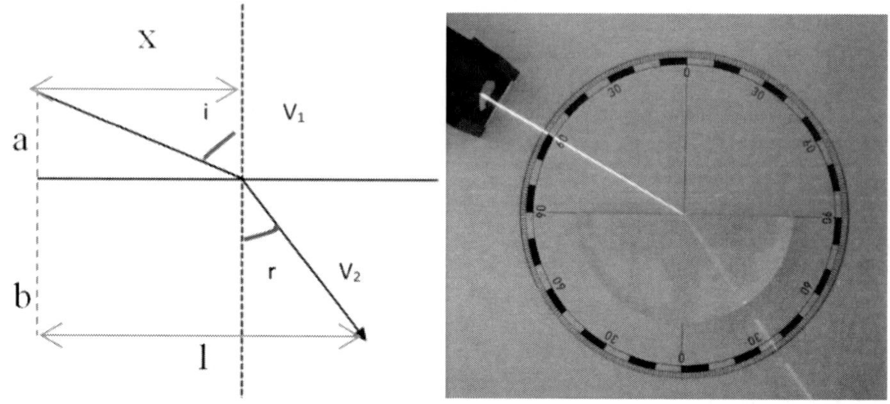

Figure 2.1. Refraction of light.

This law was first described by the Persian scientist Ibn Sahl at the Baghdad court in 984. In the manuscript *On Burning Mirrors and Lenses*, Sahl used the law to derive lens shapes that focus light with no geometric aberrations [1]. The law was rediscovered by Thomas Harriot in 1602, who however did not publish his results although he had corresponded with Kepler on this very subject. In 1621, the Dutch astronomer Willebrord Snellius (1580–1626), Snell, derived a mathematically equivalent form that remained unpublished during his lifetime. René Descartes independently derived the law using heuristic momentum conservation arguments in terms of sines in his 1637 essay *Dioptrics*, and used it to solve a range of optical problems. Rejecting Descartes' solution, Pierre de Fermat arrived at the same solution based solely on his principle of "least time." Descartes assumed the speed of light was infinite, yet in his derivation of Snell's law he also assumed the denser the medium, the greater the speed of light. Fermat supported the opposing assumptions, i.e., the speed of light is finite, and his derivation depended upon the speed of light being slower in a denser medium [2]. In effect the Snell's law can be derived from the principle of

Fermat that states "*the light travel the optical path length taking the least time path.*" Then, from Figure 2.1 the optical path length (OPL) is:

$$\mathrm{OPL} = (x^2 + a^2)^{\frac{1}{2}} + ((l - x)^2 + b^2)^{\frac{1}{2}} \tag{2.3}$$

Considering the respective speed of light the time T for taking the OPL is:

$$T = \frac{(x^2+a^2)^{\frac{1}{2}}}{v_1} + \frac{((l-x)^2+b^2)^{\frac{1}{2}}}{v_2} \tag{2.4}$$

Differentiating to minimize:

$$\frac{dT}{dx} = \frac{x}{(x^2+a^2)^{\frac{1}{2}}\, v_1} + \frac{-(l-x)}{((l-x)^2+b^2)^{\frac{1}{2}}\, v_2} = 0 \tag{2.5}$$

But, note that:

$$\mathrm{Sin}\, i = \frac{x}{(x^2+a^2)^{\frac{1}{2}}} \text{ and } \sin r = \frac{(l-x)}{((l-x)^2+b^2)^{\frac{1}{2}}} \tag{2.6}$$

Therefore:

$$\frac{dT}{dx} = \frac{\sin i}{v_1} - \frac{\sin r}{v_2} = 0 \tag{2.7}$$

$$\frac{\sin i}{v_1} = \frac{\sin r}{v_2} \tag{2.8}$$

$$\frac{\sin i}{\left(\frac{c}{n_1}\right)} = \frac{\sin r}{\left(\frac{c}{n_2}\right)} \tag{2.9}$$

And then the Snell's equation:

$$n_1 \sin i = n_2 \sin r \tag{2.2}$$

Snell's law can be derived considering the change in wavelength at the interface. When the wave goes from one material to another where the wave has a different speed v, the frequency υ of the wave will stay the same, but the distance between wave fronts or wavelength λ=v/υ will change. If the speed v is decreased, such as in the Figure 2.1, the wavelength will also decrease. With an angle between the wave fronts and the interface and change in distance between the wave fronts the angle must change over the interface to keep the wave fronts intact. From these considerations the relationship between the angle of incidence i, angle of refraction r and the wave speeds v_1 and v_2 in the two materials, the Snell's law, can be derived. Other forms for derivation Snell´s law are based from Huygens's principle (using interference of all possible paths of light wave from source to observer), from Maxwell's Equations (involving an application of the general boundary conditions of Maxwell equations for electromagnetic radiation), and from conservation of energy and momentum (see Descartes derivation) [2].

Therefore the index of refraction (n) can be calculated from the change of angle of a collimated monochromatic beam of light from vacuum into liquid using Snell's law for refraction. Using the theory of light as an electromagnetic wave, light takes a straight-line path through a medium at reduced speed (v) and wavelength (λ). The ratio v/λ is a constant equal to the frequency (ν) of the light, as is the quantized (photon) energy using Planck's constant h:

$$E = h\nu = h\frac{c}{\lambda_0} = h\frac{v}{\lambda} \tag{2.10}$$

$$\frac{c}{v} = \frac{\lambda_0}{\lambda} = n \tag{2.11}$$

Compared to the constant speed of light in a vacuum (c), the index of refraction of the medium is n = c/v.

Standard refractometers measure the extent of light refraction (as part of a refractive index) of transparent substances in either a liquid or solid state;

this is then used in order to identify a liquid sample, analyze the sample's purity and determine the amount or concentration of dissolved substances within the sample. As light passes through the liquid from the air it will slow down and create a 'bending' illusion, the severity of the 'bend' will depend on the amount of substance dissolved in the liquid. For example, the amount of sugar in water, wine or blood (Figure 2.2). In the Gladstone–Dale relation:

$$\frac{n-1}{\rho} = \sum_{1}^{n} k_i x_i = K_c \tag{2.12}$$

Where n is the index of refraction of mixture, ρ is the density of mixture, x_i is the mass fraction of component i in the mixture and k_i is the mass refractivity constant (cm^3/g) for component i and K_c the Gladstone-Dale constant of the mixture. For a binary mixture of a liquid A in water in a mass fraction α; mass is conserved on mixing, but not the volume. The plot of volume or density versus mass fraction of A in water is a quadratic curve. However, the plot of index of refraction versus weight fraction α in water is linear.

In the traditional handheld refractometer [3] a sample is placed between a measuring prism and a small cover plate (Figure 2.2). Light traveling through the sample is either passed through to the reticle or totally internally reflected. The net effect is that a shadow line forms between the illuminated area and the dark area. It is where this shadow line crosses the scale that a reading is taken. Because refractive index is very temperature dependent, it is important to use a refractometer with automatic temperature compensation. Compensation is accomplished through the use of a small bi-metallic strip that moves a lens or prism in response to temperature changes. This design was invented by Emanuel Goldberg [4]. There are many types of refractometers and the most common types are Abbe's refractometer, Pulfrich refractometer or Immersion refractometer.

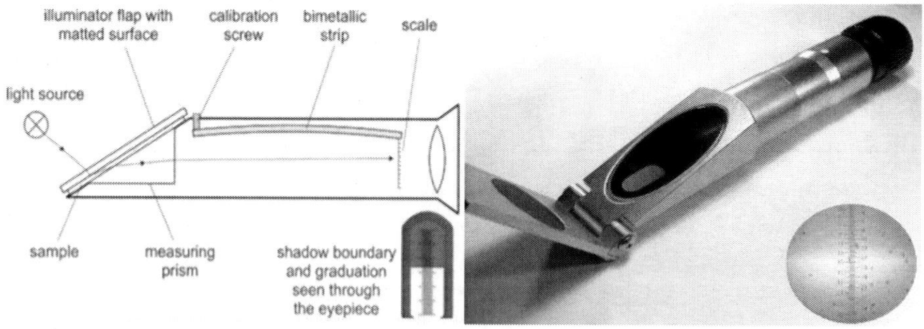

Figure 2.2. Handheld refractometer for liquid sample.

The refractive index varies with wavelength for all materials. This dispersion relation is nonlinear and is characteristic for every material. In the visible range, a decrease of the refractive index comes with increasing wavelength. In glass prisms very little absorption is observable. In the infrared wavelength range several absorption maxima and fluctuations in the refractive index appear. To guarantee a high quality measurement with an accuracy of up to 0.00002 in the refractive index the wavelength has to be determined correctly. Therefore, in modern refractometers the wavelength is tuned to a bandwidth of +/-0.2 nm to ensure correct results for samples with different dispersions.

At the atomic scale [5], the velocity of an electromagnetic wave is slowed in a material because the electric field creates a disturbance in the charges of each atom (primarily the electrons) proportional to the electric permittivity (ε) of the medium, similarly, the magnetic field creates a disturbance proportional to the magnetic permeability (μ). As the electromagnetic fields oscillate in the wave, the charges in the material will be "shaken" back and forth at the same frequency. The charges thus radiate their own electromagnetic wave that is at the same frequency, but usually with a phase delay, as the charges may move out of phase with the force driving them and the light wave traveling in the medium is the macroscopic superposition of all such contributions in the material: the original wave plus the waves radiated by all the moving charges. This wave is typically a wave with the same frequency but shorter wavelength than the original, leading to a slowing of the wave's phase velocity. Most of the radiation from oscillating material charges will modify the incoming wave, changing its velocity. However, some net energy will be radiated in other directions or even at other frequencies (scattering). Then there are several possibilities of interaction:

a) If the electrons emit a light wave which is 90° out of phase with the light wave shaking them, it will cause the total light wave to travel slower. This is the normal refraction of transparent materials like glass or water, and corresponds to a refractive index which is real and greater than 1.
b) If the electrons emit a light wave which is 270° out of phase with the light wave shaking them, it will cause the wave to travel faster. This is called "anomalous refraction," and is observed close to absorption lines (typically in infrared spectra), with X-rays in ordinary materials, and with radio waves in Earth's ionosphere. It corresponds to a permittivity less than 1, which causes the refractive index to be also less than unity and the phase velocity of light greater than the speed of light in vacuum *c* (note that the signal velocity is still less than *c*, as discussed above). If the response is sufficiently strong and out-of-phase, the result is a negative value of permittivity and imaginary index of refraction, as observed in metals or plasma.
c) If the electrons emit a light wave which is 180° out of phase with the light wave shaking them, it will destructively interfere with the original light to reduce the total light intensity. This is light absorption in opaque materials and corresponds to an imaginary refractive index.
d) If the electrons emit a light wave which is in phase with the light wave shaking them, it will amplify the light wave. This is rare, but occurs in lasers due to stimulated emission. It corresponds to an imaginary index of refraction, with the opposite sign to that of absorption.

For most materials at visible-light frequencies, the phase is somewhere between 90° and 180°, corresponding to a combination of both refraction and absorption.

Then the refractive index of electromagnetic radiation in a done material is defined by:

$$n = \sqrt{\mu_r \varepsilon_r} \tag{2.13}$$

Where ε_r the relative electric permittivity of the material and μ_r is its relative permeability.

Most naturally occurring materials are non-magnetic at optical frequencies; that is μ_r is very close to 1, therefore n is approximately $\varepsilon_r^{1/2}$. In this particular case, the complex relative permittivity $\boldsymbol{\varepsilon_r}$, with real and imaginary parts ε_r and ε'_r, and the complex refractive index n, with real and imaginary parts n and n' (the latter called the "extinction coefficient"), follow the relation:

$$\boldsymbol{\varepsilon_r} = \varepsilon_r + \varepsilon'_r i = \mathbf{n}^2 = (n + n'i)^2 \tag{2.14}$$

$$\left.\begin{aligned} \varepsilon_r &= n^2 - n'^2 \\ \varepsilon'_r &= 2nn' \end{aligned}\right\} \tag{2.15}$$

To analyze the interaction of light as an electromagnetic wave that propagates through matter in the +z direction, it is conventionally described by the equation:

$$\mathbf{A}(z, t) = \mathrm{Re}\left[\mathbf{A_0} e^{(ikz - wt)}\right] \tag{2.16}$$

Where A_o is a vector in the *x*-*y* plane, with the units of an electric field; *ω* is the angular frequency of the wave (2π/T); *k* is the angular wavenumber of the wave ((2π/λ); Re indicates real part; *e* is Euler's number.

The wavelength is, by definition.

$$\lambda = \frac{2\pi}{k} \tag{2.17}$$

For a given frequency, the wavelength of an electromagnetic wave is affected by the material in which it is propagating. The *vacuum* wavelength (the wavelength that a wave of this frequency would have if it were propagating in vacuum) is:

$$\lambda_0 = \frac{2\pi}{\left(\frac{w}{c}\right)} \tag{2.18}$$

Where c is the speed of light in vacuum. Therefore considering eq. 2.11:

$$n = \frac{\lambda_0}{\lambda} = \frac{ck}{w} \tag{2.19}$$

The intensity I(z) of the wave is proportional to the square of the amplitude, time-averaged over many oscillations of the wave, which amounts to:

$$I(z) \propto \left|\mathbf{A_0} e^{(ikz-wt)}\right|^2 = |\mathbf{A_o}|^2 \tag{2.20}$$

Note that this intensity is independent of the location *z*, a sign that this wave is not attenuating with distance. But in opaque media, it is necessary to incorporate attenuation into the mathematical description of the wave; is via an attenuation coefficient α that sometimes is called opacity:

$$\mathbf{A}(z,t) = e^{-\alpha z/2} \mathrm{Re}\left[\mathbf{A_0} e^{(ikz-wt)}\right] \tag{2.21}$$

Therefore in opaque or attenuated media, the intensity of the wave is:

$$I(z) \propto \left|e^{-\alpha z/2} \mathbf{A_0} e^{(ikz-wt)}\right|^2 = e^{-\alpha z} |\mathbf{A_o}|^2 = I_o e^{-\alpha z} \tag{2.22}$$

In general, opacity is the measure of impenetrability to electromagnetic or other kinds of radiation, especially visible light caused by absorption, reflection, and scattering (Figure 2.3).

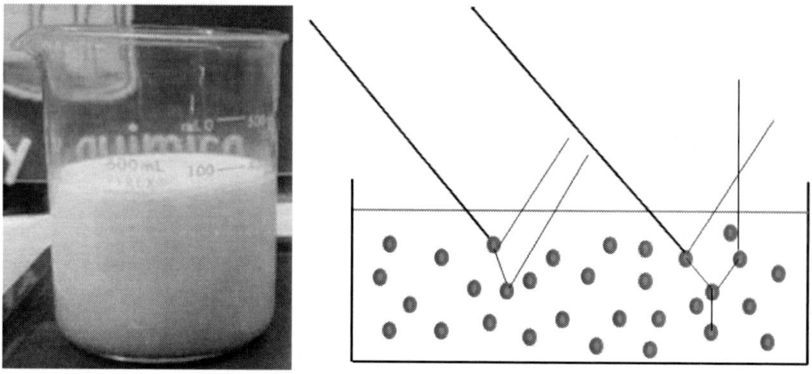

Figure 2.3. Opacifier dispersed in water and mechanism of opacification.

An opaque object is neither transparent (allowing all light to pass through) nor translucent (allowing some light to pass through). An opaque substance transmits no light, and therefore reflects, scatters, or absorbs all of it. Both mirrors (by reflection) and carbon black (by absorption mechanism) are opaque to visible light.

In the scattering [6], the light propagating through a uniform media is forced to deviate from their straight trajectory by localized "non-uniformities" in the medium through which they pass (Figure 2.3). These non-uniformities absorb light energy and re-emit light in different directions with different intensity. This phenomenon is are often called "diffuse reflection" and unscattered reflections are called "specular" (mirror-like) reflections. The non-uniformities or "scattering centers" which can cause scattering includes particles, bubbles, droplets, density fluctuations in fluids, crystallites in polycrystalline solids, defects in monocrystalline solids, surface roughness, cells in organisms, textile fibers in clothing.

When an electromagnetic wave interacts with a discrete particle, the electron orbits within the particle's constituent molecules are perturbed periodically with the same frequency as the electric field of the incident wave. The oscillation or perturbation of the electron cloud results in a periodic separation of charge within the molecule, which is called an induced dipole moment. The oscillating induced dipole moment is manifest as a source of radiation, thereby resulting in scattered light. The majority of light scattered by the particle is emitted at the identical frequency of the incident light, a process referred to as elastic scattering. In summary, the above comments describe the process of light scattering as a complex interaction between the incident electromagnetic wave and the molecular/atomic structure of the scattering object; hence light scattering is not simply a matter of incident photons or electromagnetic waves "bouncing" off the surface of an encountered object [7].

Models of light scattering can be divided into three domains based on a dimensionless "size parameter" h which is defined as:

$$h = \frac{\pi \emptyset_p}{\lambda} \tag{2.23}$$

Where $\pi\emptyset_p$ is the equivalent-circumference of the scattering centre and λ is the wavelength of incident radiation. Based on the value of h, three kinds of scattering can be defined:

a) Rayleigh scattering (small particle compared to wavelength of light); $h \ll 1$ or $h < 1/10$. In this case the particle is sufficiently small such that the particle encounters a uniform electric field at any moment, accordingly the time for penetration of the electric field is much less than the period of oscillation of the wave. Lord Rayleigh demonstrates that the formula giving the coefficient of attenuation in terms of the refraction index n and of the number of particles per unit volume μ it is inversely proportional to the number of particles per unit volume and the fourth power of the wavelength of the light [8]:

$$\alpha = \frac{32\pi^3(n-1)^2}{3\mu\lambda^4} \tag{2.24}$$

Lord Ryleigh write:

"Let us now inquire what degree of transparency of air is admitted by its molecular constitution, i. e. in the absence of all foreign matter. We may take $h=6\times10^{-5}$ cm, n-1=0.0003; whence from (eq. 2.22) we obtain as the distance x, equal to $1/\alpha$ which light must travel in order to undergo attenuation in the ratio e:1 (I/Io=1/e=1/2.7)"

$$x=1/\alpha=4.4\ 10^{-13}\,n \tag{2.25}$$

And then Lord Ryleigh comments:

"The completion of the calculation requires the value of μ. Unfortunately this number - according to Avogadro's law the same for all gases - can hardly be regarded as known" [9].

We can estimate the number of molecules under standard conditions in 19.10^{18} molec/cm^3, then $x=8.3.10^6$ cm=83 km.

Then Lord Ryleigh write:

"Although Mount Everest appears fairly bright at 100 miles (161 km) distance as seen from the neighborhood of Darjeeling, we cannot suppose that the atmosphere is as transparent as is implied in the above numbers; and of course this is not to be expected, since there is certainly suspended matter to be reckoned with. Bouguer and others estimate about 0.8 for the transmission of light through the entire atmosphere from a star in the zenith. This corresponds to 8.3 kilometres of air at standard pressure. At this rate the transmission through 83 kilometres would be $(0.8)^{10}$, or 0.11, instead of 1/e or 0.37. It appears then that the actual transmission through 83 kilometres is only less than that calculated (with the above value of μ) from molecular diffraction without any allowance for foreign matter at all. And we may conclude that the light scattered from the molecules would suffice to give us a blue sky, not so very greatly darker than that actually enjoyed."

An interesting question arises as to whether (eq. 2.24) can be applied to highly compressed gases and to liquids or solids. Since approximately (n-l) is proportional to μ, so also is α according to (eq. 2.24), Lord Rayleigh say:

> "We have no reason to suppose that the purest water is any more transparent than (eq, 2.24) would indicate; but it is more than doubtful whether the calculations are applicable to such a case, where the fundamental supposition, that the phases are entirely at random is violated."

And "hitherto we have considered the case of obstacles small compared to the wave-length."

b) Mie scattering (particle about the same size as wavelength of light, valid only for spheres); h ≈ 1. The Mie solution to Maxwell's equations describes the scattering of an electromagnetic plane wave by a homogeneous sphere. The solution takes the form of an infinite series of spherical multipole partial waves [10]. For particle sizes larger than a wavelength, Mie scattering predominates. Unlike Rayleigh's scattering, this scattering produces a pattern like an antenna lobe (Figure 2.4), with a sharper and more intense forward lobe for larger particles. Mie scattering is not strongly wavelength dependent and produces the almost white glare around the sun when a lot of particulate material is present in the air. It also gives us the white light from mist and fog [11].

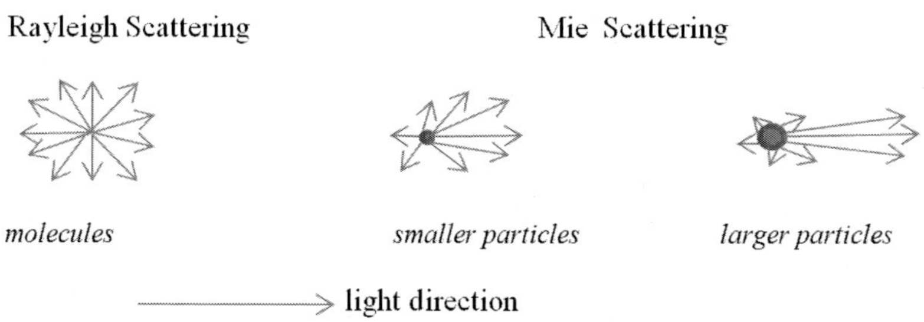

Figure 2.4. Light scattering by particles.

c) Geometric scattering (particle much larger than wavelength of light): h ≫ 1, this particles acts scattering light according to their projected area.

The scattering from molecules and very tiny particles (h<1/10 wavelength) is predominantly Rayleigh scattering. For particle sizes larger than a wavelength, Mie scattering predominates. In Figure 2.5 the monostatic radar cross section (RCS) of a perfectly conducting metal sphere as a function of frequency (computed by Mie theory) is shown.

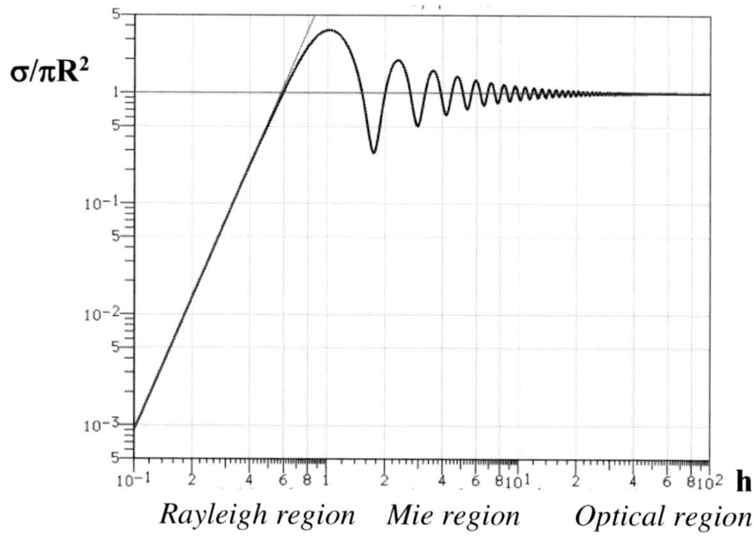

Figure 2.5. Monostatic radar cross section (RCS) of a perfectly conducting metal sphere as a function of h [11].

The x-axis is the relative frequency h, defined above as the number of wavelengths in the circumference, and equal to the angular wavenumber (k in eq. 2.16) times the radius R ($h=\pi\varnothing_p/\lambda=2\pi R/\lambda$). The y-axis is the RCS relative to the projected area of the sphere ($\sigma/(\pi R^2)$. The asymptotes are $\sigma/(\pi R^2) = 9\, h^4$ in the low frequency (Rayleigh) region ($\lambda > 2\pi R$), and $\sigma/(\pi R^2) = 1$ in the high frequency (optical) limit ($\lambda << R$) [11].

Therefore, opacity depends on the frequency of the light being considered. For instance, some kinds of glass, while transparent in the visual range, are largely opaque to ultraviolet light.

2.2. Diffuse Reflectance and Refractive Index

The optical properties of an opacifier can be predicted from the refractive index because it determines its opacity, such that the greater the refractive index of the pigment with relation to the refractive index of the medium or vehicle in which it is immersed, the greater will the opacity be. Therefore using the Mie theory [10] to predict the absorption and scattering, the refractive index of the opacifier and the medium of dispersion and the particle size, it is possible to predict the behavior of the opacifier in the considered medium [6]. The measure of the refractive index by refractometers has been described above for transparent samples (Abbe's refractometer Figure 2.2). But for other kind of samples, the measurement is complicated and the techniques to measure it are highly conditioned by the nature of the medium that it is sought to characterize (gas, liquid, or solid), by the optical characteristics (transparent, opaque, translucent), by the surface morphology (rough or smooth), etc. [12].

Specular reflection and regular transmission, the respective causes of the gloss and transparency of materials, are related to the refractive index by means of the Fresnel equations [13]: both meet a fundamental condition, which is that the incident beam, the specularly reflected beam, and the regularly transmitted beam are in the same plane. With relation to specular reflection it may readily be deduced from Snell' equation (eq. 2.2) that the incident angle and the reflected angle have the same angle. The reflection can be measured and evaluated by means of the reflectance R, the ratio between the intensity of the reflected radiation and the incident radiation. The Fresnel equations enable reflectance to be related to the refractive index according to eq. 2.26:

$$R = \frac{(n_1 \cos i - n_2 \cos r)^2}{(n_1 \cos i + n_2 \cos r)^2} \tag{2.26}$$

When the incidence is perpendicular (i=0) and the medium 1 is air (n_1=1), eq. 2.26 results:

$$R = \frac{(1-n_2)^2}{(1+n_2)^2} = \left(\frac{n_2-1}{n_2+1}\right)^2 \tag{2.27}$$

That can be written as:

$$n_2 = \frac{1+\sqrt{R}}{1-\sqrt{R}} \tag{2.28}$$

The above equations quantify the specular reflection of a material in air. But usually the light is partially reflected (a fraction R), partially transmitted T and partially absorbed A, the imbalance results:

$$R+T+A=1 \tag{2.29}$$

This equation responds to the law of conservation of energy, and it assumes that all the energy intensity of the incident radiation is reflected, transmitted or absorbed. As it is discussed above, absorption can be explained from the complex refraction index (eq. 2.14):

$$\mathbf{n}^2 = (n + n'i)^2 \tag{2.14}$$

The real part n, corresponds to the change of speed that light undergoes when it travels through a medium (the classic or real Thomas Young's refractive index, eq. 2.1) and the imaginary part n', called the coefficient of extinction, is linked to the absorption of the material. It is related to this by means of its coefficient of absorption α according to:

$$R = \frac{\alpha\lambda}{4\pi} \tag{2.30}$$

The Fresnel equation (eq. 2.26) is known to be highly suited for obtaining the refractive index from the reflectance measurement of materials that reflect specularly, its application is generally restricted to polished surfaces. In gloss surfaces with highly reflecting, other techniques can be used to measure, such as ellipsometry, which even allows the complex part of the refractive index to be measured, techniques that involve measurement of the limit angle, or techniques that involve measurement of the Brewster angle [13, 14] which is related to the polarization of reflected light. According to Brewster's law, if the incident light is polarized in the plane of incidence (*p*–polarization), at some angle of incidence reflectivity is close to zero (R=0); this angle is called the Brewster's angle θ and is associated with the refractive index n_2 of the material (n_1 is the incident medium, usually air) by eq. 2.31:

$$\theta_B = \arctan(n_2/n_1) \tag{2.31}$$

But usually, in the case of powders or matte paints or ceramics, at macroscopic level the reflection that occurs is no longer specular and the reflected radiation in the foregoing conditions is termed diffuse radiation and is characterized by diffuse reflectance (Figure 2.6).

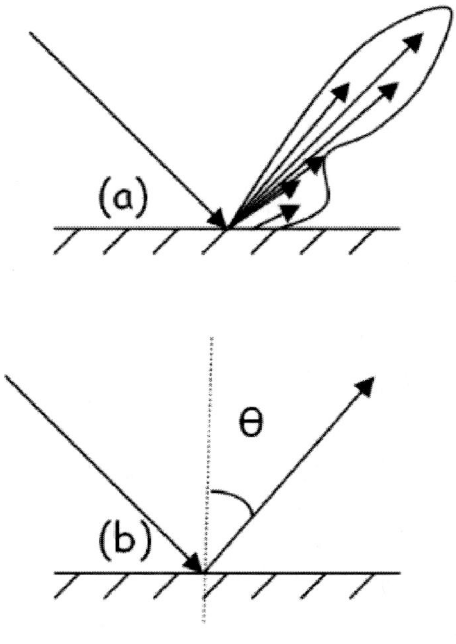

Figure 2.6. (a) Diffuse reflectance, (b) specular reflectance.

If the scattering is null in all directions except for the reflecting direction θ=i where i is the incident angle, the reflectance is specular (Figure 2.6.b). In the case of diffuse reflectance (Figure 2.6.a) the gloss in a direction θ (angle between the normal and the interface, Figure 2.6), is defined by eq. 2.31:

$$B = \frac{I_\theta}{A_\theta} \tag{2.32}$$

where B is the gloss in the θ direction, I_θ is the reflected intensity in the θ direction and A_θ elemental normal area in the θ direction (i.e., $A_o \cos\theta$, with A_0 the elemental area in the θ direction). The reflectance is called diffuse if the scattering of the light follow the cosine law or Lambert's law (eq. 2.33):

$$I_\theta = I_0 \cos\theta \tag{2.33}$$

where I_0 is the intensity of scattered light at θ=0, therefore the gloss in a done direction θ is constant:

$$B = \frac{I_o \cos\theta}{A_o \cos\theta} = \frac{I_o}{A_o} = \text{constant} \tag{2.34}$$

Freshly deposited snow or carbon conforms to the cosine law presenting ideal diffuse reflectance behavior. The diffuse reflectance can be measured with a spectrophotometer equipped with an integrating sphere. A simplified scheme of an Ulbricht integrating sphere is shown in Figure 2.7.

Light enters the integrating sphere forming an angle of θ with relation to the vertical of the substrate to be measured. The light diffused by this substrate is collected inside the sphere, which has a highly reflecting white lining (i.e., using MgO coating). After multiple reflections the light exits the sphere and reaches the detector.

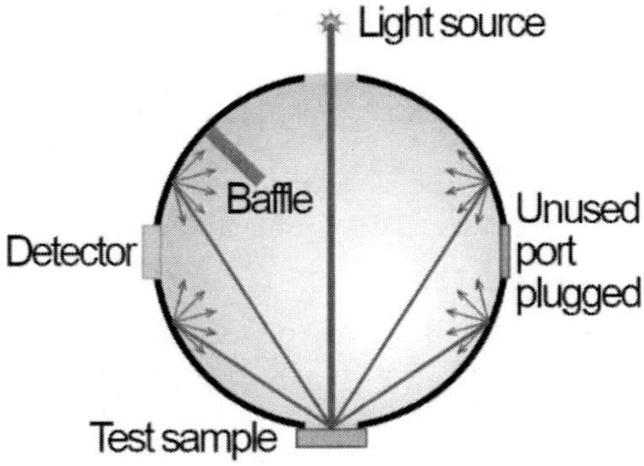

Figure 2.7. Simplified scheme of an integrating sphere or Ulbricht sphere (https://es.qaz.wiki/wiki/Integrating_sphere).

In order to measure the refractive index of rough matte surfaces such as the powders of opacifiers, the starting assumption was made that as a result of their surface irregularity, the specular radiation energy that would be observed if the surface was perfectly sintered and polished, which would obey Snell's Law and the Fresnel equations, is distributed across all the material surface, since the impinging light is reflected in every direction and plane perpendicular to that surface. Since an integrating sphere enables the total energy spread over the entire surface, reflected by this, to be measured, the total of this diffuse radiation energy may be assumed to be very similar to the energy reflected specularly if the material was polished, since the refractive index is an intrinsic property of the material and not of its surface [14, 15]. Hence, to determine the refractive index, the spectral dependence of the reflectance depending on the angle of incidence θ of *p*–polarized light beam for the desired wavelength is checked to find the angle corresponding to the minimum value of reflectance and calculate n using eq. 2.28 as if the diffuse radiation were specular [14, 15].

Solids with high n appear more brilliant than those with a low refraction index. Some solids are singly refractive but usually are doubly refractive showing two different refractive indices. When a beam of light enters a doubly refractive material, it is split into two beams, each travelling at a different speed and on a different path through the crystal. Birefringence (δ) measures the difference between the two refractive indices in materials that are doubly refractive, and it ranges from 0.003 to 0.287. Very few solids are singly refractive; in fact, the only well-known with that property are diamond, spinel, garnet or cerianite. Therefore, according to the number of refractive indices, we have three types of materials: I (Isotropic, with a single refractive index), U (Uniaxial, two refractive indices), B (Biaxial, three refractive indices) (Table 2.1).

The appearance of a solid can be described using terms such as brilliance, luster or dispersion that are technical terms with fairly precise meanings, but they are often misused. Brilliance refers to the light reflected from the interior of a solid. Thus only a transparent solid can be said to have brilliance. Luster, on the other hand, refers to the light reflected from the surface. If someone describes an opaque star ruby as 'brilliant,' for example, they are misusing the term. They need to talk about the ruby's luster instead.

2.3. Comparative Analysis of the Main Historical Opacifiers in Ceramics

As discussed in Chapter 1, both scheelite (n_ω = 1.918–1.921, n_ε = 1.935–1.938) and zircon (n_ω = 1.967–2.015, n_ε = 1.920–1.960) show high refractive index with birefringence δ = 0.017 and δ=0.047–0.055 respectively. Done the high refractive index of the crystalline phase as well as its reticular stability, scheelite can be used as a bleaching and opacifying additive for ceramic glazes. The structural and optical characteristics of the main historical opacifiers in ceramics described in Chapter 1(section 1.2.2) are summarized on Table 2.1 [17].

The diffuse reflectance of several classic oxide opacifiers (industrial powders) is shown in Figure 2.8.a together the calculation of the refractive index following the methods described above using the minimum of diffuse reflectance at θ =8° [14] by eq. 2.28 (Figure 2.8.b).

Table 2.1. Structural and optical characteristics of the main historical opacifiers in ceramics: I (Isotropic), U (Uniaxial), B (Biaxial)

Opacifier	ρ(g/cm[3])	Cryst. System	Space Group	melt. point (ºC)	Refractive indices (400 nm)			
$Ca_2Sb_2O_7$	6.58	Cub.	Fd-3m		B	2.01	2.10	
SnO_2 cassiterite	6.95	Tet.	$P4_2/mnm$	1630	U	1.97	2.02	-
TiO_2 rutile	4.25	Tet.	$P4_2/mnm$	1850	U	2.87	2.58	-
ZrO_2 baddeleyite	5.75	Mon.	$P2_1/c$	2710	B	2.13	2.19	2.20
$ZrSiO_4$ zircon	4.65	Tet.	$I4_1/amd$	2550	B	1.92	1.97	-
CeO_2 cerianite	7.22	Cub.	Fm3m	2400	I	2.13		
$CaWO_4$ scheelite	6.01	Tet.	I41/a	1580	B	1.92	1.94	

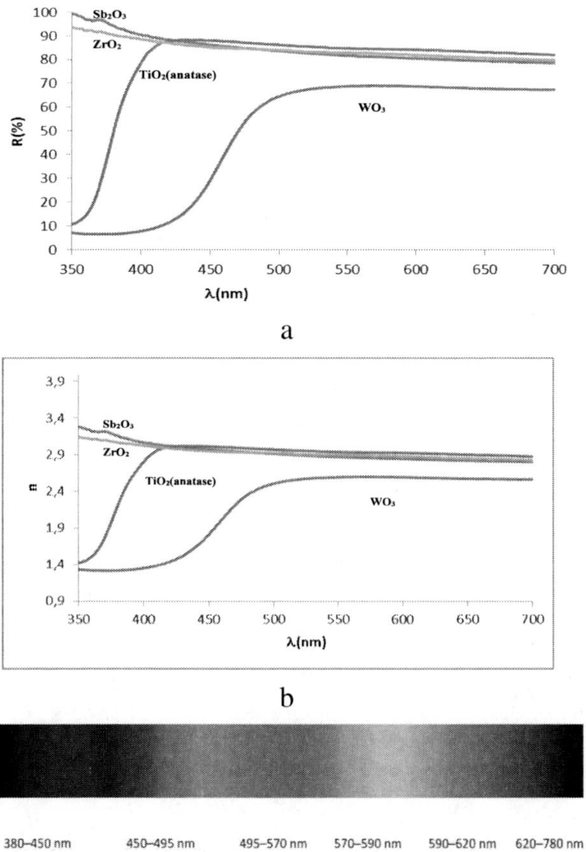

Figure 2.8. (a) Diffuse reflectance spectra of several oxide opacifiers, (b) its calculated refractive index by eq. 2.28 at θ =8º (minimum of diffuse reflectance).

From Table 2.1 it is observed that the classic oxide opacifiers show high symmetry (cubic or tetragonal crystalline system) and high thermal stability (all shows melting points above 1500ºC). For the analyzed classic oxide opacifiers in θ =0º (Figure 2.9.a) the diffuse reflectance of Sb_2O_3 and ZrO_2 (both white colored) shows high reflectance in the whole interval of wavelength analyzed (always above 80%), in the case of TiO_2 (white colored) and WO_3 (yellow colored) show a charge band transfer at 425 and 450 nm respectively (with a band gap associated of 3.5 and 2.9 eV respectively). The associated refractive index obtained from eq. 2.28 indicates a variation between of 2.9-2.7, 2.8-3.1, 2.7-3.2 and 2.6-1.3 in the visible range analyzed (400-700 nm) for Sb_2O_3, TiO_2 (anatase), ZrO_2 and WO_3 respectively, similar to the trend observed with other optical methods: TiO_2(anatase)> ZrO_2 >Sb_2O_3>WO_3.

The evolution and technology of the opacifiers is very close to the glass technology. Historically, calcium antimonate-lead antimonate, cassiterite, rutile-anatase, zirconia-zircon, cerianite and scheelite have been used successively. The ancient opacifiers are discussed below.

3.2.1. Calcium Antimonate-Lead Antimonate: The Egyptian and Roman Opacifier

The first opacifier in the history is the lead antimony and calcium antimony oxides. It is generally accepted that the first artificial glass was obtained in the Mesopotamian area at the beginning of the Bronze Age (3300-2100 BC) as a by-product of metallurgical activities [17]; the treatment of copper minerals is known to produce vitreous colored and opaque slags. Quartz melts at 1670ºC, consequently glass can only be produced as long as elements capable of reducing it are added (fluxes) which in Mesopotamia consisted in alkaline salts obtained from plant ashes. Metal ions such as copper, cobalt or iron dissolved in the glass or lead antimony and calcium antimony oxides forming small crystalline precipitates were also incorporated giving color to the glass. The earliest glass objects, turquoise blue beads were found in the excavations of the cemetery of Ur

(2500 BC). The oldest treatise on the fabrication of glass was found in the Palace of Assurbanipal (668-626 BC), a set of tablets in cuneiform script containing recipes for the production of both colorless and color glass. Phoenician glass production became particularly important since the 8th century BC and during the pre-Roman time [17].

As Boschetti et al. describe [18] opacity of glass implies either the addition of mineral pigments to the base glass or the in situ precipitation of crystals during the thermal cycle. Calcium antimonate, a synthetized pigment which does not exist in nature, is attested in the form of hexagonal ($CaSb_2O_6$) or cubic ($Ca_2Sb_2O_7$) crystals to generate opaque white glass. The use of antimony as opacifier is documented since the very beginnings of glass making in the Late Bronze Age, both in Mesopotamia as well as in Egypt [19]. Calcium antimonate was used not exclusively for making white glass, but in combination with strongly colored glasses; it also achieves a number of cool shades, for instance blue, aqua-green, turquoise, grey, and light purple. During the same period, lead antimonate ($Pb_2Sb_2O_7$) served to produce opaque yellow and shades of green. Unlike calcium antimonate, lead antimonate occurs in nature as bindheimite or lead antimony oxide hydroxide ($Pb_2Sb_2O_6$ [O,OH]) and was probably used by glass artisans both in its mineral and in its synthetized form. The appearance of both calcium antimonate and lead antimonate as pigments seem to be related to the origins of glass making. No use of either calcium or lead antimonate prior to the invention of vitreous materials has been reported. Calcium antimonate-based pigments continued to be used as opacifiers during the Hellenistic and the Roman period, and have been discussed in the literature particularly in connection with Hellenistic mosaic tesserae and vessels as well as Roman mosaic tesserae and opus sectile panels. For reasons not well understood, the use of calcium antimonate as an opacifying agent for glass ceased towards the end of the Western Roman Empire, when it was replaced by calcium stannate [20].

By the end of the second millennium BC Syria became also an important glass production centre and later between approximately 1500 and 1300 BC (18th dynasty), glass was also produced in Egypt reaching its maximum splendor at the glass workshop of Tell-el-Amarna. It is worth to mention that during this period Syria had already been conquered by Thutmose III who took the best artisans from the conquered area. The Egyptian glass was of the type high soda-high lime type made of natron (a highly hydrated variety of sodium carbonate found in the valley of Wadi al Natrum in Egypt). Glass was mainly used to produce ornamental objects and for the production of precious small colorful flask and bottles. The Egyptian supremacy in glass production lasted until the resurgence of the Assyrian glass industry in the last millennium BC [17].

During the Roman Empire many technological innovations were introduced and glass reached the highest quality standards. Dichroic glass, such as the Lycurgus cup (later discussed, see chapter 4, Figure 4.16), red in transmission and green in reflection, was obtained by gold and silver colloidal metallic particles. Lead oxide is also a flux, and during Roman times was directly applied on ceramic wares to produce a high lead glaze. Those glazes were mainly used to waterproof and protect the ceramic objects. The colors obtained were yellow-green due to the presence of Fe^{3+} and Fe^{3+} dissolved in the glaze. However, alkaline based glazes were not applied on ceramics during Roman times. In ancient times, glass vessels were made by dipping a core of animal dung and clay on the end of a metal rod into the pot containing the batch. After the glass had cooled, the rod and core were removed (Figure 2.9). Glass blowing, however, is thought to have developed in the era of Caesar, mid first century BC. Glass in the ancient world was considered a very valuable commodity available only to the extremely wealthy, and in small sizes for cosmetic, not practical purposes. With the invention of glass blowing, glass became generally available for inclusion in various public works and to the general population.

The studies over both Egyptian and Roman yellow glasses, a significant proportion of the lead antimonate particles exhibit an irregular, ragged morphology and a clumped distribution, whereas other particles exhibit euhedral morphologies suggesting that they were formed during cooling from the melt. It is therefore argued that preformed lead antimonate, rather than separate lead and antimony compounds, was added to the glass melt. The lead antimonate then suffered partial dissolution in the molten glass, resulting in both surviving clastic particles and euhedral particles that had crystallized from the melt [17].

Figure 2.9. Yellow roman glass unguetarium opacified by $Pb_2Sb_2O_7$ (https://www.antiquities.co.uk/roman-glass-yellow-unguentarium).

Therefore, as proposed by Shortland [19] the lead antimonate pigment used in the production of New Kingdom Egyptian yellow glass was most probably produced by roasting a mixture of galena (PbS) and stibnite (Sb_2S_3) containing excess lead. The result would have been the formation of lead and antimony oxides which then combined to produce lead antimonate. Because of the refractory nature of zinc oxide (m.p. 1975°C), the loss of zinc in this process would have been limited. It seems reasonable to assume that the Egyptian glass makers would have been able to select and combine these two metal ores since they were also being used for quite different purposes. Thus stibnite was being used in the production of calcium antimonate white glasses and occasionally as a kohl (a black eye cosmetic traditionally made by grinding sulphides such as stibnite (Sb_2S_3) or galena (PbS)), and galena was being used in the production of both lead and silver glasses as well as being used as a kohl. Furthermore, the two ores were most probably being obtained from different places, the galena, from Gebel Zeit in Egypt, and the stibnite from the Caucasus where antimony mines that were active in the Late Bronze Age have been discovered [19].

Conversely, because of the absence of zinc from the Roman yellow glasses, antimonial litharge resulting from the cupellation of silver is a possible source of the lead antimonate used in the production of these glasses. In the cupellation process of metal ores (usually smelted lead ores) or alloyed metals, the mixture is melted in the cupel (made of absorbent bone ashes or clay) by a gas-fired burner, and oxygen is blown through the molten bath of metal. The process is based on the principle that precious metals do not oxidize or react chemically, unlike the base metals, so when they are heated at high temperatures, the precious metals remain apart, and the others react, forming slags where the stibnite and litharge (PbO) are content. However, there are some doubts as to the extent to which cupellation litharge containing sufficient antimony would have been available. The alternative would have again been to roast a mixture of galena and stibnite, but in this case, zinc must have been absent from the galena chosen.

3.2.2. Cassiterite: The Islamic Opacifier

As above discussed, antimony-based opacifiers (lead antimonate yellow and calcium antimonate white) were used from the beginnings of glass production in the Near East and Egypt around 1500 BC through into the Roman period. Towards the end of the Roman period (i.e., from about 4th century AD onwards), lead stannate replaced lead antimonate in the production of opaque yellow glasses [20] and it was not until the late 15th century AD that lead antimonate started to be used again both in glass production in Venice [21].

Actually, from a chemical point of view, probably the transition from white ($CaSb_2O_6$) to the "lead-tin-yellow type I" (orthorhombic Pb_2SnO_4), the oldest classic yellow pigments known in Europe since at least the thirteenth and its Si-doped modification called "lead-tin-yellow type II" (cubic pyrochlore $PbSn_xSi_{1-x}O_3$) and after the called Naples's yellow or lead (II) antimonate $Pb_2Sb_2O_7$ (cubic pyrochlore, space group *Fd3m*) is a simple compositional evolution: in ancient glazes, as discussed above, the opacification (pigmenting) starts with the roasting a mixture of galena (PbS) and stibnite (Sb_2S_3) containing excess lead which reacts to give the $Pb_2Sb_2O_7$ yellow pyrochlore, and adding $CaCO_3$ the white calcium antimonate; in order to produce a better yellow, romans introduce tin oxide (Pb_2SnO_4), and after silica (pyrochlore $PbSn_xSi_{1-x}O_3$), but surprisingly avoiding the presence of CaO in glazes, they rediscovery the called Naples' Yellow which arrives to Venice in Italy from the Byzantine Empire [22] and as a yellow pigment in Italian Renaissance paintings [23,24] and in Italian majolica production [20].

In effect, as Tite describes [20], the introduction of lead stannate yellow in northern and central Europe in the second and first centuries BC was probably the result of glassworkers not having access to sufficient supplies of imported lead antimonate yellow glass. Possible reasons for the switch from antimony to tin-based opacifiers within the Mediterranean region during the fourth century AD include either a breakdown in the supply of antimony or the transfer of this new technology from India. In either case, the introduction of tin-based opacifiers in the Germanic kingdoms and Scandinavia most probably represented an independent event. The choice of tin-based opacifiers almost certainly reflected the ready availability of tin, through its use in metal production. Furthermore, metalworking processes, involving the use of both lead and tin (e.g., leaded bronzes, soft solder, pewter), could have provided the context in which the lead stannate yellow colorant was first discovered.

Concurrently, in the Islamic world, tin oxide started to be used in the production of white opaque glazes in Abbasid Iraq from the ninth century AD [26]. From there, tin-opacified glazes spread throughout the Islamic world, including Islamic Spain (Figure 2.11) [27] from where the technology ultimately reached Italy in the 13th century AD, resulting in the beginnings of Italian majolica [28]. Finally tin-based opacifiers were used in the production of yellow and white enamels applied to Islamic glasses from about the 12th century AD) and Venetian glasses from about the 13th century AD [29].

Therefore the introduction of Islamic tin-opacified glazes (Figure 2.10) was probably the result of technological transfer from Byzantine glass producers. However, the Islamic potters went on to make significant developments to the opaque glaze technology, and seem to have taken precedence in the use of tin opacifiers over Islamic glassworkers.

Sometime during the fourth century AD, tin-based opacifiers started to replace antimony-based opacifiers in glass produced from the eastern Mediterranean through into northern Europe. In the eastern Mediterranean, the earliest published examples are the yellow and flesh-colored glass pieces in opus sectile panels (Thomas Panel and panel from Kenchreai in Greece) that were probably produced in Egypt and date from the fourth century AD. Then, from the fifth century AD, tin-based opacifiers were used in the production of mosaic tesserae in Roman Spain (identified by XRD), and in Byzantine contexts in Israel and Greece. Tin-based opacifiers were also used in Italy from the fifth century AD onwards, but at the same time, antimony-based opacifiers continued to be used, or reused, until the 13th century [20, 25].

Figure 2.10. White of cassiterite and green of copper glazed tiles of Mudejar Torre de San Pedro (Teruel, Spain, XIIIth century).

References

[1] Rashed, R., "A pioneer in anaclastics: Ibn Sahl on burning mirrors and lenses." *Isis*. 81 (3) (1990): 464–491. doi: 10.1086/355456.

[2] Kwan, A., Dudley, J., Lantz, E., "Who really discovered Snell's law?." *Physics World*, 15 (4) (2002): 64. doi: 10.1088/2058-7058/15/4/44.

[3] Khan J., Farooqui M., Quadri S. H., "Verification of the molar refraction as an additive and constitutive property of binary liquid mixtures of water-ethanol and benzeneethanol," *Rasyan J. Chem.*, 4(4) (2011) 944-946.

[4] Goldberg, Emanuel, Goldberg, Herbert: 1956. US. Pat. US000002768553A. *Refractometers*.

[5] Fox M., *Optical Properties of Solids*, (Oxford: Volum 3 de Oxford Master Series in Physics, 2010)6.

[6] Bhoren C. F., Huffman D. R., *Absorption and Scattering of Light by Small Particles* (Wiley, New York 1998).

[7] Kerker, M., *The scattering of light and other electromagnetic radiation*. (Academic, New York, 1969).

[8] Lord Rayleigh F. R. S., "XXXIV. "On the transmission of light through an atmosphere containing small particles in suspension, and on the origin of the blue of the sky," *The London, Edinburgh, and Dublin Philosophical Magazine and Journal of Science,* 47(1899) 375-384.

[9] Maxwell J. C., "Molecules," *Nature*, VIII (1873)440.

[10] Mie, G. "Beiträge zur Optik trüber Medien, speziell kolloidaler Metallösungen [Contributions to the optics of cloudy media, especially colloidal metal solutions]." *Annalen der Physik*. 330(3) (1908) 377–445.

[11] Greenler, R., *Rainbows, Halos, and Glories*, (Cambridge University press, Cambridge, 1980).

[12] Singh, S., "Refractive Index Measurement and its Applications," *Physica Scripta*, 65(2002)167-180.

[13] Born, M., Wolf, F., *Principles of Optics*, (Cambridge University Press, Cambridge, 1999).

[14] Kozlova, N. S., Busanov, O. A., Zabelina, E. V., Kozlova, A. P., Kasimova, V. M., "Optical properties and refractive indices of $Gd_3Al_2Ga_3O_{12}$:Ce^{3+}crystals," *Crystallography Reports*, 61, 3 (2016): 474-478.

[15] Tolosa, A., Alcón, N., Sanmiguel, F., Ruiz, O., "Method of determining the optical properties of ceramics and ceramic pigments: measurement of the refractive index," *Qualicer 12*, O23(2012)1-10.

[16] Dondi, M., Eppler, R. A., *Ceramic colorants*, (Ullmann's Encyclopedia of Industrial Chemistry, 2000, 1-18).

[17] Fernández-Navarro, J. M., *El Vidrio* [*Glass*], CSIC. Sociedad Española de Cerámica y Vidrio, 2003.

[18] Molina, G., *Colour and Technology in historic decorated glazes and glasses*, Ph. Thesis, Barcelona, 2014.

[19] Boschetti, C., Leonell, C., Rosa, R., Romagnoli, M., Valero, M. A., Schibille, N., "Preliminary Thermal Investigations of Calcium Antimonate Opacified White Glass Tesserae," *Heritage*, 3 (2020): 549–560.

[20] Shortland, A. J., "The use and origin of antimonate colorants in early Egyptian glass." *Archaeometry* 44, (2002):517–530.

[21] Tite, M., Pradell, T., Shortland, A., "Discovery, production and use of tin-based opacifiers in glasses, enamels and glazes from the late iron age onwards: A reassessment." *Archaeometry*, 50(2008): 67–84.

[22] Biringuccio, V., *The Pirotechnia of Vannoccio Biringuccio* (C. S. Smith, M. T. Gnudi, Trans.). The MIT Press, Cambridge, MA., 1966.

[23] Cerro, S., Llusar, M., Gargori, C., Monrós, G., "Cool and photocatalytic yellow ceramic pigments; from lead-tin to Cr doped scheelite pigments," *Ceramics International* 45 (2019) 4613–4625.

[24] Dik, J., *Scientific analysis of historical paint and the implications for art history and art conservation: the case studies of Naples yellow and discoloured smalt*, Thesis, Faculty of Sciences, University of Amsterdam, (2003)115–120.

[25] Dik, J., Hermens, E., Peschar, R., Schenk, H., "Early production recipes for lead antimonate yellow in Italian art," *Archaeometry*, 47 (2005):593-607.

[26] Fiori, C., Vandini, M., and Mazzotti, V., 2004, *I colori del vetro antico – Vetro musivo Bizantino* [*The colors of ancient glass - Byzantine mosaic glass*], Il prato Ed., Saonara (PD), 2004.

[27] Mason, R. B., Tite, M. S., The beginnings of tin-opacification of pottery glazes, *Archaeometry*, 39 (1997):41–58.

[28] Molera, J., Pradell, T., Salvadó, N., and Vendrell-Saz, M., "Evidence of tin oxide recrystallization in opacified lead glazes," *Journal of the American Ceramic Society*, 82 (1999): 2871–5.

[29] Tite, M. S., Technological investigations of Italian Renaissance ceramics, in *Italian Renaissance pottery* (ed.T. Wilson), 280–5, British Museum Press, London, 1991.

[30] Freestone, I. C., and Stapleton, C. P., Composition and technology of Islamic enamelled glass of the thirteenth and fourteenth centuries, in *Gilded and enamelled glass from the Middle East* (ed. R. Ward), 122–8, British Museum Press, London, 1998.

Chapter 3

CERAMIC PIGMENTS: COOL AND PHOTOCATALYTIC PIGMENTS

3.1. CONCEPT OF CERAMIC PIGMENT

The so-called eco-friendly cool pigments are being studied in recent years, although their evaluation remains in many cases ambiguous and debatable because not always an intense colored powder becomes a "pigment." This kind of pigments must fulfil two fundamental characteristics: a) to behave as true pigments, being powders with acceptable colorimetric values when introduced in polymeric, vitreous or ceramic matrices, b) to show high reflectance in the NIR region of the spectrum (700-2500 nm).

3.1.1. Pigment Definition and Color Evaluation

In the case of pigment characteristics, colored powders are classified by CPMA in three categories [1, 2]:

- *Category A* deals with pigments suspended in glass matrices which require the highest degree of heat stability and chemical resistance to withstand the attack of molten glass.
- *Category B* deals with pigments suspended in plastics and other polymers, which require only moderate heat stability.
- *Category* C deals with pigments suspended in liquid vehicles, which require little, or no heat stability.

Highly colored powders are often unstable and classified in Category C indicating a poor pigmenting capacity. On the other hand, color is a subjective perception due to the sensitivity of the cone photoreceptors in human eye to the light (L long cones sensible to red color, M middle to green and S short to blue light). Physically, we can characterize a color by its spectral reflectance distribution (its optical spectra), but sometimes we can reproduce a color using different spectral distributions if the cone response is the same as the original distribution (this phenomenon is called metamerism).

In 1853, Grassmann published a theory of how colors mix; its three color laws are still taught, as Grassmann's laws [3]. A "color model" is an abstract mathematical model describing the way colors can be represented as tuples of numbers (e.g., triples in RGB or quadruples in CMYK); however, a color model with no associated mapping function to an absolute color space is a more or less arbitrary color system with no connection to any globally understood system of color interpretation. In effect, a “color space” is a specific organization of colors in combination with color profiling supported by various physical devices, and supports reproducible representations of color, whether such representation entails an analog or a digital representation. A color space may be arbitrary; i.e., with physically realized colors assigned to a set of physical color swatches with corresponding assigned color names (including discrete numbers) as in the Pantone collection. In this sense Hex numbers are the hexadecimal notation for the combination of Red, Green, and Blue color values of the RGB model. The first two numbers are red, second two are green, and last two are blue. The two numbers designate a strength of value from 0 to 255: 00 being black (no color), and FF being white (full color). Hence, #000000 is pure black and #FFFFFF is pure white. #FF0000 is hex red, #00FF00 is hex green, and #0000FF is hex blue, cyan (mixture of blue and green lights) is #00FFFF.

As above cited, the main color models are the RGB and CMYK (Figure 3.1.a):

- RGB model uses additive color mixing (Red, Green, and Blue); his means that this color model is based on adding and mixing light, when Red, Green, and Blue lights are mixed, a pure white is obtained. It is understood that white is truly the combination of all colors and black is the absence of color. RGB stores individual values for red, green and blue. RGBA is RGB with an additional channel, alpha, to indicate transparency. Common color spaces based on the RGB model include sRGB, Adobe RGB, ProPhoto RGB, scRGB, and CIE RGB.
- CMYK uses subtractive color mixing (Cyan, Magenta, and Yellow); a subtractive color model works exactly opposite from an additive color model (RGB), it works by partially or entirely masking certain colors from white. So instead of adding light to achieve a color, CMYK is using ink to subtract brightness from white. Therefore, Cyan, Magenta, and Yellow combined is black, but Key (Black) is being used to additionally subtract brightness in printing (producing deeper darker tones) and heavily reduced ink usage of the other tones. CMYK stores ink values for cyan, magenta, yellow and black. There are many CMYK color spaces for different sets of inks, substrates, and press characteristics (which change the dot gain or transfer function for each ink and thus change the appearance).

There are different methods for an "objective" measurement of color such as the developed by CIE ("Comission International de l´Eclairage") [3]. Color Matching Functions (CMF) such as RGB or XYZ CIE color functions deal with the accurate reproducibility of color; they tell us if one color looks like another but not what they look like. Color Appearance Models (CAM) seek to solve the much more difficult problem of quantifying colors in dimensions that are meaningful and intuitive to humans. They use dimensions like hue, saturation and brightness.

The CIEL*a*b* [3] colorimetric method referred to a standard lighting (e.g., D65) and an observer's field of view (*e.g.* 10º) uses three parameters: L* measures the lightness (100=white, 0=black) and a* and b* the chroma (-a*=green, +a*=red, -b*=blue, +b*=yellow).

Similarly to CIEL*a*b*, the CIEL*C*h color space (correlates well with human eye color perception. It has the same diagram as the L*a*b* color space but uses cylindrical coordinates instead of rectangular coordinates. In this color space, L* indicates lightness like in CIEL*a*b* model, C* represents chroma, and h* is the hue angle. The value of chroma C* is the distance from the lightness axis (L*) and starts at 0 in the center. Hue angle starts at the +a* axis and is expressed in degrees (e.g. 0° is +a* axis, or red, and 90° is +b axis, or yellow). The values of C* and h* can be estimated from a* and b* parameters by the equations 3.1 and 3.2 respectively:

$$C^* = (a^2 + b^2)^{1/2} \tag{3.1}$$

$$h^* = \arctan (b^*/a^*) \tag{3.2}$$

Color Appearance Models (CAM) based on hue, lightness and relative chroma first appeared in the early 19th century, but the key concept of absolute chroma was devised by the American artist Albert Munsell [4]. Munsell gives a classification of colors of objects according to dimensions of hue, lightness value (or greyscale value) and chroma or relative chroma (often loosely referred to as "saturation"):

- Hue refers to the circular scale of "pure" or "saturated" colors formed by the colors seen in the spectrum (red, orange, yellow, green, cyan, blue and violet), together with the non-spectral colors like magenta, seen when the two ends of the spectrum are mixed.

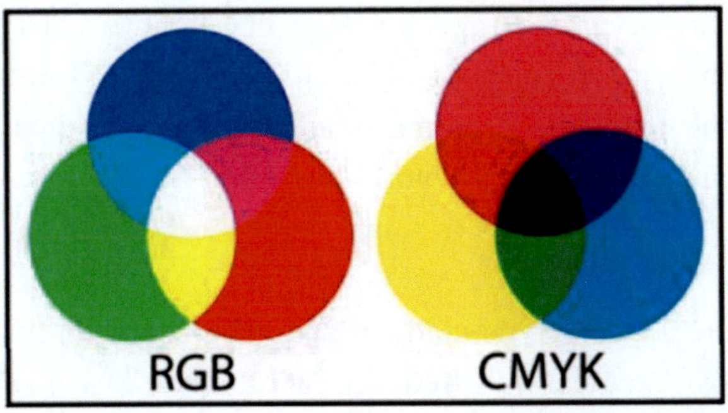

a

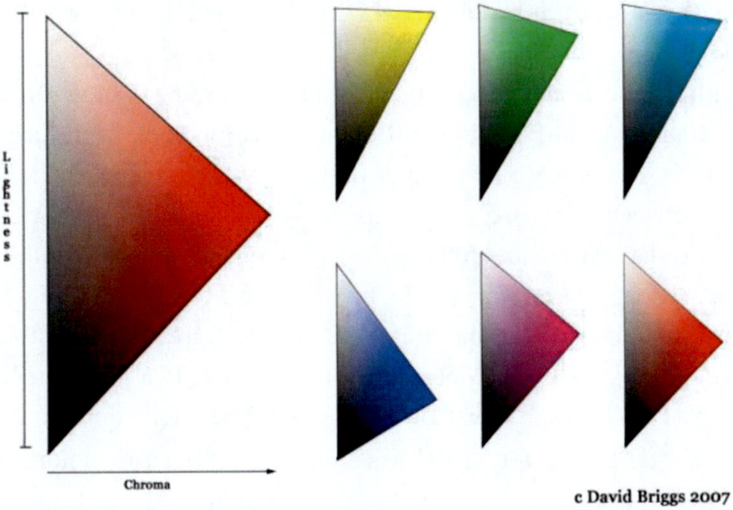

b

Figure 3.1. (a) Color models, (b) Diagram of all possible variations of a single hue of colors, showing variations in the range of possible chroma with lightness [5] (http://www.huevaluechroma.com/082.php).

- Lightness refers to the scale from black to white; tone value and greyscale value are synonyms or very closely related.
- Chroma or color strength refers to the amount of visual difference from a grey of the same value. Because the maximum chroma attainable with any set of paints varies with value, a hue page that shows all variations of value and absolute chroma attainable for a given hue has an irregular right margin (Figure 3.1) [4], and because

> this margin varies for each hue, a hue-lightness-absolute chroma space has an irregular, tree-like shape. Simpler systems arbitrarily treat the maximum chroma attainable for each hue as uniform, resulting in a circular "color wheel" and a cylindrical color space [5].

From Figure 3.1 can be pointed out that the optimal chroma is obtained when Lightness L* (measured by the L*a*b* method) is high for yellow (L*=80-90), cyan (L*=75-85) and green (L*=70-80) hues, middle for magenta (L*=50-60) and red (L*=45-55) hues and finally low for blue (L*=25-35) hues.

Finally, another advanced topic is that regarding Color Difference Models (CDM), quantifying how different two colors look from each other, so that color spaces with perceptual uniformity are created. In these, the distance between any two colors in the coordinate space mirrors the perceived visual difference in the colors. One of these models is the CMC tolerance system based on CIELCh which provides better agreement between visual assessment and measured color difference. CMC tolerance was developed by the Color Measurement Committee (CMC) of the Society of Dyers and Colorists in Great Britain and became public domain in 1988. The tolerance ΔE^*_{CMC} for visual assessment between two colors is a 3-dimensional ellipsoid with axes corresponding to hue (Δh*), chroma (ΔC*) and lightness (ΔL*) that is larger for chromatic colors (high C*>10), while in the orange area of color space the ellipsoids are longer and narrower than the broader and rounder ones in the green area. The CMC tolerance system is specified in the ISO International Standard 105-J03 (part J03: calculation of color differences) [6] but other standards consider the tolerance ΔE* based on L*a*b* parameters:

$$\Delta E^* = \sqrt{\Delta L*^2 + \Delta a*^2 + \Delta b*^2} \quad (3.3)$$

ISO 12642-25 (referred to graphic technology to characterize 4-color printing process) considers unacceptable a difference ΔE*>5. Instead, the International Standard ISO 10545-16 for ceramic tiles stablishes a limit of

$\Delta E^*<0.75$, concerning the determination of small color differences between plain colored ceramic tiles, which are designed to be of uniform and consistent color. The several and usual color spaces such as Lab, Rgb, Cmy, Cmyk, Hsl, Xyz, Lab, Lch and Yxy can converted using color converter systems (e.g., http://colormine.org/color-converter).

In summary, in order to consider a practical evaluation of color based on CIEL*a*b* and Munsell considerations it can be assumed that:

a) The yield of color should be high; then values L* around 85 for yellow-cyan hues, 75 for green, 55 red-brown-magenta and 30 for blue-black hues can be considered optimal. A low L * value indicates a dark color and a high one a pale color.
b) The hue should be measured by high positive a* for red hue, high negative a* for green hue, high positive b* for yellow hue and high negative a* for blue hue.
c) The chroma C* (eq. 3.1) should be high: e.g., an L*a*b*=45/20/15 is a dark brown color.

3.2. Cool Pigments

3.2.1. Cool Powders

Cool pigments show high reflectance in the NIR range (700-2500 nm). When a light falls on an object, both diffuse and specular reflection operate simultaneously by quite distinct processes, since the specular reflection typically retains the color of the light source, while the diffuse reflection is often colored by the material (see section 2.2).

Diffuse reflectance is considered to result from subsurface scattering; that is, light enters a surface layer, is scattered multiple times, and then exits in a random direction [7] (Figure 2.4). The exact proportion of the two kinds of reflection depends essentially on the type of material (metals exhibit high specular and no diffuse reflection), and on the angle of incidence (light striking at a low angle to a surface is more likely to be reflected specularly), but not on the roughness of the surface. Polishing a rough surface will make a "fuzzy" specular reflection more concentrated by reducing variation in microfacet orientation, but will not necessarily increase the total amount of specular reflection unless it involves removal of a diffuse-reflecting coating.

The spectrometers for collection of UV-Vis-NIR diffuse reflectance spectra (see Chapter 2) give the data in absorbance (A) using arbitrary units or in reflectance (R) using the Kubelka-Munk model. The Kubelka-Munk (K-M) model has a particularly simple solution in the case of semi-infinite samples such as in reflection of radiant energy at boundary surfaces of matte (dull) scattering surfaces. All the geometric peculiarities of an inhomogeneous sample are condensed into a single parameter, the scattering coefficient s. The diffuse reflectance $R_\propto$ is given as [8]:

$$R_\propto = 1 + \frac{k}{s} - \sqrt{\frac{k}{s} - (2 + \frac{k}{s})} \qquad (3.4)$$

Where k is the absorption coefficient of the sample. This relatively simple form is easily solved for k/s yielding the familiar K-M remission function:

$$A = \frac{k}{s} = \frac{(1 - R_\infty)^2}{2R_\infty} \qquad (3.5)$$

The K-M transform of the measured spectroscopic observable is approximately proportional to the absorption coefficient and hence is approximately proportional to the absorption, and then to the concentration of absorbent species which absorb light associated to charge transfer and d-d transitions of metal ions in lattice environments or also to stretching and deformation modes of functional groups, vibrations of supported species [9]. The scattering coefficient was introduced into the theoretical description of diffuse reflection as a semi-empirical parameter to account for the internal scattering processes. The scattering coefficient s is, in fact, dominated by particle size and refractive index of the sample. It is not a strong function of the wavelength or the absorption coefficient, so the K-M model considers it a constant.

Therefore in order to achieve cool pigments, the following factors should be considered for powders:

a) Absence of absorbent species in the NIR range (700-2500 nm.): e.g. Co^{2+} in tetrahedral environment shows a strong absorbance in the 1200-1600 nm (due to the coupling of d-d transitions bands $^4A_2(F) \rightarrow ^4T_1(F)$ at 1400 nm and $^4A_2(F) \rightarrow ^4T_2(F)$ at 1600 nm) which difficult its application as NIR reflector [10]. In this sense, a good NIR reflector is a semiconductor with relatively high band gap such as TiO_2 (2.5 eV) or CdS, ZnO and $SrTiO_3$ (3.4 eV), with an intense charge transfer band located at UV range that allows the pigment block the UV radiation and protect the organic binders.
b) High thermal emittance. Thermal emittance is the measure of a material ability to release heat that it has absorbed and therefore its temperature is moderated [11]. The emittance index of solids is high (e.g. ceramics 0.9, asphalt 0.88) except in the case of conductive metals (e.g., Ag 0.02, Al 0.03, Cu 0.04).

3.2.2. Cool Coatings with Cool Pigments

Obviously a pigment should have high pigmenting yield when dispersed in the matrices (polymers, glass, ceramics…). In the interaction pigment-medium it is important to note the following characteristics of the pigment:

a) High refractive index that allows achieving high scattering efficiency of light when the pigment is added to a matrix with very different refractive index. The scattering efficiency increases with the difference of the refractive index pigment-medium. The polymorphs of TiO_2 anatase and rutile show high refractive index (2.55 and 2.73 respectively), then pigments based on rutile lattice such as ochre of Cr-Sb rutile or yellow of Ni-Sb rutile are two of the best classical cool pigments. Related pigments such as Cr-pseudobrookite (Cr-Fe_2TiO_5) [12], Cr-armalcolite (Cr-

($MgFe_2Ti_3O_5$) [13] or Ni-geikielite (Ni-$MgTiO_3$) [14] based on titanates also show high NIR reflectance.

b) Particle size should be about half the wavelength of the light, then for NIR light (700-2500 nm) the particle diameter should be between 350-1250 nm. But it is important to note that, for a similar weight loading of a pigment with high particle size in the range of NIR spectrum (500-1500 nm) and of a fine pigment with lower particle size (100-500 nm), the surface area and number of particles is higher for the fine pigment, which probably compensates the low scattering of dissimilar size, and therefore it can show higher NIR reflectance. Usually, reducing the particle size to a size equivalent to the wavelength is said to decrease the proportion of specular reflection of light and maximize the scattering efficiency. The particle shape and packing status are other important factors, in addition to the particle size.

In summary, opacity or hiding power of paints is the ability to hide the surface of an object. The hiding power of paint measures its ability to obscure a background of contrasting color of the coating. The opacity of an ideal coated cool pigment should be complete, namely the transmittance of the coating should be naught. Opacity is dependent of the above characteristics: absence of absorbance bands in NIR range, high scattering associated to the high miss-match between refractive index of pigment and medium and adequate particle size of the opacifier pigment. Besides, the opacity increases with the thickness of the coating and, thus, the thickness should be similar when comparing the reflectance of coatings.

Therefore, high opacity of a coating is associated with low transmittance of the light through the coating layer. A simple way to evaluate the utility of a pigmented coating for "cool'' applications is to consider its NIR absorbance and NIR transmittance. If the NIR absorbance is low, the pigment is cool. However, a cool pigment that has high NIR transmittance will require a NIR-reflective background (typically white or metallic) to produce a NIR-reflecting coating. In this context the nature of the background of the coating modifies the reflectance index of the coating; *e.g.*

in 7% in volume in alkyd paint the rutile white pigment changes from R=0.88 over a white background to 0.43-0.64 over a black background, the nickel titanate yellow from 0.77-0.85 to 0.42-0.64 and the chromium oxide green from 0.50-0.57 to 0.33-0.40 [15]. Materials that are opaque even on a microscopic scale have no diffuse reflection, and appear either black, sub-metallic (*e.g.* graphite) or metallic, depending on the amount of specular reflection they exhibit. The amount of specular reflection generally appears to be greater on dark objects than on white objects of similar composition, but this is an illusion caused by swamping of the specular reflection by the greater diffuse reflection from the latter.

According to Akbary et al. [15] the classical pigments analyzed with high NIR reflectance (measured in films of the order of 11-30 μm thick in PVC applications) are: a) White: TiO_2 rutile, a strongly scattering, weakly absorbing, stable, inert, nontoxic, inexpensive, and hence extremely popular white pigment, b) Black: inorganic black of Cr-hematite solid solution and organic Perylene black pigment that absorbs strongly in the visible and very weakly in the NIR (unusable in ceramic matrices and its nature as PAH could induce negative environmental effects), c) Brown: Fe_2O_3 hematite, Fe_2TiO_4 and Cr-$ZnFe_2O_4$ (franklinite) spinels and Mn,Sb-TiO_2 manganese antimony titanium buff rutile, d) Blue: ultramarine blue ($Na_{7.5}Si6Al_6O_{24}S_{4.5}$) that owe the blue color to the stabilization of isolated S_3 molecules (the metal oxide skeleton forms an open clathrate sodalite structure that stabilizes S^{-3} ions in cages to form the chromophores), copper phthalocyanine, blue cobalt aluminate $CoAl_2O_4$ and cobalt chromite $Co[Al,Cr]_2O_4$ spinels (despite strong absorbance in the 1200-1600 nm), e) Green: chlorinated copper phthalocyanine green, Cr_2O_3 and cobalt titanate green $CoTiO_3$, f) Red-orange: organic red pigments such as acra burnt orange, acra red, monastral red, naphthol red and inorganic cadmium sulfoselenide $Cd(S_{1-x}Se_x)$ red (high x) or chrome titanate ochre Cr,Sb-TiO_2, g) Yellow: organic yellows such as Hansa yellow and diarylide yellow and inorganic $Cd(S_{1-x}Se_x)$ yellow (low x) or nickel titanate yellow Cr,Sb- TiO_2. Many of these pigments are not stable in ceramic matrices and show high toxicity associated to presence of toxic ions (Cr(VI), Cd(II), Se^{2-}, Pb(II)…).

3.2.3. Solar Reflectance R and SRI (Solar Reflectance Index)

The total solar reflectance R or the solar reflectance in the NIR range R_{NIR} are evaluated from UV-Vis-NIR spectra through diffuse reflectance technique as the integral of the measured spectral reflectance and the solar irradiance divided by the integral of the solar irradiance in the range of 350 to 2500 nm for R or 700 to 2500 nm for R_{NIR}, as in the equation 3.6:

$$R = \frac{\int_{350}^{2500} r(\lambda)i(\lambda)d\lambda}{\int_{350}^{2500} i(\lambda)d\lambda} \qquad (3.6)$$

Where, (λ) is the spectral reflectance (Wm^{-2}) measured from UV-Vis-NIR spectroscopy and $\boldsymbol{i}(\lambda)$ is the standard solar irradiation ($Wm^{-2}nm^{-1}$) according to the American Society for Testing and Materials (ASTM) Standard G173-03.

Buildings are responsible of about the 36% of primary energy consumption and 30% of greenhouse gas emissions in U.S. Therefore, strategies such as Leadership in Energy and Environmental Design (LEED) certification, promoted by the U.S. Green Building Council, analyze the sustainability of buildings considering factors associated with their design and construction such as the content of recycled material in their composition, the use of regional raw materials for their manufacture, innovation and design (functionality) and the heat island effect (HIE). Regarding the heat island effect, the determination of the Solar Reflectance Index (SRI) for opaque materials (wall, roof and floor) aids the LEED certification. SRI reveals the ability of a surface to reflect solar radiation and to emit thermal radiation, reducing the increase in temperature caused by the incidence of solar radiation on the surface. The SRI of a specific surface changes with the solar reflectance and thermal emittance of the materials discussed in section 3.2.1, and also according to certain environmental conditions, such as the solar flux, the convection coefficient, the air temperature and the sky temperature (ASTM E1980 2011). In the case of roof tiles for buildings, the SRI value required for a specific material is dependent on the slope of the roof (SRI≥39 for steep slope; SRI≥82 for low slope). The maximum temperature reached on a sun-irradiated surface T_s is also used to estimate the reflectance power of this surface [15].

3.2.4. Rare Earths and Cool Pigments

Electronic configuration of the valence layer of the trivalent rare earth ions, $[Xe]4f^n$, involves the activity of internal f electrons, strongly protected by 5s and 5p electrons. Because of the shielding effect of the outer 5s and 5p shell electrons, the crystal-field interaction with inner 4f electrons is weak and can be treated as a perturbation (Stark effect) of the free-ions states. Accordingly, the energies of the corresponding levels of $4f^n$ configuration are only weakly sensitive to the type of the crystal host. Therefore, the resulting crystalline field caused by their interaction with neighboring ions is very low and usually optical spectra of lanthanide compounds present weak and profuse bands. On the other hand, the 5d electrons are not effectively shielded by other electrons, and the crystal field influence on the energy levels of $4f^{n-1}5d^1$ electronic configuration is strong. Accordingly, crystal field splitting of 5d levels is large and the energies of levels within $4f^{n-1}5d$ electronic configuration can strongly differ for different crystal hosts.

In the case of rare earth based pigments, the mechanism of the color is based on the shift of the charge transfer band of a semiconductor to higher wavelengths, by introducing an additional electronic level through doping with a chromophore rare earth ion. For example, CeO_2 crystallizes in the fluorite structure, where FCC packed Ce^{4+} ions are surrounded by eight oxide ions, occupying alternate centers of tetrahedral cavities in the FCC lattice. The 4f valence shell of Ce^{4+}([Xe]) in cerianite is empty, and that of O^{2-} ($[He]2s^2\ 2p^6$) is full: adjacent Ce^{4+} ions are virtually in contact in the fluorite lattice and, as a result, 4f orbitals overlap in a cationic conduction band; similarly, overlap of 2p orbitals of oxygen ions gives to an anionic valence band. The band gap between the anionic band and the cationic band is 2.76 eV [16]. Pr^{4+} doping into cerianite introduces an additional conduction band and its associated charge transference band absorbs from 570 nm (band gap 2.2 eV) inducing a red color; since all wavelengths above

570 nm are reflected, except the weak and sharp absorption due to 4f transitions; therefore $Pr^{4+}:CeO_2$ is a good NIR reflector material.

Recently, the industrial utilization of lanthanides has increased rapidly because of their low toxicity, although some hazardous effects have been detected in the literature [16, 17]; consequently, a large number of rare earth based NIR reflective pigments have been proposed as alternative to traditional pigments:

a) For blue pigments Smith et al. [18, 19] synthesized by ceramic route at 1200ºC the solid solution Mn-$YInO_3$ with perovskite structure ($YIn_{0.8}Mn_{0.2}O_3$ powder shows L*a*b*=34.1/11.7/-44.4 and R=34%, and 20% in vinyl paint shows 77,9/-2,9/-19,3 with R_{NIR}=62%). Sheethu et al. and Zhang et al. [20, 21] describe blue pigments based on the solid solutions of La,Li and Eu respectively in the ancient based Egyptian blue of wesselsite $SrCuSi_4O_{10}$, ($Sr_{0,3}La_{0,7}Cu_{0,3}Li_{0,7}Si_4O_{10}$ powder shows L*a*b* = 49.6/10.4/-51.7 and R_{NIR}=66% obtained by ceramic route at 950ºC/16h, while $Sr_{0,8}Eu_{0,2}Cu_{0,3}Si_4O_{10}$ shows L*a*b*=46.6/-4.6/-28.1 and R_{NIR}=72.3 synthesized by citrate route at 900ºC/2h).
b) For magenta-brown shades García et al. [16, 17] describe the application of brown-red pigments based on $Ce_{1-x}M_xO_2$ solid solutions (M=Pr,Tb) showing R_{NIR}=82% for x=0.05. Vishnu et al. [22] report brown pigments by solid state reaction at 1500ºC/18h based on the solid solution Pr-$Y_2Ce_2O_7$ with fluorite structure ($Y_2Ce_{1,7}Pr_{0,3}O_7$ powder shows L*a*b*=32.6/18.1/12 and R_{NIR}=58%), and yellow pigments from Mo-$Y_2Ce_2O_7$ solid solutions ($Y_2Ce_{1,5}Mo_{0,5}O_7$ powder shows L*a*b*=90.2/-4.5/62.4 and R_{NIR}=81%); later, Raj et al. [23] report a brown pigment doping with Tb the same structure. Jovani et al. [24] report a brown pigment based on $Y_{2-x}Tb_xZr_{2-y}Fe_yO_{7-\delta}$ (L*a*b*=64/34/33 for 4wt% addition in a conventional frit of powder x=0.35).
c) For yellow pigments, Gargori et al. [25] report the preparation of a yellow pigment based on the codoped V^{5+}, Ca^{2+}-$Y_2Sn_2O_7$ pyrochlore ($Ca_xY_{2-x}V_ySn_{2-y}O_7$) solid solutions at 1200ºC/6h (the powder

x=y=0.16 shows L*a*b*= 85.7/1.1/34 and 5wt% enameled in a double firing frit at 1000ºC shows L*a*b*=79.3/2.6/35 with R_{NIR}=81%). Lili et al. [26] develop by citrate route at 700ºC/6h high NIR reflectance yellow pigments based on Al-$LaFeO_3$ ($LaAl_xFe_{1-x}O_3$) perovskite solid solutions: at low x the color is brown and at x>0.7 yellow (L*a*b*=71.2/2.8/10.0 with R_{NIR}=57%). Although the authors do not evaluate the NIR reflectance, the $(Ca_{1-x-y}Eu_xZn_y)_2Al_2SiO_{7+\delta}$ ($0\leq x\leq 0.15$; $0\leq y\leq 0.07$) solid solutions with gehlenite structure ($Ca_2Al_2SiO_7$) show high NIR reflectance $(Ca_{0.87}Eu_{0.10}Zn_{0.03})_2Al_2SiO_{7+\delta}$ pigment exhibited the most vivid yellow color (L*a*b*=74/0/76) [27]. The origin of the yellow color is the 4f-5d transition of divalent Eu^{2+} ion, and codoping with Zn enhances the yellowness by increasing the Eu^{2+} amount in the samples. Schildhammer et al. report a yellow pigment based on $Yb_6Mo_2O_{15}$ crystallizing in a trigonal structure by ceramic method at 1350ºC/20h with the color coordinates L*a*b*=88.7/6.6/55 and R_{NIR}=94%. Color variation from pale yellow to brick-red was realized by partial substitution of Yb^{3+} or Mo^{6+} with various metal ions [28].

d) Obtaining black pigments based on rare earths as chromophores is very complicated given that the origin of the color, as discussed above, is based on 4f-5d transitions or on the generation of additional charge transfer bands induced by entrance of rare earths ions in the network, because the absorptions associated with 4f transitions are sharp and weak. Then, rare earth chromophores only absorb in the blue-green range producing yellow or brown colors; blues are due to traditional chromophores in rare earth network such as Mn^{3+} in trigonal bipyramid [18] which could be obtained in other free-rare earth structures [20], or to weakly allowed bands associated to Cu^{2+} [20, 11] that connects with ancient pigments such as the Egyptian Blue Family [30, 31]. Therefore, to obtain a whole absorbent pigment of the visible range is virtually impossible using rare earths as chromophores, but there are black or dark brown pigments involving rare earth lattices: Gargori et al. describe a black pigment based on perovskite $CrNdO_3$, a green pigment that turns into black when it is modified by fluorides [32], probably associated to the high concentration of vacancies introduced by the dopants [33] (adding 2 wt.% BaF_2 and 8 wt.% MgF_2 at 1100ºC/3h the powder shows L*a*b*=43.4/2.4/4.0 but low R_{NIR}=4%). Likewise, the undoped $YInO_3$ shows a black color although the authors [28, 29] only measure the NIR reflectance of the Mn^{3+} doped blue pigment.

3.3. Photocatalytic Pigments

3.3.1. Pigments, Photocatalysis and Photolysis

Besides high tinctorial capacity and durability, pigments must show high opacity in order to efficiently cover the colored surface. In the case of paints, titania stands out by its high refractive index (around 2.73 for rutile and 2.55 for anatase) showing a pure withe color. Anatase was used as opacifier and bleaching agent since 1918 replacing classical bleaching agents such as white lead $(PbC0_3)_2Pb(OH)_2$ or the white lithopone (a equimolecular coprecipitate of $BaSO_4$ and ZnS). In 1940, rutile displaces anatase as opacifier in paints due to its higher scattering capacity, associated to its higher refractive index and above all the durability of the polymers of paints with rutile due to its lower photocatalytic activity than anatase.

In effect, photocatalysis is an advanced oxidation process (AOP) that use chemical agents (e.g., hydroxyl radical $HO^{\cdot}$) generated in the surface of a semiconductor in wet environments to oxidize organic pollutants. However, the UV component (300-380 nm) associated with other environment factors such as oxygen, ozone and other industrial pollutants produces the depolymerization of plastics by photochemical oxidation or photolysis. Both anatase and rutile absorbs the UV component or solar radiation protecting paints from its photolytic degradation but not from the photocatalytic degradation associated to semiconductor performance of anatase. In this sense, an efficient photocatalytic pigment for paints must to degrade the simple volatile organic compounds (VOCs) and/or nitrogen oxides (NO_x) but protecting the polymers from the photolytic degradation (UV blocking effect); this effect is not necessary when pigments acts in glass or ceramic matrices.

In the photocatalytic process, when a photocatalyst (with large absorption coefficients and high photochemical stability) is irradiated, chemical agents are generated in interaction with the solution medium that reacts with organic compounds in this medium. If photocatalytic agent is solved in the media the photocatalysis is called homogeneous photocatalysis, if the photocatalyst is in different phase than the media (usually a solid photocatalyst in liquid or gas media) it is called heterogeneous [34].

In the heterogeneous catalysis [35] the light is absorbed for the catalyst which shows a semiconductor behavior, promoting electrons from the valence band to conduction band and creating electron-vacant pairs (e^--v^+) that interacts with the medium producing highly reactive chemical species. The energy difference between the valence band and conduction band is called band gap (E_g): in photocatalysis, when a photon shows an energy hν that exceeds E_g, an electron e^- is promoted from the valence band to the conduction band leaving a vacancy v^+ in the valence band: in the case of anatase this energy is around 3.2 eV (Figure 3.2):

$$TiO_2 + h\nu \rightarrow TiO_2\,(e^- \text{-} v^+) \qquad (3.7)$$

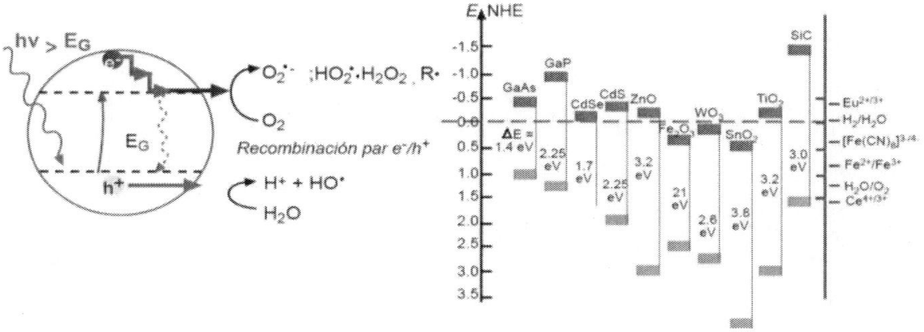

Figure 3.2. Heterogeneous photocatalysis in a semiconductor, active chemical agents production and band gap range of the usually photocatalytic oxides [35].

These electron-vacancy pair, reacts with molecules of the medium in contact (adsorbed) with the surface of the catalyst. The vacancies act as oxidant agents and the electrons as reducing agent over the medium molecules generating high reactive species such as hydroxyl radical ($HO^{\cdot}$), superoxide radical ($O_2^{\cdot -}$)...:

$$TiO_2\ (v^+) + H_2O\ (ads.) \rightarrow TiO_2 + HO^{\cdot}(ads.) + H^+ \quad (3.8)$$

$$TiO_2\ (e^-) + O_2\ (ads.) \rightarrow TiO_2 + O_2^-(ads.) \quad (3.9)$$

For example, the hydroxyl radical ($HO^{\cdot}$) or superoxide radical ($O_2^{\cdot -}$) generated in the photocatalysis process can oxidize the NO_x gases to soluble nitrates that are washed by rain acting as fertilizer:

$$HO^{\cdot}(ads.) + NO(ads.) \rightarrow HNO_2^{\cdot}(ads.) \quad (3.10)$$

$$HNO_2^{\cdot}(ads.) + HO^{\cdot}(ads.) \rightarrow NO_2(ads.) + H_2O \quad (3.11)$$

$$NO_2(ads.) + HO^{\cdot}(ads.) \rightarrow HNO_3 \quad (3.12)$$

$$NO(ads.) + 3HO^{\cdot}(ads.) \rightarrow HNO_3 + H_2O \quad (3.13)$$

or

$$NO(ads.) + O_2^{\cdot -}(ads.) \rightarrow NO_3^-(ads.) \quad (3.14)$$

In order to modulate the color and catalyst activity, modifiers and photosensitizer agents are added to the pigment structure. Modifiers introduced by solid solution alter the crystal structure of the pigment (they are added in molar proportion between 0.01 y 0.2 moles per formula weight. Photosensitizers modify the optical spectra of the pigment increasing its NIR reflectance or modifying its band gap.

3.3.2. Methods of Evaluation

For the evaluation of the catalyst activity above described, it is necessary to follow detailed assay's methods for its measurement, such as the following.

3.3.2.1. ISO 22197-1-2007

(part 1 NOx degradation, part 2 Removal of acetaldehyde and part 3 Removal of toluene) and ISO 10678:2010 (degradation of methylene blue) methods.

The norm from the International Organization for Standardization ISO 22197-1-2007 for fine ceramics (advanced ceramics, advanced technical ceramics) [36], is a test method for air-purification performance of semiconducting photocatalytic materials which consist in the introduction of a controlled flux of NO gas over standard plane tiles irradiated with UV light. The test analyzes the content of NO in the evolved flux when the light is on and when it is off.

The main experimental conditions of the test are summarized in Table 3.1 and the essential procedure in the test follows the following steps:

1) Placement of standard tiles (100x50 mm) in the reactor to be irradiated with the UVA radiation (λ=300-400 nm and irradiation power > 10 $W \cdot m^{-2}$) during at least 5 hours, then the tiles are washed with deionized water and dried to constant weight.
2) Verification of the uniform irradiation by the measure of the irradiation power in three different points of the tiles (the difference in the measurements must be lower than 0.5 $W \cdot m^{-2}$).
3) Verification of the flux of NO (1 ppm) during 10 min.
4) Flow the NO current without irradiation during at least 30 min.
5) Irradiation of the sample during at least 5h and record of $NO/NO_2/NO_X$ concentrations in the evolved flow.
6) Turn out the irradiation and registry $NO/NO_2/NO_X$ concentrations in the evolved flow.
7) Results are expressed as %NO eliminated during the test.

On the other hand the ISO standard 10678:2010, called "Determination of photocatalytic activity of surfaces in an aqueous medium by degradation of methylene blue" specifies a method for the determination of the photocatalytic activity of surfaces by degradation of the dye molecule methylene blue (MB) in aqueous solution using artificial ultraviolet (UV) radiation, and characterizes the ability of photoactive surfaces to degrade dissolved organic molecules on ultraviolet radiation [37].

Table 3.1. Main experimental conditions of the test ISO 22197 [36]

Characteristic	
Width of standard tile (mm)	49,5± 0,5
Length of standard tile (mm)	99,5± 0,5
NO concentration in air flow (ppm)	1.00± 0.05
Relative moisture at 25ºC ± 2.5ºC	50%
UVA (300-400 nm) irradiation power ($W\cdot m^{-2}$)	10,0
Distance between tile surface and glass window (mm)	5.0 ± 0-5
Total air flow (l/min)	3.00

The test method specified is also applicable to evaluation of the specific photocatalytic self-cleaning activity of surfaces covered with respective coatings. But this method is not applicable to characterizing the photoactivity of surfaces on visible illumination, regarding direct soiling, degradation of gaseous molecules and the determination of antimicrobial photoactivity of surfaces.

3.3.2.2. Methods of Elution: Nitrates Deposition and Orange II Photodegradation Test

Although the above described tests are the approved by ISO in order to measure the photocatalytic activity of tiles, there are several problems associated whit its application [38].

a) The used concentration of NO in the air flow (1 ppm) is excessive compared with the usual values of NO_x ($NO+NO_2$) in urban areas: e.g., EU stablish an NO_2 average hour maximum limit for annual periods of 40 $\mu g\ m^{-3}$ (i.e., 22 ppb) and the level of NO in town centre such as London present annual average of only 67 ppb.
b) The gas flow over the samples is relatively slow (0.2 ms^{-1}), therefore it is possible that the mass transference mask the photonic efficiency of the catalyst.

An alternative test to estimate the photocatalysis activity on NOx by oxidation to nitrates for air quality studies is the elution method. In this method the photocatalytic powder is added in an alkyd paint in the ratio paint:water:powder=7:2:5 which is brushed on an aluminum grid with a fixed blade with 15° of slope. The slide is placed over a funnel and irradiated by a UVA lamp (365 nm, 25W, 230V), then an air current of 12 l/min tested for NO-NO_2 [39] is suctioned during 48 h. (Figure 3.3.a). The slide is washed with deionized water and the first 20 ml of washing water are collected. Nitrates are finally measured in this washing water by spectrophotometry at 205 nm and the efficiency of nitrates deposition E(%) is estimated.

In the case of described ISO standard 10678:2010 for the determination of photocatalytic activity of surfaces in an aqueous medium by degradation of methylene blue (MB), the main underlying assumptions considered by the standard are, namely [40]: (i) dye purity, (ii) adsorption and pH, (iii) light source, (iv) stirring and diffusion, (v) reaction mechanism and (vi) kinetics. Possible sources of errors arising from these assumptions are identified and changes to the standard's protocol suggested. The major suggested changes are:

i. A source of MB of known and proven on site high purity (>90%) should be used to make up the standard test solution, which should have a referenced absorbance, 0.74, at 665 nm.

ii. The conditioning solution should be the same concentration as the standard test solution.

iii. The initial pH of the reaction solution should lie in the range: 5.5–6.0.

iv. A BLB (Black Light Blue) UVA light with a europium-doped strontium fluoroborate or borate phosphor should be used.

v. The reaction solution should be vigorously and continuously stirred if possible.

vi. The solution height should be the minimum recommended value of 2 cm, and the photoreactor cylinder 4.7 cm.

vii. After the 3 h irradiation the reaction solution should be left stirring overnight to check that no dye photoreductive bleaching has occurred.

Application of most of these suggestions should improve considerably the standard's reported current poor percentage errors for repeatability within a lab (9.2%) and reproducibility between labs (30.6%) [40].

As well the photocatalytic activity of powders can be estimated by the well-known degradation of the azo dye Orange II in aqueous solutions [35]. The photocatalytic tests were carried out adding a catalyst loading of 0.5 g/L to an Orange II solution of 0.6 10^{-4} mol/l, which has been prepared dissolving an amount of Orange II ($C_{16}H_{11}N_2NaO_4S$) in a pH 7.42 phosphate buffer media ($NaH_2PO_4.H_2O$ 3.31 g and $Na_2HPO_4.7H_2O$ 33.77 g solved in 1L of water). The UV irradiation source was a mercury lamp of 125 W emitting in lines at 254, 313 and 365 nm in a refrigerated quartz reactor (Figure 3.3.b). Before measuring, the suspensions were first stirred in the dark during 15 min to reach sorption/desorption equilibrium. The evolution of reaction was followed taking samples every 15 min. Orange II concentrations in the different sample were determined by means of UV-Vis spectroscopy at 485 nm. The commercial TiO_2 Degussa P25 was used as reference of comparison with the home prepared samples [41].

(a) (b)

(c)

Figure 3.3. (a) Assembly for NO_x photocatalytic test by oxidation to nitrates, (b) Quartz reactor for the Orange II degradation test, (c) Evolution of the intensity color of irradiated Orange II sample with the time of irradiation.

The resulting photodegradation curves are analyzed following the Langmuir-Hinshelwood model that follows the equation 3.15:

$$\ln \frac{c}{c_0} = kKt = K_{app} t \tag{3.15}$$

Where t is the time of irradiation, C is the concentration of colorant and C_0 the initial concentration of the colorant. The model follows a first order kinetic for the degradation with a kinetic constant K_{app}. The half-life time ($t_{1/2}$) can be calculated by equation 3.16:

$$\ln\frac{c_0/2}{c_0} = K_{app}t_{1/2} \qquad t_{1/2} = -\frac{\ln 2}{K_{app}} \tag{3.16}$$

3.4. Cool and Photocatalytic Pigments: Characteristics

From the above discussion, photocatalyst are semiconductors of wide band (E_g~1.5-3 eV): incident light with enough energy promotes electrons from the valence band of the semiconductor to its conduction band generating electron-vacant pairs that acts as reductor-oxidant generating active oxidants such as superoxide $O_2^{\cdot-}$ or $OH^{\cdot}$ radicals. The semiconductor appearance and the measure of the band gap E_g can be stand up by the UV-Vis-NIR absorbance spectra by diffuse reflectance [8] of the materials following the Kubelka-Munk model by Tauc method [42].

As described above, in literature are reported numerous examples of cool pigments or pigments with high NIR reflectance, but some have low possibilities of use in glazes and ceramics due to its low thermal stability. They are sometimes based on crystal structures that are not considered as pigment structures in the classic literature [1, 2]; they are calcined polyphase mixtures of semiconductor metal oxides (Cr_2O_3, CeO_2, TiO_2, MoO_3 ...), and rare earth elements are often used (Y, Ce, Eu ..) as structural or doping elements, which make materials more expensive without major SRI developments. Some companies have developed a line of cool pigments [43, 44]. However, there are classic pigments that have interesting characteristics such as cool pigments with photocatalytic activity. Gargori et al. [32] analyzed two CMY families of classical pigments, those based on zircon and rutile, as well as a non-classical family based on magnesium geikielite $MgTiO_3$. From the obtained results, the best photocatalytic and cooling pigments enameled in a conventional double firing glaze of the $ZnO-SiO_2$-CaO system that matures at 1080ºC is the CMY palette based on vanadium turquoise-zircon (total solar reflectance R=48% in the Vis-NIR range and half-life period $t_{1/2}$ = 117 min for Orange II), iron-ilmenite magenta (R=67% and $t_{1/2}$ = 493 min) or iron-zircon (Rnir=65% and $t_{1/2}$ = 145 min.) and yellow of Ni,Sb-rutile (Rnir=66% and $t_{1/2}$ = 431 min) or Pr-zircon (Rnir=78% and $t_{1/2}$ = 184 min.) respectively. Only the iron-doped form of ilmenite geikielite $MgTiO_3$ obtains comparable results to classical zircon or rutile pigments, but in this case it is actually $MgTi_2O_5$ pseudobrookite with residual rutile.

In general, and according to the literature, photocatalytic pigments with high reflectance capacity shows the following characteristics.

3.4.1. Cool Pigments Are Inorganic Pigments

Figure 3.4 shows some organic pigments (copper phthalocyanine, acra red, diarylide yellow) and carbon black (10wt% in alkyd paint), which show high NIR reflectance except carbon black. However, inorganic pigments [1, 2] give very uniform colorations in paints and glazes; they are very stable against sunlight, maintaining the color in outdoor applications, also against temperature (which makes them suitable, in general, for glazed matrices and ceramics). They also allow special applications such as anti-corrosion paints (zinc molybdate or barium metaborate, ecological alternatives to the basic zinc or lead chromates). Finally, it is not possible to obtain pure whites or blacks with organic pigments. Instead titanium oxide (in its anatase or rutile polymorphs) is widely used in white paints given its chromatic purity, its modular particle size to very powders fine and its high opacity and staining power, which allows high performances in covering surfaces paints [45, 46, 47]. Pure black is obtained with inorganic carbon black or spinels such as $(CoMn)(FeCr)_2O_4$.

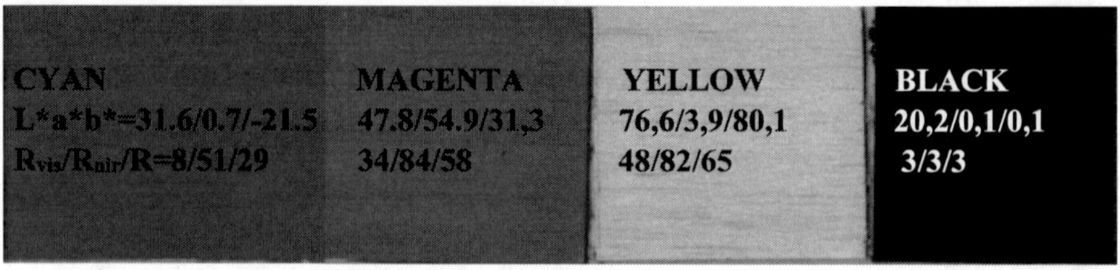

a

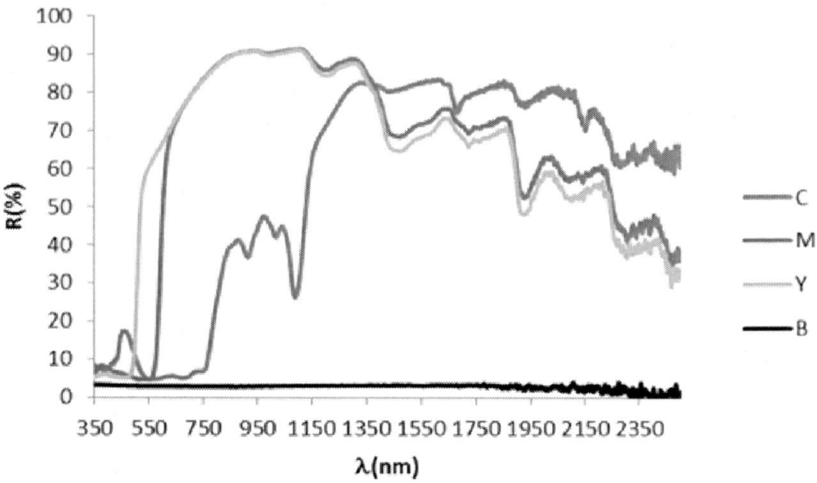

b

Figure 3.4. Organic-based pigments and carbon black powder: (a) in 10% alkyd paint with its CIEL*a* b* values, (b) UV-Vis-NIR diffuse reflectance spectra of samples (a).

3.4.2. Composite Effect

Cool pigments have a composite nature at a submicron level, with synergistic interactions between the components (composite effect). Obtaining this effect makes it essential to calcine the components and grind the sintered material to a suitable particle size. The use of non-conventional methods such as MOD or Sol-Gel is very interesting in obtaining this effect. Figure 3.5 shows the absorption spectrum of a cool pigment based on the pink coral of iron in zircon. X-ray diffraction indicates that in this material the phase is not unique and both quartz and residual monoclinic zirconia appear with relevant peaks, with reflectance capacity in NIR range.

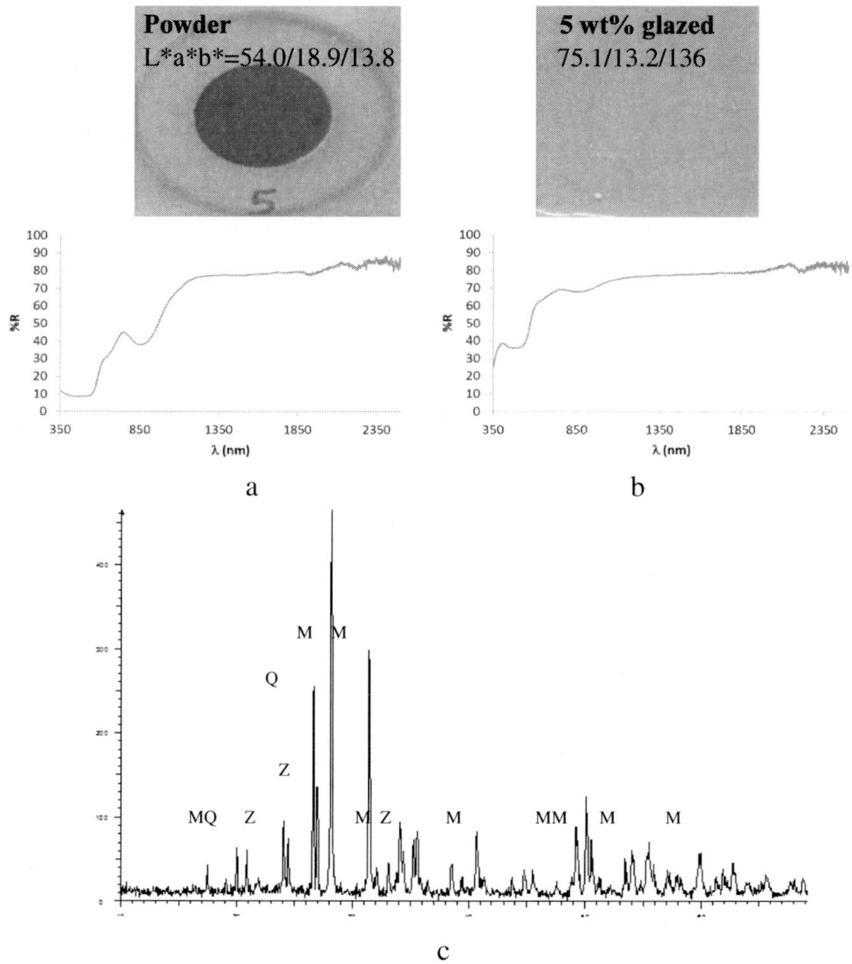

Figure 3.5. Absorption spectrum of a cool pigment based on the pink coral of iron in zircon ((a) powder and (b) glazed in a double firing glaze) and (c) X-ray diffraction of the powder: CRYSTALLINE PHASES Z (zircon), M (monoclinic zirconia), Q (quartz).

3.4.3. Pigments with Photocatalytic Reflective Properties Are Wide Bandgap Semiconductors (E_g between 3.4 and 2.0 eV)

Obviously, the photocatalytic capacity requires the semiconductor nature of the material, in addition, the need to act as a UV blocker requires an intense charge transfer in the UV zone (100-400 nm, although fundamentally in the higher UVA range of 315 to 400 nm present in solar radiation). In addition to the stability of the material, for development of the photocatalytic capacity, they are ceramic semiconductors with a relatively wide forbidden band (E_g between 3.4 and 2.0 eV). However, photosensitization with absorption shoulders in blue may be interesting to generate photocatalytic activity.

Figure 3.6 shows a yellow Ni-doped ilmenite ($Mg_{0,5}Ni_{0,5}TiO_3$) pigment obtained by the traditional CE ceramic method, by an organic decomposition method MOD with citric acid and a hyper-mineralization method by adding excess NaCl to the MOD sample (molar ratio ilmenite:NaCl=1:1) fired at 1000°C/3h.

The UV-Vis-NIR absorption spectra show a shoulder in the blocking UV zone and also one in the band gap blue zone around 532 nm (Eg=2.3 eV). The Vis-NIR reflectance spectrum shows a relative high reflectance R=52% (SRI = 61, T_s = 59°C) lower than the yellow of Ni,Sb-rutile or the Pr-zircon yellow. But under these conditions, its photocatalytic capacity over Orange II is high, with half-live time $t_{1/2}$ of 60-125 min. These values are very similar to those obtained with commercial photocatalytic anatase (the reference P25 Degussa under these conditions has half-live of 35 min). The SEM microscopy of the ceramic sample (left) and the non-conventional one (right) did not indicate a composite effect in this case, with very fine and similar particle sizes (around 0.5 μm for CE sample and 1-3 μm and aggregated in the hyper-mineralized sample.

3.4.4. Case of the "Cool" Black Pigment

Black has little interest in buildings but instead it has great interest in public works such as asphalt coating of roads and cities. On the market, coatings based on the commercial pigment Paliogen® Black L 0086 [43] are commercialized; a powder with a chemical nature based on perylene, a polycyclic aromatic hydrocarbon HAP, which according to the producer has a Total Solar Reflectance (R) value of 42%, admitting that the coating matrix shows an emittance of the order of 0.9, its SRI would be of the order of 48 (compared to 5 for carbon black widely used in paints), with T_s = 64ºC.

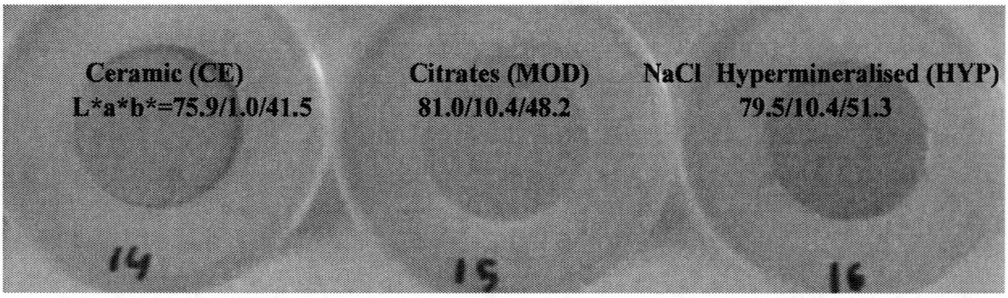

a

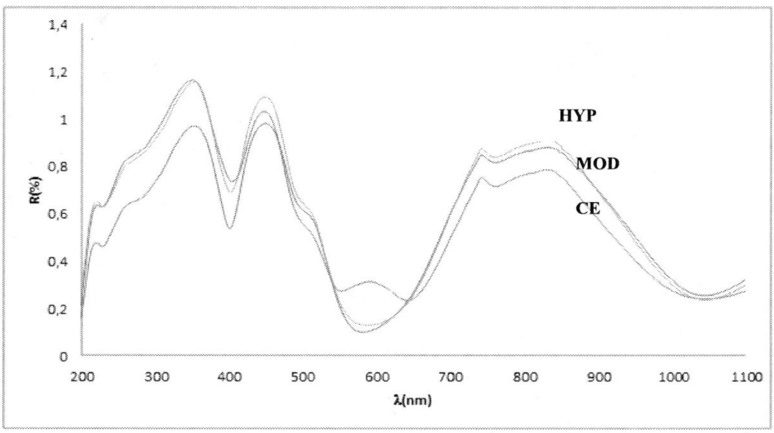

b

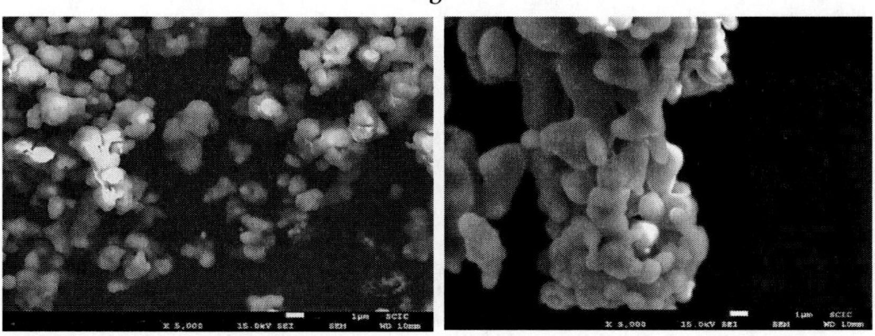

c

Figure 3.6. (a) Yellow Ni-doped ilmenite-based pigment obtained by the traditional CE ceramic method, MOD citrates and a hypermineralization method with NaCl, (b) UV-Vis-NIR absorption spectra, (c) SEM of CE and HYP.

Its organic nature makes it unusable in ceramic matrices and its nature as PAH could induce negative environmental effects [48]. Figure 3.7 shows an alternative with black inorganic pigment based on the solid solution $(FeCr)_2O_3$ applied at 5% in a double-firing frit (1050ºC). The R value is 23% in powder and 20% in the glaze.

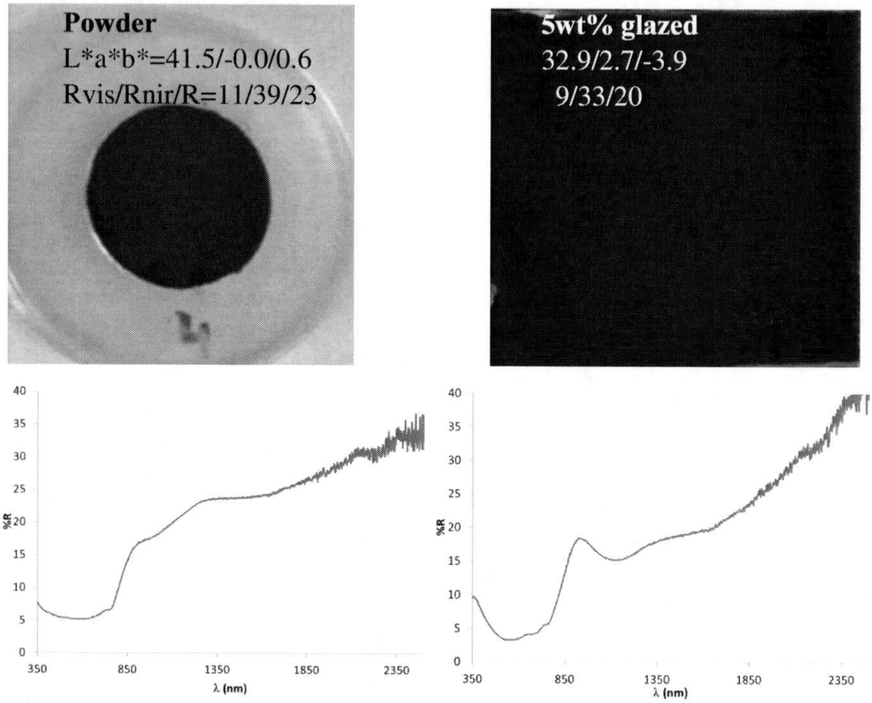

Figure 3.7. Black inorganic pigment $(FeCr)_2O_3$ 5 wt.% applied on conventional vinyl transparent paint and enameled on a double firing frit (1050ºC).

References

[1] *CPMA Classification and chemical description of the complex inorganic color pigments*, fourth ed. Alexandria, Dry Color Manufaturers Association, 2010.

[2] Monrós, G., Pigment, Ceramic (in *Encyclopedia of Color Science and Technology*, Ronnier Luo ed., Springer; 2014. ISBN 978-1-4419-

8070-0. ON LINE ISBN 978-3-642-27851-8. http://www.springerreference.com/docs/html/chapterdbid/348055.html.

[3] CIE Comission International de l´Eclairage, *Recommendations on Uniform Color Spaces, Color Difference Equations, Psychometrics Color Terms*. Suplement nº2 of CIE Pub. Nº15 (E1-1.31) 1971, Bureau Central de la CIE, Paris (1978).

[4] Munsell, A. H., *Atlas of the Munsell color system* (Malden, Mass., Wadsworth, Howland & Co., Inc., Printers, 1915).

[5] Briggs, D. J. C, *Modern color theory for traditional and digital painting media* http://www.huevaluechroma.com/082.php.

[6] ISO International Standard 105-J03, *Part J03: calculation of color differences*, 2009.

[7] Hanrahan, P., Kruger, W., SIGGRAPH '93, *Proceedings of the 20th annual conference on Computer graphics and interactive techniques*, 165-174, 1993.

[8] Hecht, H. G., "The Interpretation of Diffuse Reflectance Spectra," *Journal of research of the National Bureau of Standards-A. Physics and Chemistry* 80A, 4(1970): 567-583.

[9] Kortüm, G., *Reflectance Spectroscopy, Principles, Methods, Applications*, (Springer; 1969, ISBN 978-3-642-88071).

[10] Llusar, M., Bermejo, T. Primo, J. E., Gargori, C., Esteve, V., Monrós, G., "Karrooite green pigments doped with Co and Zn: Synthesis, color properties and stability in ceramic glazes," *Ceram Int* 12,43(2017): 9133-9144.

[11] ASTM C1371, *Standard Test Method for Determination of Emittance of Materials Near Room Temperature Using Portable Emisometers*, (2004).

[12] Cerro, S., Gargori, C., Llusar, M., Monrós, G., "Orthorhombic (Fe_2TiO_5)-monoclinic (Cr_2TiO_5) solid solution series: synthesis by gel routes, coloring and NIR reflectivity evaluation," *Ceram Int* 44, 11(2018):13349-13359. doi.org/10.1016/j.ceramint.2018.04.167.

[13] Gargori, C., Cerro, S, Fas, N., Llusar, M., Monrós G., "Red-brown ceramic pigments based on chromium doped ferrian armalcolite, effect

of mineralizers," *Ceram Int* 43(2017): 5490-5497. doi.org/10.1016/j.ceramint.2017.01.065.

[14] Gargori, C., Cerro, S., Fas, N., Llusar, M., Monrós, G., "Study of the photocatalytic activity and cool characteristics of a novel palette of pigments," *Bol Soc Esp Ceram Vidr* 56(2017)166-176. http://dx.doi.org/10.1016/j.bsecv.2017.01.001.

[15] Akbari, H., *The cool colors Project*, Berkeley National Laboratory http://coolcolors.lbl.gov/.

[16] García, A., Llusar, M., Calbo, J., Tena, M. A., Monrós, G., "Low-toxicity red ceramic pigments for porcelainised stoneware from lanthanide-cerianite solid solutions," *Green Chem* 3(2001): 238.

[17] Rim K. T., Koo K. H., Park, J. S., "Toxicological Evaluations of Rare Earths and Their Health Impacts to Workers: A Literature Review," *Saf Health Work*, 4(2013):12-26.

[18] Smith, A. E., Mizoguchi, H., Delaney, K., Spaldin, N. A., Sleight, A. W., Subramanian, M. A., "Mn^{3+} in trigonal bipyramidal coordination: a new blue chromophore," *J Am Chem Soc* 131(2009):17084-6.

[19] *Materials with trigonal bipyramidal coordination and methods of making the same*, US Patent 8,282,728 B2.

[20] Sheethu, J., Mundlapudi, L. R., "Lanthanum strontium copper silicates as intense blue inorganic pigments with high near-infrared reflectance," *Dyes Pigm* 98(2013):540-546.

[21] Zhang, Y., Zhang, Y., Zhao, X., Zhang, Y., "Sol–gel synthesis and properties of europium–strontium copper silicates blue pigments with high near-infrared reflectance," *Dyes Pigm* 131(2016):154-159.

[22] Vishnu, V. S., Reddy, M. L., "Near-infrared reflecting inorganic pigments based on molybdenum and praseodymium doped yttrium cerate: Synthesis, characterization and optical properties," *Solar Energy Materials & Solar Cells* 95(2011)2685–2692.

[23] Athira K. V. Raj, P. Prabhakar Rao, S. Sameera, S. Divya, "Pigments based on terbium-doped yttrium cerate with high NIR reflectance for cool roof and surface coating applications," *Dyes Pigm* 2015; 22:116-125.

[24] Jovani, M., Sanz, A., Beltrán-Mir, H., Cordoncillo E., "New red-shade environmental-friendly multifunctional pigment based on Tb and Fe doped $Y_2Zr_2O_7$ for ceramic applications and cool roof coatings," *Dyes Pigm* 133(2016):33-40.

[25] Gargori, C., Galindo, R., Cerro, S., García, A., Llusar, M., Monrós, G., "Synthesis of a new $Ca_xY_{2-x}V_xSn_{2-x}O_7$ yellow pigment," *Physics Procedia* 8(2010):84-87.

[26] Lili, L., Aijun, H., Mingquan, Y., Zhao, Z., "Synthesis and characterization of Al^{3+} doped $LaFeO_3$ compounds: A novel inorgànic pigments with high near-infrared reflectance," *Solar Energy Materials & Solar Cells* 131(2015)377–384.

[27] Bae, B., Tamura, W. S., Imanaka, N., "Novel environmentally friendly inorganic yellow pigments based on gehlenite-type structure," *Ceram Int* 42, 13(2016):15104-15106. https://doi.org/10.1016/j.ceramint.2016.06.106.

[28] Schildhammer, D., Fuhrmann, G., Petschnig, L., Weinberger, N., Schottenberger, H., Huppertz, H., "Synthesis and characterization of a new high NIR reflective ytterbium molybdenum oxide and related doped pigments," *Dyes Pigm* 138(2017):90-99.

[29] Kim, T., Kim, S., Lin, C., Liu, R., Chan, T., Im, S., "Melilite-type blue chromophores based on Mn^{3+} in a trigonal-bipyramidal coordination induced by interstitial oxygen," *J Mater Chem* 37,1(2013):5843–5848. doi: 10.1039/c3tc30795f.

[30] Pozza, G., Ajò, D., Chiari, G., De Zuane, F., Favaro, M., "Photoluminescence of the inorganic pigments Egyptian blue, Han blue and Han purple," *J Cult Heritage* 1(2000)393-398.

[31] Kendrick, E., Kirk C. J., Dann, S. E., "Structure and color properties in the Egyptian Blue Family, $M_{1-x}M'_xCuSi_4O_{10}$, as a function of M, M' where M, M'=Ca, Sr and Ba," *Dyes Pigm* 73(2007):13-18.

[32] Gargori, C., Cerro, S., Fas, N., Llusar, M., Monrós, G., "Study of the photocatalytic activity and cool characteristics of a novel palette of pigments," *Bol Soc Esp Ceram Vidr* 56(2017):166-176. http://dx.doi.org/10.1016/j.bsecv.2017.01.001.

[33] Seitz, G., Penin, N., Decoux, L., Wattiaux, A., Duttine, M., Gaudon, M., "Near the Ferric Pseudobrookite Composition (Fe_2TiO_5)," *Inorg Chem*, 55(2016):2499–2507, doi: 10.1021/acs.inorgchem.5b02847.

[34] Hoffmann, N., "Photochemical reactions as key steps in organic synthesis," *Chem. Rev.* 2008, 108, 3(2008): 1052−1103.

[35] Konstantyinou, I. K., Albanis T. A., "TiO_2-assisted photocatalytic degradation of azo dyes in aqueous solution: kinetic and mechanistic investigations. A review," *App. Catalyst B: Environmental* 49(2004):1-14.

[36] International Standar ISO 22197-1 2007 *Fine ceramics (advanced ceramics, advanced technical ceramics, Test method for air-purification performance of semiconducting photocatalytic materials-* Part 1: removal of nitric oxide, Switzerland, 2007.

[37] International Standar ISO 10678:2010, *Fine ceramics (advanced ceramics, advanced technical ceramics, Determination of photocatalytic activity of surfaces in an aqueous medium by degradation of methylene blue*', Switzerland, 2010.

[38] Mills, A, "An overview of the methylene blue ISO test for assessing the activities of photocatalytic films," *Applied Catalysis B: Environmental* 128 (2012) 144–149.

[39] Mills, A., Hill, C., Robertson, P. K. J., "Overview of the current ISO tests for photocatalytic materials," *Journal of Photochemistry and Photobiology A: Chemistry* 237 (2012): 7–23.

[40] Glasius, M., M. F. Carlsen, T. S. Hansen, C. Lohse, "Measurements of nitrogen dioxide in funen using diffusing tubes," *Atmospheric Environment*, 33(1999)1177-1185.

[41] Galindo R., Gargori C., Badenes J., Llusar M., Tena M. A., Monrós G., "Photocatalytic degradation of orange II azo dye by low titania doped sol-gel glasses," *XIVth Internacional Sol-Gel Conference*, Abstracts book, Montpellier (France) 2-7 September, 2007.

[42] Tauc, J., Grigorovici, R., Vancu, A., "Optical Properties and Electronic Structure of Amorphous Germanium," *Phys tatus solidi*, 15, 2, (1966): 627–637.

[43] BASF, *Pigments for solar heat management in paints*: https://www.dispersions-pigments.basf.com/portal/streamer?fid=560474.

[44] FERRO, *Cool Colors™ Pigments Infrared (IR) Heat and Energy Saving Pigments*: http://www.ferro.com/NR/rdonlyres/7814B923-21C7-4E44-90B5-438F6AECE6A9/0/CoolColorsBrochure_EMEA_March2014.pdf.

[45] Clark, M. D. T., Surface Coatings. Volume 1 - *Raw Materials and Their Usage*; Oil and Colour Chemists' Association, Australia, 1983.

[46] Montagner, C., Bacci, M., Bracci, S., Freemand, R., Picollo, M., "Library of UV–Vis–NIR reflectance spectra of modern organic dyes from historic pattern-card coloured papers," *Spectrochimica Acta Part* A 79 (2011) 1669–1680.

[47] Aceto M., Agostino A., Fenoglio G., Idone A., Gulmini M., Picollo M., Ricciardie P., Delaney J. K., "Characterisation of colorants on illuminated manuscripts by portable fiber optic UV-visible-NIR reflectance spectrophotometry," *Anal Methods*, 6(2014): 1488–1500. https://doi.org/10.1039/C3AY41904E.

[48] Xue, W., Warshawsky, D., "Metabolic activation of polycyclic and heterocyclic aromatic hydrocarbons and DNA damage: A review." *Toxicology and Applied Pharmacology*, 206, 1(2005):73–93. doi: 10.1016/j.taap.2004.11.006.

Chapter 4

CHEMICAL MECHANISMS OF COLOR IN CERAMIC PIGMENTS

The classical complex inorganic pigments based on mixed metal oxides are based on transition metals and the mechanism of color is associated to its broad and intense absorption bands due to d-d transitions that fall within the visible range: however, transition metals such as Co, Cr and Ni are toxic and there is a need to develop new high NIR reflectance inorganic pigments that are less hazardous to the environment; therefore the analysis of other color mechanisms is of great importance to be able to develop these objectives [1, 2].

The color origin in inorganic pigments is associated to the following main atomic mechanisms.

4.1. CRYSTAL FIELD MECHANISM (CFM)

In this case the color is due to the degeneration of d orbitals of structural transition metal ions by the influence of the anion coordination environment in the crystal lattice, therefore when pigment is illuminated, the d-d electronic transitions produce absorption bands associated to transition metals integrated in the solid lattice of the pigment following the Crystal Field Theory: e.g., Co^{2+} in willemite Zn_2SiO_4 (trigonal, *R*-3H) replaces Zn^{2+} in the tetrahedral site ($(Co_xZn_{2-x})SiO_4$ solid solution), showing d-d absorbance bands in the range 450-700 reflecting the blue component (Figure 4.1.a). In effect, for blue Co(II) d^7 in a tetrahedral ligand field, the spin allowed transitions are (Figure 4.1.a):

υ_1 : ${}^4A_2(F) \rightarrow {}^4T_2(F)$ in NIR region (1600 nm)
υ_2 : ${}^4A_2(F) \rightarrow {}^4T_1(F)$ in NIR region (1400 nm)
υ_3 : ${}^4A_2(F) \rightarrow {}^4T_1(P)$ splitting in three bands (green-red range)

The two first L-S spin-orbit Russel-Saunders of υ_2 : ${}^4A_2(F) \rightarrow {}^4T_1(F)$ in NIR at 1400 nm and υ_1 : ${}^4A_2(F) \rightarrow {}^4T_2(F)$ in 1600 nm fall in the infrared range. Only the third is present in the red spectral region of the visible region giving to a triple band centre at 540, 590, 640 nm of the ${}^4A_2(F) \rightarrow {}^4T_1(P)$ transition split by a Jahn-Teller effect of tetrahedral vacancies (Figure 4.1.a).

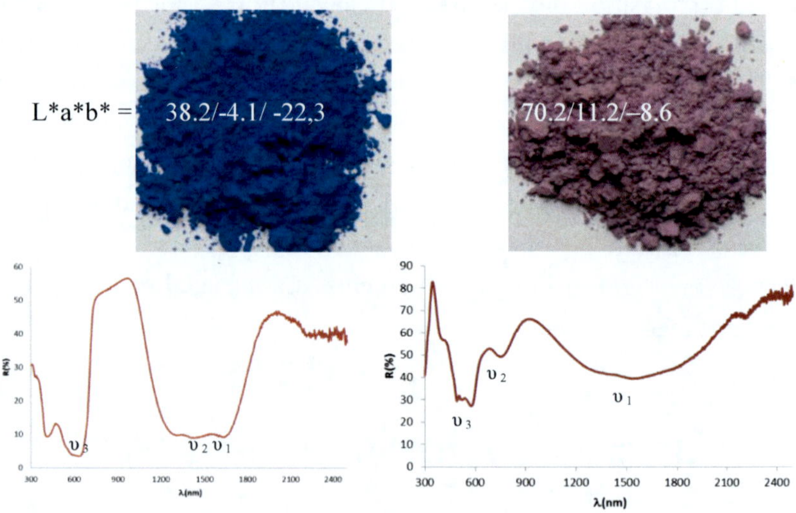

a b

Figure 4.1. Co-Zn_2SiO_4 willemite (blue) and Zn- Co_2SiO_4 olivine (violet) solid solutions.

If the cobalt amount x increases in the mixture $(Co_xZn_{2-x})SiO_4$ crystallizes the solid solution of Zn^{2+} in olivine Co_2SiO_4 (orthorhombic, *Pbmn*) which consists of a hexagonal closed-packed oxygen array in which half of the octahedral sites are occupied by Co^{2+} ions (with two sites M1 and M2) that are replaced by Zn^{2+}, showing that d-d absorbance bands shift and it is reflecting the violet component. In the case of violet Co(II) in octahedral ligand field, the spin allowed transitions are (Figure 4.1.b).

υ_1 : $^4T_1(F) \rightarrow ^4T_2(F)$ in NIR region (double band at 1300 and 1350 nm to both sites)

υ_2 : $^4T_1(F)) \rightarrow ^4A_2(F)$ yellowness-orange (double band 660 and 700 nm)

υ_3: $^4T_1(F) \rightarrow ^4T_1(P)$ blue-violet (400-600 nm multiple band due to rhombic distortion by the Jahn–Teller effect).

Besides, it appear other spin forbidden transitions associated to the singlet terms $^2A_2(2F)$,$^2A_1(2G)$, $^2T_2(2G)$, $^2T_1(2G)$ (from highest to lowest energy) [3].

Both Co-willemite (blue) and Zn-olivine (violet) solid solutions results unstable in glazes and Co(II) dissolves in the glazes imparting a blue color associated to Co(II) in tetrahedral coordination, a behavior known as "cobalt leaching" in a term coined by Eppler [4].

The discussed CFM mechanism is associated with a many ceramic pigments based on spinel structure, a wide family of ceramic pigments [1].

4.2. METAL-OXYGEN CHARGE TRANSFER MECHANISM (MOCTM)

This mechanism of color is based on a wide band semiconductor whose charge transfer band is shifted by the entrance of a metal in solid solution: e.g., the rutile based pigments use this mechanism.

Figure 4.2 shows the Ni,Sb-rutile (yellow) and Cr,Sb-rutile (orange-brown) solid solutions 3 w.% enameled in a double firing glaze (1000ºC) and its UV-Vis-NIR reflectance spectra. Both samples show a semiconductor feature with a clear sigmoid aspect of their UV-Vis-NIR spectra. In the case of Ni doped rutile, UV-Vis-NIR spectrum shows an absorbance edge at 450 nm and an inflection point at 475 nm (associated band gap energy E_g=2.61 eV), but the maximum of reflectance in the visible range is centered at 580 nm (yellow). In the case of Cr doped rutile the absorbance edge is at 680 nm and the inflection point at 500 nm (associated band gap energy E_g=2.48 eV), the maximum of reflectance in the visible range is centered at 680 nm (orange-red).

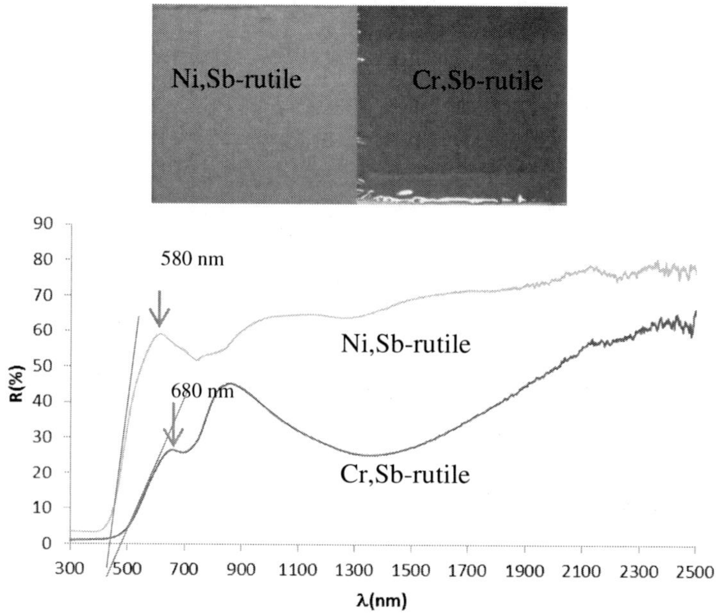

Figure 4.2. Ni,Sb-rutile (yellow) and Cr,Sb-rutile (orange-brown) solid solutions 3 wt.% enameled in double firing glaze (1000ºC) and its UV-Vis-NIR reflectance spectra.

4.3. Band Theory Mechanism (BTM)

The BTM color mechanism is based on the inclusion of crystals of a vivid colored semiconductor (guest) in the lattice of a transparent ans stable host crystal. This inclusion of crystals of a colored semiconductor in a hight stable crystal host (inclusion or encapsulated pigments) are also known as "hetermorphic pigments" [5]: e.g., pigment $ZrSiO_4$-$Cd(S_xSe_{1-x})$ based on the encapsulation of crystals of cadmium sulfoselenide into zircon crystals varies from yellow for high x to red for high selenium content [6].

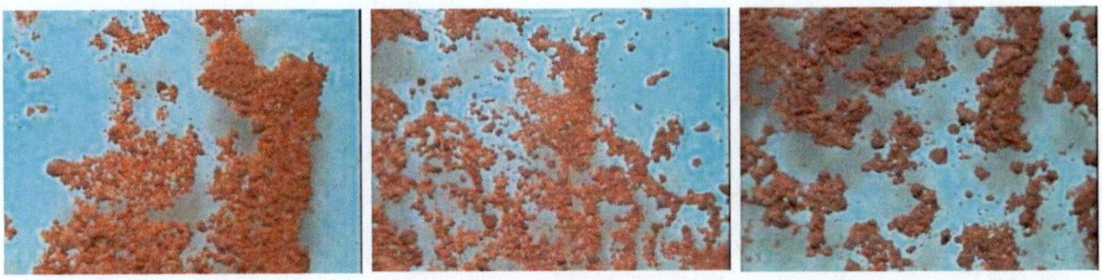

a

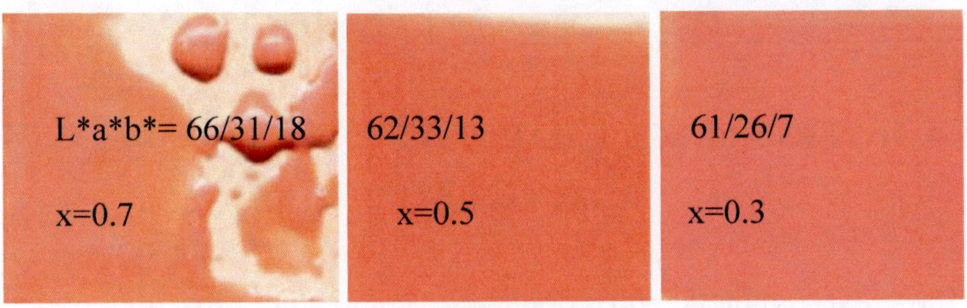

b

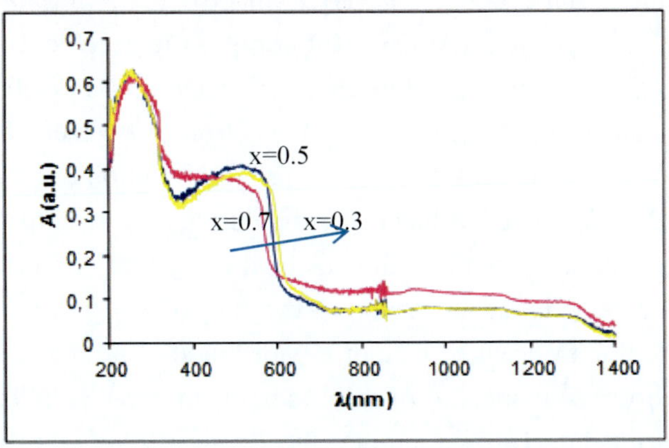

c

Figure 4.3. (a) $ZrSiO_4$-$Cd(S_xSe_{1-x})$ zircon-cadmium sulfoselenide pigments: powders (magnification x40), (b)3 wt.% enameled in single firing glaze (1080°C), (c) UV-Vis-NIR absorbance spectra.

Figure 4.3 shows powders image (magnification x40) and 3 wt.% enamelled samples in a single firing glaze (1080°C) of zircon-cadmium sulfoselenide pigments $ZrSiO_4$-$Cd(S_xSe_{1-x})$ for x=0.7, 0.5 and 0.3. When Se increase in the composition, the pigment darknens (L* decreases), yellow tone b* decreases and red chroma a* shows a maximum for x=0.5.

UV-Vis-NIR spectra show a band of high absorbance at 250 nm associated with zircon metal-oxygen transfer band and a lower one in intensity at 500 nm associated with $Cd(S_xSe_{1-x})$ semiconductor band transfer. The inflection point of the above band appears at 610 nm for x=0.7 (E_g=2.03 eV), at 625 nm for x=0.5 (E_g=1.98 eV) and at 630 nm for x=0.3 (E_g=1.97 eV) in agreement to the observed color which shifts towards dark red.

4.4. Intervalence Charge Transfer Mechanism (IVCTM)

The first synthetic mixed-valence compound was the so-called "Prussian blue," $[KFe^{II}Fe^{III}(CN)_6.xH_2O$, made by mixing the colorless solutions of $Fe(CN)_3$ and $K_4Fe(CN)_6$. Not only is the deep color of Prussian blue at first sight perplexing, but its chemical constitution is left ambiguous by the fact that a substance having an electronic spectrum and powder pattern identical to those of formed by mixing solutions of $Fe(CN)_2$ and $K_3Fe(CN)_6$. Prussian blue has been prized industrially for centuries as an ink and dye-stuff. This infinite three-dimensional cubic array of alternate Fe(II) and Fe(III) centers linked by bridging cyanide (CN^-) groups exhibits an intense blue coloration which cannot be attributed to a combination of the individual absorption spectra of the constituent ions [7]. The color is explained by the intervalence charge: the Prussian blue is an iron(II,III)–cyanide complex in which there is an iron(II) atom surrounded by six carbon atoms of six cyanide ligands bridged to an iron(III) atom by their nitrogen ends. An electron-transfer reaction occurs via the cyanide ligands to give iron (III) associated with an iron (II)-cyanide complex (with a transfer distance by 3 Å):

Fe(II) – C – N – Fe(III) ↔ Fe(III) – C – N – Fe(II)
1.85 1.18 Å

There is some localization of the distinct valences, but there is low activation energy for their interconversion. Some activation is required to induce electron transfer from one site to another via the bridge. These species exhibit an intense "Intervalence charge transfer" (IVCT) band; a broad intense absorption in the infrared or visible part of the spectrum, centered at 680 nm in the case of Prussian blue reflecting an intense blue color, and also exhibit magnetic exchange coupling at low temperatures, in the case of Prussian blue the measured moment would correspond to five unpaired spins per two iron ions (5.72 B.M. per molecule of $KFe_2(CN)_6$), if orbital contributions were small or non-existent. The degree of interaction between the metal sites can be estimated from the absorption profile of the IVCT band and the spacing between the sites.

This IVCT mechanism is observed in the Cr-doped cassiterite (SnO_2) and is object of controversy in similar chromium based pigments such as: Cr-$CaTiO_3$ perovsquite, Cr-$CaSnSiO_5$ malayaite and its isomorphic sphene Cr-$CaTiSiO_5$ titanite. At the beginning of the 19th century, the British industry had developed a pink pigment to color under-glazes on earthenware. The synthesis being secret, Brongniart asked Malaguti, chemist at the Manufacture de Sèvres in 1838, to find the composition of this pigment. Malaguti succeeded to obtain the chemical composition and to identify the elements necessary to retrieve the pink color, as well as unneeded constituents which were added on purpose to mask the real pigment elements. He identified the key role of calcium, tin, silica and chromium in malayaite $CaSnSiO_4$ pink pigment. His researches also led him to synthesize a new lilac pigment, by simply calcining tin oxide and chromium oxide, leading to Cr-doped cassiterite (SnO_2) [8].

Chromium speciation in the malayaite and cassiterite structures is less straightforward than in the corundum structure. Authors agree on the presence of Cr^{4+} in both structures, mainly substituting for Sn^{4+} in octahedral symmetry [9. 10]. In the case of malayaite, a very small amount of Cr^{4+} in tetrahedral symmetry in substitution for Si^{4+} was also detected by optical spectroscopy [10, 9]. Furthermore, the presence of Cr^{3+} substituting for Sn^{4+} with the formation of oxygen vacancies to accommodate the charge was evidenced by optical spectroscopy and X-ray and neutron diffraction [10].

The distribution of chromium between tetrahedral and octahedral sites and its valence variability are certainly the reasons why the color of pigment composed of Cr-doped malayaite is known to be hardly reproducible, and strongly dependent on synthesis conditions such as thermal treatment, used fluxes or precursors [11, 12].

In the case of Cr-malayaite $CaSnSiO_4$ (monoclinic, space group *C2/c*), the lattice consists of corner-sharing SnO_6 octahedra, forming slightly kinked chains parallel to a axis, which are mutually connected via SiO_4 tetrahedra, so within the SnO_6–SiO_4 framework irregular CaO_7 polyhedra appears; the off-centered position of Cr^{4+} at the octahedral site is able to justify the anomalously high optical density of its bands, which is allowed by its non-centrosymmetric location, which relaxes Laporte's selection rule [13, 14].

Crucciani et al. [10] consider that in the case of malayaite the anomalous optic efficiency cannot be accounted for an intervalence charge transfer mechanism (IVCT, e.g. $Cr^{3+}\leftrightarrow Cr^{4+}$) because the metal–metal distances in malayaite (approximately 3.6 Å for Sn–Sn and 3.3 Å for Si–Sn) are too long with respect to usual distances observed in IVCT (2.6–3.1 Å) [15].

In Figure 4.4 is depicted the color evolution of powders with the chromium content of $Cr_xCaSi_{1-x}SnO_5$ solid solutions (x40), 5wt.% enameled samples in a double firing glaze (1000ºC)) and optical UV-Vis-NIR absorbance spectra of samples. Likewise in Figure 4.5 XRD of $Cr_xCaSi_{1-x}SnO_5$ x=0.05 mineralized solid solution (showing malayaite and free SnO_2 associated to peak at 26.6 º2θ and α-cristobalite to peak at 22 º2θ) and SEM images of x=0.05 without and with mineralizer addition. The powders were synthesized by solid state ceramic method from carbonates and oxides fired at 1250ºC/6h.

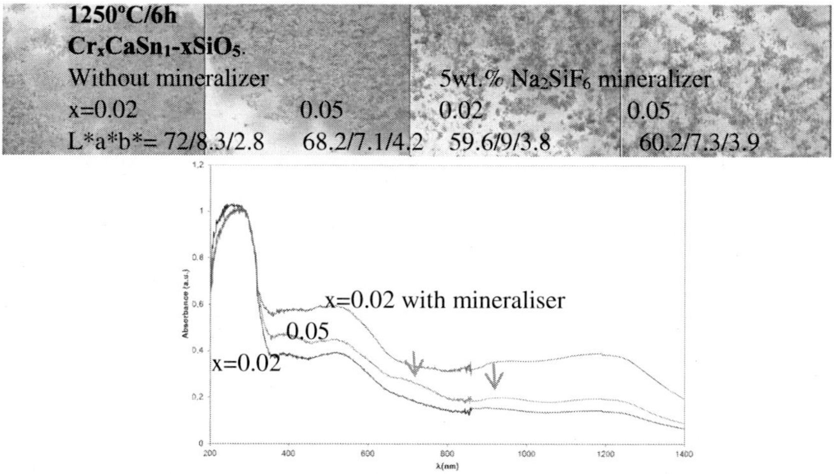

a

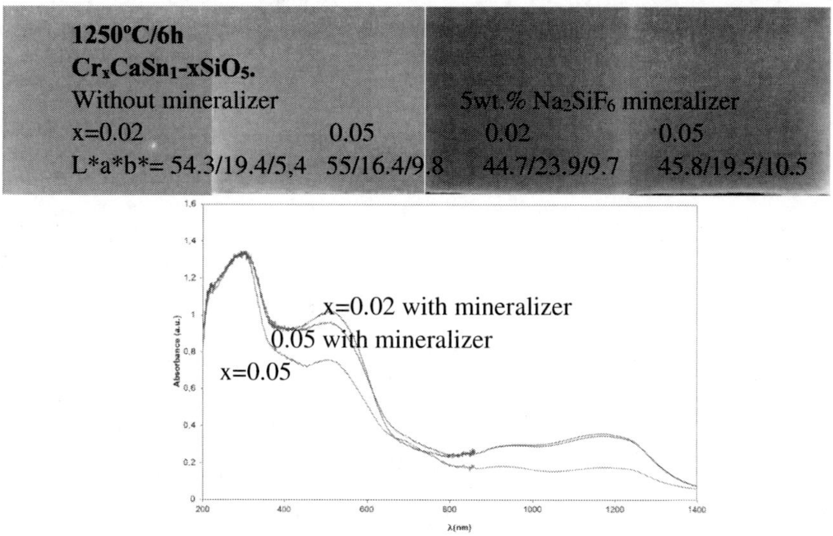

b

Figure 4.4. (a) Color evolution of powders with the chromium content of $Cr_xCaSn_{1-x}SiO_5$ solid solutions (x40), (b) 5wt.% enameled in a double firing glaze (1000ºC), with its respective optical UV-Vis-NIR absorbance spectra.

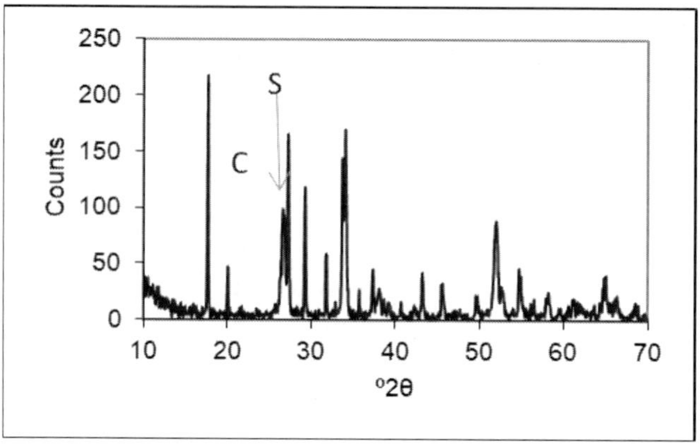

a

b

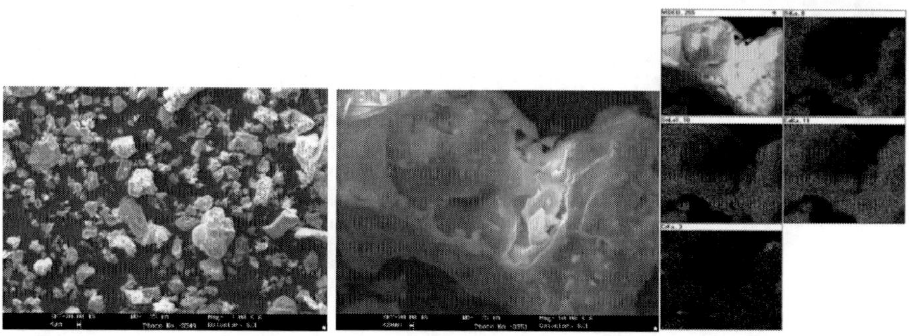

c

Figure 4.5. (a) XRD of $Cr_xCaSn_{1-x}SiO_5$ x=0.05 mineralized solid solution (showing malayaite and free SnO_2 (S) associated to peak at 26.6°2θ and α-cristobalite (C) to peak at 22°2θ) and SEM-EDX images of x=0.05 without (b) and with mineralizer addition (c).

From the results shown in Figure 4.4 and Figure 4.5 in the case of powders, it can be pointed out:

i. The entrance of chromium in the malayaite lattice for achieve an effective color is relatively low in agreement with literature [9,10]: the no-mineralized powders with x=0.02 shows a better red shade (a*) than x=0.05 but with higher brightness L* and the UV-Vis-NIR absorption spectra of these unmineralized samples (Figure 4.4.a) are similar but with higher general absorption (darkness) for the x=0.05 sample.

ii. The UV-Vis-NIR spectra of powders show the following absorbance bands (Figure 4.4.a): 260 nm (sharp and strong), 400 (weak), 550 (strong), 730 (medium), 920 (weak), 1180 (medium) and 1200 (medium).These bands can be associated to [10]:
   - The cation-anion charge transfer band at 260 nm.
   - Cr^{4+} (d^2) in octahedral coordination that has a ground state $^3T_{1g}(3F)$ with two Laporte-forbidden and spin-allowed transitions:
     - $^3T_{1g}(3F)$ to $^3T_{2g}(3F)$ at 550 nm
     - $^3T_{1g}(3F)$ to $^3T_{1g}(3P)$ at 400 nm.

The band at 1200 nm can be associated to the overlapping contributions of spin-forbidden transitions $^1E_g(1D)$ and $^1T_{2g}(1D)$.

- Cr^{3+} (d^3) in octahedral coordination that has a ground state $^4A_{2g}(4F)$ with two intense, parity-forbidden and spin-allowed transitions:
  - $^4A_{2g}(4F)$ to $^4T_{2g}(4F)$ at 560 nm (overlap with the above for Cr^{4+})
  - $^4A_{2g}(4F)$ to $^4T_{1g}(4F)$ at 430 nm (overlap with the above for Cr^{4+})

Furthermore, two tiny bands at 920 and 730 nm associated to doubly forbidden transitions $^2E_g(2G)$ and $^2T_{1g}(2G)$, respectively.

- Cr^{4+} in tetrahedral coordination shows a ground state $3A_{2g}(3F)$ with three spin allowed transitions;

- $3A_{2g}$(3F) to $^3T_{2g}$(3F) (split in two bands at 1100 and 960 nm which overlaps with the above for Cr^{3+}).
- $3A_{2g}$(3F) to $^3T_{1g}$(3F) (at 700 nm overlapping with above band for Cr^{3+})
- $3A_{2g}$(3F) to $^3T_{1g}$(3P) (at 470 nm that overlaps with above of for Cr^{4+} and Cr^{3+}).

Further two weak bands, at energies close to 910 nm (overlaps with the above for Cr^{3+}) and 625 nm (not observed in Figure 4.4.a) associated to spin-forbidden transitions $^1A_{1g}$(1G) and $^1T_{1g}$(1G).

In conclusion it is very difficult to distinguish the presence of $Cr^{3+}$and Cr^{4+} in octahedral-tetrahedral sites on malayaite structure in UV-Vis-NIR spectra of samples containing the ions in these positions simultaneously, because the bands are overlapped to a high extent. However the intense band associated to $3T_{1g}$(3F) to $^3T_{2g}$(3F) transition at 560 nm that dominates the spectrum of the powders indicates the presence of Cr^{4+} in octahedral coordination with its characteristic absorption band [18]. Furthermore the two tiny bands at 920 and 730 nm (see the marks in the spectra of Figure 4.4.a), associated to doubly forbidden transitions ^2E_g(2G) and $^2T_{1g}$(2G) respectively of octahedral Cr^{3+}, increase in intensity in the most concentrated sample (x=0.05) and practically disappears in mineralized samples x=0.02, indicating an higher content in octahedral coordinated Cr^{4+} and low presence of Cr^{3+} in the more intense red (a*) samples. The entrance of Cr^{3+} associated with oxygen vacancies in the lattice, induces the color darkens and loss of red shade (a* parameter). The absence of specific bands associated to Cr^{4+} in tetrahedral coordination does not confirm the existence of this kind of substitution in malayaite lattice.

iii. In agreement with literature, the color of Cr-doped malayaite is known to be hardly reproducible, and strongly dependent on synthesis conditions such as thermal treatment [10]; in effect, the addition of 5 wt.% of Na_2SiF_6 as flux or mineralizer (which from 500°C slowly decomposes in HF and silica), clearly improves the lilac shade of the powder increasing the a* color parameter between

3-5 points, reaching a*=23.9 in the case of mineralized x=0.02 powder.

iv. XRD of powders in all cases shows the presence of malayaite with unreacted SnO_2 as it is usually in the literature [10].

v. For no mineralized x=0.05 sample, SEM micrographs of powders (Figure 4.5.b) show a low particle size with rounded grains between 2-6 μm covered with fine particles that appear richer in Sn at EDX analysis. In the case of mineralized sample, Figure 4.5.c shows a sintered appearance of particles with a size between 5-25 μm which allows it to better resist the attack of the glazes, which helps to improve the coloration of the enamel.

When the powders were 5 wt.% enameled in a double firing glaze (1000ºC) (Figure 4.4.b), the burgundy shade shows higher intensity in mineralized samples and in the lower chromium content sample x=0.02, both in non-mineralized and mineralized samples. In agreement with the observed color intensity, the UV-Vis-NIR absorbance spectra of mineralized samples show an intense band centered at 560 nm associated to $^3T_{1g}(3F) \rightarrow$ $^3T_{2g}(3F)$ transition of Cr^{4+} in octahedral coordination, likewise the specific bands at 920 and 730 nm associated to doubly forbidden transitions $^2E_g(2G)$ and $^2T_{1g}(2G)$ respectively in Cr^{3+}, practically disappears in mineralized samples x=0.02, corroborating the association of lilac color with higher content in Cr^{4+} in octahedral coordination and low presence of Cr^{3+}; therefore the intervalence charge transfer mechanism (IVCT) between $Cr^{3+} \leftrightarrow Cr^{4+}$ should be discarded in this pigment.

But in the case of Cr-cassiterite (tetragonal, space group *P42/mnm*) Serment et al. [22] study the low content incorporation of chromium (≤5 mol%) into SnO_2 cassiterite by a coprecipitation route in a basic medium, followed by an annealing step under an O_2 flow at T = 800°C and 1000°C. UV–Vis and EPR spectroscopy investigations show the coexistence of isolated Cr^{4+} and Cr^{3+} ions as well as ferromagnetic Cr^{4+}–Cr^{3+} and antiferromagnetic Cr^{3+}–Cr^{3+} interactions. The strong purple hue is related to the isolated Cr^{4+} ions stabilized in a distorted octahedral site. This is thanks to the second-order Jahn–Teller (SOJT) effect with a crystal field splitting

10Dq value around 2.4 eV, whereas the 10Dq value is around 2 eV for isotropic Cr^{3+} ions, partially substituted for Sn^{4+} ions in cassiterite.

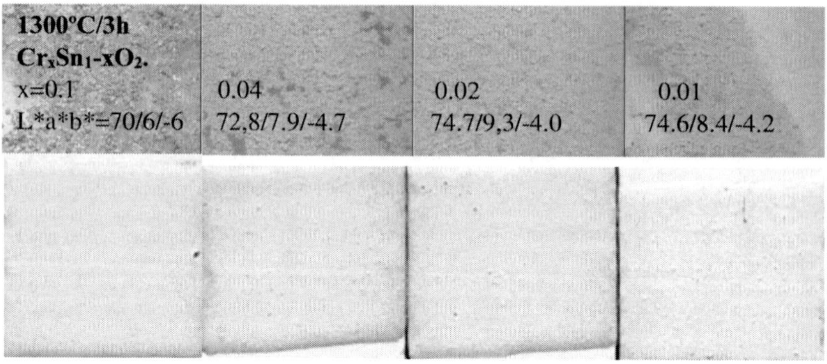

Figure 4.6. Color evolution with the chromium content x (powder and 3 wt.% enameled in a porcelain stoneware single fire glaze (1200ºC)) of samples $Cr_xSn_{1-x}O_3$ synthesized by solid state reaction starting from SnO_2 and Cr_2O_3.

Just after the coprecipitation process, only Cr^{3+} species are stabilized in the cassiterite network with a poor crystallinity. The isolated Cr^{4+} content remains high after annealing at 800°C for 2 days especially for the highest Cr rate (2 and 5 mol %), leading to a darker purple color, but unfortunately the Cr^{3+} content also increases for a higher Cr concentration. A lighter purple hue can be reached after calcination at a higher temperature (T = 1000°C) for a shorter time (4 h) but with a lower Cr content to avoid Cr clusters. This is due to stabilizing a high content of isolated Cr^{4+} species and limiting the Cr^{4+}–Cr^{3+} ferromagnetic interactions, which are optimal for 2% Cr content and also cause the color to darken. The key roles of the Cr^{4+} rate and the Cr^{4+}–Cr^{3+} clusters create local defects whose concentration strongly varies with a total Cr content, which have then been demonstrated to strongly influence the optical and magnetic properties [22].

Figure 4.6 shows the color evolution with the chromium content x of the powder and 3 wt.% glazed sample in a porcelain stoneware single fire glaze (1200ºC) of samples $Cr_xSn_{1-x}O_3$. In agreement with Serment et al. [22], it is observed that the best pink color is detected for the incorporation of 2 mol% of chromium (x=0.02) to cassiterite lattice; although the improvement of the color is not significant compared with x=0.01, it is clear that color bluewish and darkens when the chromium content increases. The concentration of isolated Cr^{4+} species is optimal at x=0.02 limiting the Cr^{4+}–Cr^{3+} interactions that cause the color to darken at higher chromium content.

Although Cr-cassiterite is a pigment with a marginal importance in the conventional technologies it has been revitalized by the advent of digital decoration, even though handicapped by the high cost of tin oxide due to its rather good performance in digital decoration and its peculiar violet color on glazes [23].

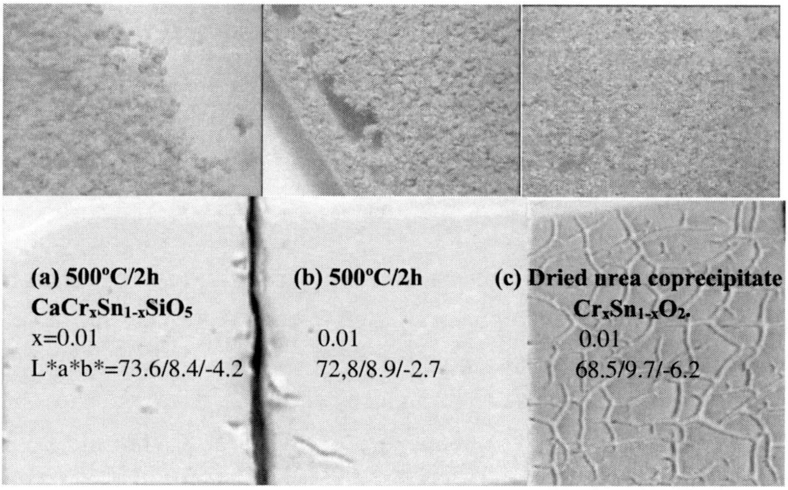

Figure 4.7. Color of 3 wt.% enameled in a porcelain stoneware single fire glaze (1200°C) of samples $Cr_xSn_{1-x}O_3$ x=0.01 synthesized by citrate route compared with Cr-malayaite ($CaCr_xSn_{1-x}SiO_5$ x=0.01).

In order to illustrate that the pigment is hardly reproducible, and strongly dependent on synthesis conditions, Figure 4.7 shows the color of 3 wt.% enameled in a porcelain stoneware single fire glaze (1200°C) of samples $Cr_xSn_{1-x}O_3$ x=0.01 synthesized by citrate route (Pechini method) compared with Cr-malayaite ($CaCr_xSn_{1-x}SiO_5$ x=0.01).

The sample a in Figure 4.7 is a Pechini gel obtained from solving $Ca(NO_3)_2.4H_2O$, $SnCl_2.2H_2O$, ludox (colloidal silica), $Cr(NO_3)_3.9H_2O$ for produce 0.5M of final composition in 200 ml of water. Then 200 ml of citric acid 0.5M is added to the continuous stirred solution and after diethylene glycol (molar ratio citric:diethylene glycol=1:2) and the mixture was dried at 110ºC in order to obtain the corresponding polyester resin. The sample b is the Cr-cassiterite x=0.01 obtained from the same precursors and Pechini procedure.

Finally sample c is a urea coprecipitated gel of Cr-cassiterite x=0.01: the precursors $SnCl_2.2H_2O$ and $Cr(NO_3)_3.9H_2O$ for 4 g. of final product were solved in 100 ml of diethylene glycol, then 4 moles of urea per mol of product were added to stirred solution that was refluxed during 1 hour, finally the precipitated was separated by centrifugation, washed with ethanol and dried at air.

The both polyester resins (a) and (b) were fired ("charring" treatment) at 500ºC/2h, and the obtained powders were 3 wt.% glazed in a porcelain stoneware single fire glaze (1200ºC). In the case of urea coprecipitated sample (c) the dried powder was directly 3 wt.% glazed in a porcelain stoneware single fire glaze (1200ºC).

Figure 4.7 shows the peculiar violet color of Cr-cassiterite pigment for sample (c); but sample b shows a very similar pink to the obtained by the polyester of Cr-malayaite pigment (Figure 4.7.a) and the ceramic samples of the Cr-cassiterite showed in Figure 4.6.

Similar pink colors where the intervalence charge transfer mechanism (IVCT) between $Cr^{3+}\leftrightarrow Cr^{4+}$ has been proposed are: a) chromium-$CaTiO_3$ perovskite and b) chromium-$CaTiSiO_5$ titanium sphene (titanite).

a) Chromium-$CaTiO_3$ perovskite pink.

Gargori et al. [24] study $Ca(Cr_xTi_{1-x})O_3$ x = 0.025, 0.05, 0.1 and 0.2 compositions obtained with and without addition of NH_4Cl as mineralizer (Figure 4.8). Pink color in glazes at low x was obtained, but $CaCrO_4$ crystallizes when x increases, producing pin-hole defects and green colors in glazed samples. The crystallization of chromates can be avoided using NH_4Cl as mineralizer. Chromium doped calcium titanate only produce a complete solid solution in presence of NH_4Cl: the $Ca(Cr_xTi_{1-x})O_3$ perovskite produces pink color in glazes at low x and dark blue shades at high x. Coprecipitated samples produce blue colors at lower x and at lower temperature (1000ºC instead 1200ºC for CE sample). Cr^{4+} and Cr^{3+}, both in octahedral coordination, replaces Ti^{4+} in titanate lattice: Cr^{4+} ion acts as the red chromophore agent associated with its 520 nm absorption band in red-pink samples, but at higher x values (blue samples), Cr^{3+} ion affects the color with its absorption band at 600 nm.

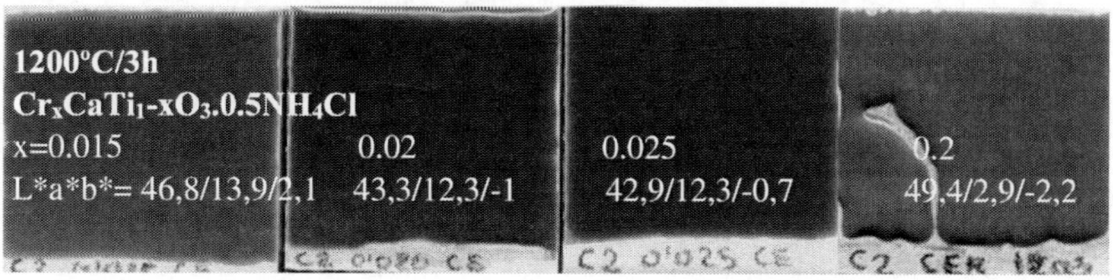

a

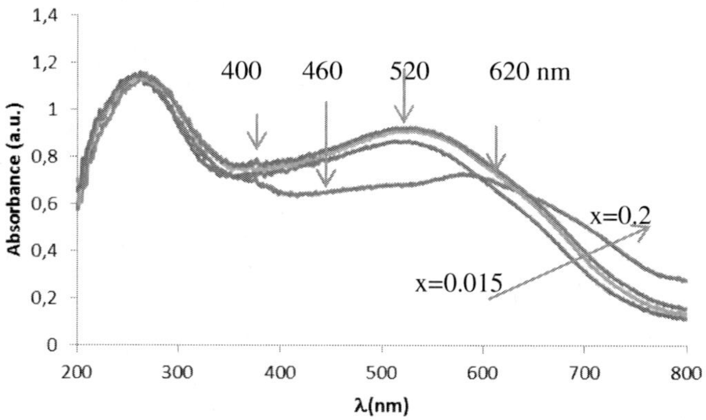

b

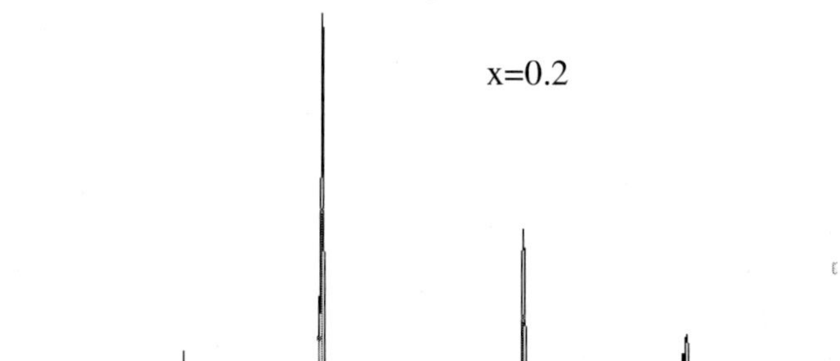

c

Figure 4.8. (a) Color evolution with chromium content of $Cr_xCaTi_{1-x}O_3$ (with $0.5NH_4Cl$ addition as mineralizer) perovskite solid solutions (5wt.% enameled in a double firing glaze (1080°C)), (b) optical UV-Vis-NIR absorbance spectra, (c) XRD of powder of $Cr_{0.2}CaTi_{0,8}O_3$ sample. CRYSTALLINE PHASES: perovskite $CaTiO_3$ (red peaks).

Figure 4.8 shows the color evolution with chromium content of $Cr_xCaTi_{1-x}O_3$ (with 0.5NH_4Cl addition as mineralizer) perovskite solid solutions (5wt.% enameled in a double firing glaze (1080ºC)), the UV-Vis-NIR optical absorbance spectrum and the XRD of powder of $Cr_{0.2}CaTi_{0.8}O_3$ sample. Samples were prepared by a solid state reaction starting from calcium carbonate, eskolaite Cr_2O_3, anatase, and NH_4Cl: the mixtures were grinded in a ball mill in acetone medium and fired at 1200ºC/3h.

It can be pointed out, in agreement with literature, that the pink color darkens (L* diminish) and bluewish (b* is more negative) with chromium increase. At high chromium concentration (x=0.2) the sample is opaque, dark (L*=49) and blue (b*=-2.2). The XRD shows a single phase of perovskite indicating that the solid solution of chromium is reached.

The UV-Vis spectra of enameled samples (Figure 4.8) show a feature compatible with Cr^{4+} (d^2) in octahedral coordination with the spectra dominated by its main Laporte-forbidden and spin-allowed transition $^3T_{1g}$(3F) to $^3T_{2g}$(3F) at 520 nm (the transition $^3T_{1g}$(3F) to $^3T_{1g}$(3P) at 400 nm only is slightly observed for the x=0.2 sample). The band position shifts to higher wavelengths with chromium content, in this way for x=0.2 the band is located at 610 nm: this shift can be explained by the entry of a smaller ion (Cr^{4+}, Shannon-Prewit radius 0.69 Å in octahedral coordination) replacing Ti^{4+} (Shannon-Prewit radius 0.745 Å in octahedral coordination) relaxing the crystal field over the Cr^{4+} (d^2) ion.

In addition a band at 620 nm that shifts at higher values with x is compatible with the $^4A_{2g}$(4F) $\rightarrow ^4T_{2g}$(4F) transition in Cr^{3+} (d^3) in octahedral coordination (however the $^4A_{2g}$(4F) to $^4T_{1g}$(4F) transition at 460 nm is only perceptible in x=0.2 sample and the forbidden transition to $^2T_{1g}$(2G) at 730 nm is not detected).

Therefore the pink color of Cr-perovskite pink solid solution show a mechanism in which both Cr^{4+} and Cr^{3+} are involved (although, as in the case of Cr-cassiterite pigment, the unquestionable pink chromophore agent is Cr^{4+}). Indeed, for Cr-perovskite pink pigment, the intervalence charge transfer mechanism (IVCT) should be considered as in the case of Cr-cassiterite.

b) Chromium-$CaTiSiO_5$ titanium sphene (titanite).

Stoyanova Lyubenova et al. [25] study the Cr-doped titanite ($CaTiSiO_5$) pigments which were synthesized through spray pyrolysis of aerosols generated from aqueous solutions containing colloidal silica, calcium chloride, titanium (IV) oxychloride and chromium (III) nitrate. This process yielded amorphous powders with spherical morphology and broad size distribution (<10 μm) after thermal decomposition at 600 C. The titanite phase was obtained by further calcination at 800ºC without any addition of flux agents.

The brown color of the pigments can be attributed mainly to the existence of Cr (IV) ions occupying both, octahedral positions of Ti (IV) and tetrahedral position of Si (IV), together with a small amount of Cr (III) present as Cr_2O_3. The optimum pigment obtained by this method corresponded to a Cr/titanite mole ratio of 0.04.

Figure 4.9 shows the evolution of color with chromium content of $Cr_xCaTi_{1-x}SiO_5$ titanium sphene (titanite) solid solutions (5wt.% enameled in a single firing glaze (1080ºC)), with XRD and optical UV-Vis-NIR absorbance spectrum of sample $Cr_{0.06}CaTi_{0,94}O_3$. It can be observed, in agreement with literature, that the pink color darkens (L* diminishes) and greenish (a* decreases) with the increase of chromium. At high chromium concentration (x=0.3) the sample is dark (L*=48.5) and green (b*=-3). The optimal color is for the x=0.06 sample which XRD shows intense peaks associated to titanite with residual quartz and Ca-Ti perovskite.

The UV-Vis spectra of enameled samples show a feature compatible with Cr^{4+} (d^2) in octahedral coordination with the spectra dominated by its main Laporte-forbidden and spin-allowed transition $^3T_{1g}$(3F) to $^3T_{2g}$(3F) at 500 nm and the transition $^3T_{1g}$(3F) to $^3T_{1g}$(3P) at 400 nm. The band position shifts to lower wavelengths respect the isomorphic Cr-$CaSnSiO_5$ malayaite (560 nm) and Cr-$CaTiO_3$ perovskite (520 nm).

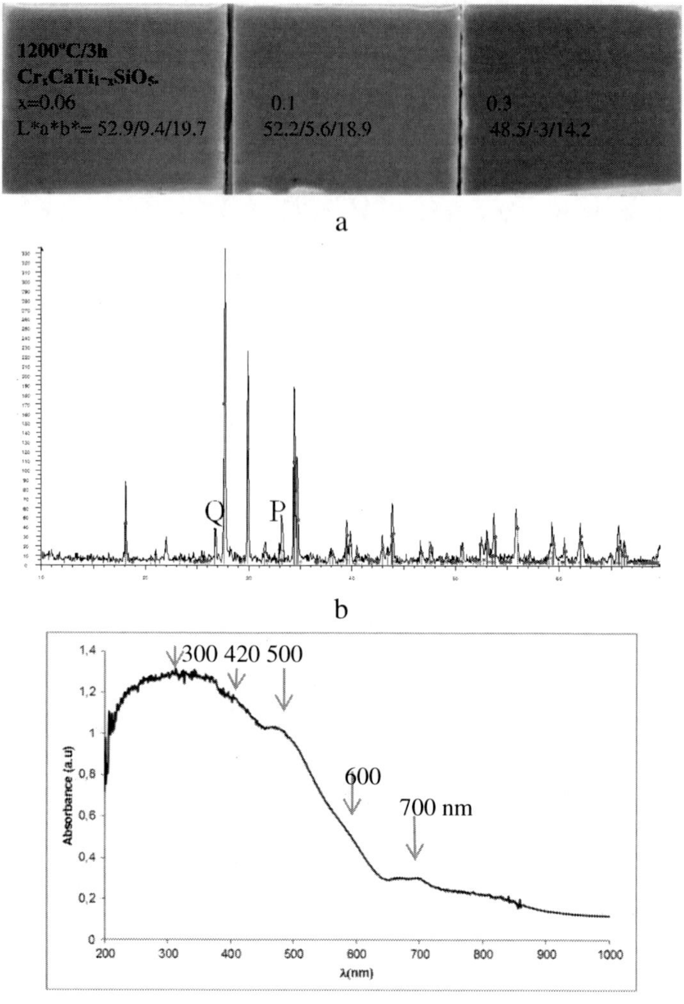

Figure 4.9. (a) Evolution of color with chromium content of $Cr_xCaTi_{1-x}SiO_5$ (NH_4Cl mineralized) titanium sphene (titanite) solid solutions (5wt.% enameled in a double firing glaze (1080ºC)), XRD of powder (b) and optical UV-Vis-NIR absorbance spectrum (c) of sample $Cr_{0.06}CaTi_{0,94}O_3$, CRYSTALLINE PHASES: titanite or Cr-Ti sphene $CaTiSiO_5$ with marked peaks of residual quartz (Q) and Ca-Ti perovskite ($CaTiO_3$).

The ion Cr^{4+} (Shannon-Prewit radius 0.69 Å in octahedral coordination) replaces Ti^{4+} (Shannon-Prewit radius 0.745 Å in octahedral coordination) in titanite and Sn^{4+} (Shannon-Prewit radius 0.830 Å in octahedral coordination) in malayaite; therefore the crystal field over the Cr^{4+} (d^2) ion is more relaxed in malayaite and the transition appears at higher wavelength. In the case of titanite the ion replacement are the same but the presence of $SiO_4^{=}$ in the lattice of titanite would intensify the crystal field over Cr^{4+}.

UV-Vis-NIR spectrum in Figure 4.9 shows also bands compatible with Cr^{3+} (d^3) in octahedral coordination: the two intense, parity-forbidden and spin-allowed transitions, $^4A_{2g}(4F) \rightarrow {}^4T_{2g}(4F)$ at 600 nm and $^4A_{2g}(4F) \rightarrow {}^4T_{1g}(4F)$ at 420 nm, and furthermore the tiny band at 700 nm associated to doubly forbidden transition to $^2T_{1g}(2G)$. Therefore, as in the case of Cr-cassiterite, Cr-$CaTiO_3$ or Cr-malayaite pigments, the unquestionable pink chromophore agent is Cr^{4+}, but as in the case of Cr-perovskite the presence of Cr^{3+} is involved in the best lilac colors, therefore the intervalence charge transfer mechanism (IVCT) cannot be discarded in this pigment.

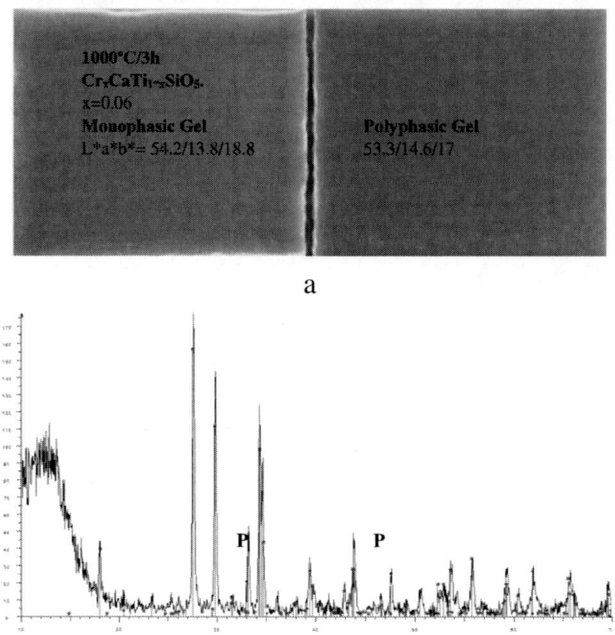

a

b

Figure 4.10. (a) Color of $Cr_xCaTi_{1-x}SiO_5$ x=0.06 titanium sphene (titanite) solid solution from coprecipitated gels and NH_4Cl addition fired at 1000°C/3h (5 wt.% enameled in a double firing glaze (1080°C)), (b) XRD of monophasic gel: CRYSTALLINE PHASES: titanite or Cr-Ti sphene $CaTiSiO_5$ with marked peaks of Ca-Ti perovskite $CaTiO_3$ (P).

Figure 4.10 shows color of $Cr_xCaTi_{1-x}SiO_5$ x=0.06 titanium sphene (titanite) solid solution from coprecipitated gels and NH_4Cl addition as mineralizer(5wt.% enameled in a single firing glaze (1080ºC)).

The gels $Fe_{1-x}Cr_xTiSiO_5.0,5NH_4Cl$ (monophasic and polyphasic) were prepared starting from TEOS (tetraethylorthosilicate), titanium isopropoxide, calcium nitrate and NH_4Cl. At 100 ml of ethanol was successively added, at room temperature and continuous stirring, the calcium nitrate, the acetylacetone (acac) (molar ratio acac:Ti=1:1), the titanium isopropoxide and finally TEOS. The mixture was maintained in stirring 1 hour and the monophasic gel is obtained by drying the solution under IR lamp, the polyphasic gel was prepared by slow dropwise of a solution of ammonia to the stirred solution until pH 8 was reached, then the sample was dried under IR lamp. NH_4Cl was added by ball milling in acetone medium to the two dried gels. The final mixtures were fired at 1000ºC/3h.

The results showed in Figure 4.10 stands out the improvement of the pink color with the gel samples that give values of a* around 14 versus a*=9 for ceramic samples (Figure 4.9), lightness L* and b* parameter are similar in both samples but gels are fired at only 1000ºC and ceramic powders at 1200ºC, showing the efficiency of the gel method in order to achieve a good pink color.

Compared with the pink color obtained in the isomorphic $Cr\text{-}CaSnSiO_5$ malayaite (best pink color in double fire frit at 1000ºC L*a*b*=44.7/23.9/9.7 for x=0.02 mineralized in Figure 4.4) and $Cr\text{-}CaTiO_3$ perovskite (best pink color in single fire frit at 1080ºC L*a*b*=46.8/13.9/2.1 for x=0.015 mineralized in Figure 4.8), the pink color is similar to the obtained with $Cr\text{-}CaTiO_3$ perovskite and with lower a* than in Cr-malayaite, but the firing frit used is less aggressive in malayaite case.

4.5. Cooperative Transitions (COT)

In this mechanism the chromophore enters in solid solution replacing an isovalent cation but with dissimilar size, if the chromophore is bigger than the replaced ion it becomes "compressed" even in small concentration even more if the incorporation of the chromophore ions M occurs with increasing covalence and polarizability of the M-O bond. M-O distance decrease and the high tetragonal distortion of d orbitals of the transition ion enhance the crystalline field over de ion. Consequently absorption bands shift to higher frequencies and the color shifts resulting to red. If the chromophore is small than the replaced ion it becomes "relaxed" in the lattice site and the crystalline field over the d orbitals goes down, therefore the absorption bands shift to lower wavelengths. In short, in this mechanism there are a cooperative transition between the chromophore and the replaced ion that modifies the crystalline field over the chromophore ion. The following pigments show this mechanism.

a) Chromium-Al_2O_3 corundum pink.

The solid solution of chromium in corundum is done in the Ruby (red) or in Sapphire (pink) depending the chromium content in the solid solution [26]. The solid solution $Cr_xAl_{1-x}O_3$ x=0.02 to 0.2 obtained from eskolaite (Cr_2O_3) and AlO(OH) as starting precursors, fired at 1200°C/2h, is followed by XRD and CIEL*a*b* in Table 4.1.

Table 4.1. XRD (intensity of peak at 35.1°2θ in counts/s) and CIEL*a*b* of solid solutions $Mn_xAl_{1-x}O_3$ x=0.002 to 0.2 obtained from eskolaite (Cr_2O_3) and AlO(OH) as starting precursors fired at 1200°C/2h (5wt.% enameled in a double firing glaze (1000°C)): CRYSTALLINE PHASES C(corundum α-Al_2O_3)

x	Phase (counts/s)	L*a*b*
0.02	C(810)	85.2/3/10.6
0.05	C(760)	76.8/5.3/12.8
0.1	C(550)	68.7/6.2/14.8
0.2	C(355)	57.1/4.3/11.1

From data on Table 4.1 is observed that the XRD intensity of corundum decreases as x increases, the pink powder darkens with x and both red a* and yellow b* parameters of 5wt.% enameled sample in a double firing glaze (1000ºC) show a maximum at x=0.1 that is considered the optimal composition. Probably at low x, the Cr^{3+} ions remain isolated but progressively clustered in the corundum lattice and the color darkens and saturates.

Figure 4.11 shown the evolution of color with temperature (5wt.% enameled in a double firing glaze (1000ºC)), XRD and optical UV-Vis-NIR absorbance spectrum of sample $Cr_{0.2}Al_{1.8}O_3$ obtained from eskolaite (Cr_2O_3) and AlO(OH) as starting precursors.

At 900ºC/2h the sample has not reacted and XRD shows the peaks of eskolaite that gives a pure green color when 5 wt.% enameled in the double fire glaze (1000ºC). At 1100ºC/2h only very weak residual peaks of eskolaite are detected by XRD and the enameled sample shows a pink color (a*=5). Finally at 1200ºC the reaction is completed and only corundum is detected by XRD and the glazed sample shows a lighter pink (a*=6.2).

UV-Vis-NIR absorption spectra on Figure 4.10 show that the unreacted sample at 900ºC shows a feature associated with Cr^{3+} (d^3) in octahedral coordination in eskolaite lattice: a cation-anion charge transfer band at 250 nm and the above described bands for Cr^{3+} (d^3) in octahedral coordination with two intense, parity-forbidden and spin-allowed transitions:

$^4A_{2g}(4F) \rightarrow {}^4T_{2g}(4F)$ at 610 nm.
$^4A_{2g}(4F) \rightarrow {}^4T_{1g}(4F)$ at 460 nm.

Furthermore, the tiny band at 730 nm associated to the forbidden transition at $^2T_{1g}(2G)$ is also detected but not the transition to $^2E_g(2G)$ at 920 nm.

When the solid solution is reached (1100 and 1200ºC) the bands shift to lower wavelengths: 590 and 410 nm respectively. This shift can be explained by the entrance of a bigger ion (Cr^{3+}, Shannon-Prewit radius 0.755 Å in octahedral coordination) replacing Al^{3+} (Shannon-Prewit radius 0.675 Å in octahedral coordination) enhancing the crystal field under the Cr^{3+} (d^3) ion. Therefore the pink color of Cr-corundum pink solid solution can be explained by the crystal field theory with a cooperative transition of Cr^{3+} influenced for Al^{3+} and the intervalence charge transfer mechanism (IVCT) is discarded [27].

b) Manganese-Al_2O_3 corundum pink.

The manganese-Al_2O_3 corundum pink is a solid solution of Mn^{3+} in the corundum lattice replacing Al^{3+} [28]: as the Cr-corundum pigment is rather stable in most applications. Color variants are obtained by co-doping Mn with Ca or P. Both pigments are not used in digital decoration; they are not suitable for quadrichromie and cost of micronization (due to the hardness of corundum). Both show easy synthesis by solid state, possible also in absence of mineralizers [29] and by mechanosynthesis [30].

The solid solution $Mn_xAl_{1-x}O_3$ x=0.02 to 0.2 obtained from eskolaite (Cr_2O_3), AlO(OH) and Mn_2O_3 as starting precursors, fired at 1200°C/2h, is followed by XRD and CIEL*a*b* in Table 4.2.

Table 4.2. XRD and CIEL*a*b* (5wt.% enameled in a double firing glaze (1000°C)), of solid solution $Mn_xAl_{1-x}O_3$ x=0.002 to 0.2 obtained from eskolaite (Cr_2O_3) and AlO(OH) as starting precursors fired at 1200°C/2h: CRYSTALLINE PHASE C(corundum α-Al_2O_3, 35.1°2θ peak), H(hausmanite, Mn_3O_4, 36.1°2θ peak)

x	Phase (counts/s)	L*a*b*
0.02	C(890)	79.4/13.3/10.5
0.05	C1(740)	71.9/13.05/11.5
0.1	C1(640) H(40)	67.6/12.3/11.7
0.2	C1(480) H(45)	60.4/11.2/10.8

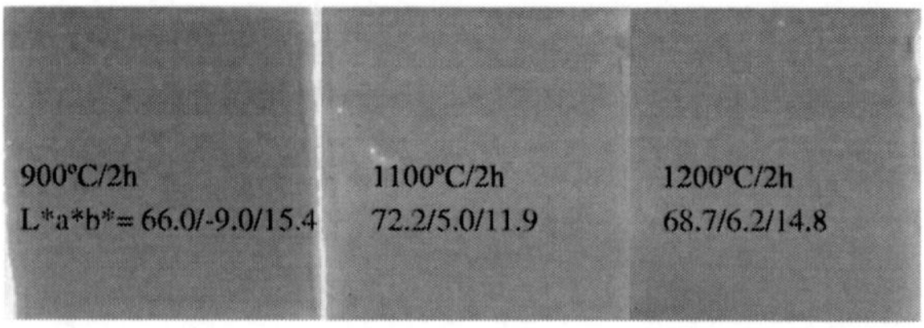

a

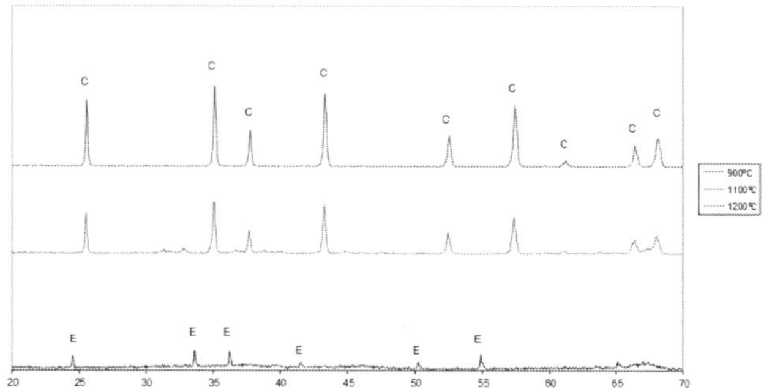

b

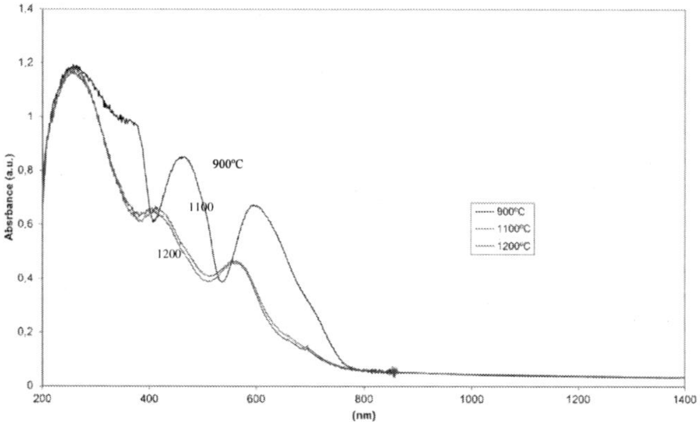

c

Figure 4.11. (a) Color evolution with temperature of color of sample $Cr_{0.2}Al_{1.8}O_3$ obtained from eskolaite (Cr_2O_3) and AlO(OH) (5wt.% enameled in a double firing glaze (1000°C)), (b) XRD and (c) UV-Vis-NIR absorbance spectrum of sample $Cr_{0.2}Al_{1.8}O_3$. CRYSTALLINE PHASES: E (eskolaite), C (corundum).

From data on Table 4.2 is observed that samples x=0.02 and 0.05 show corundum as single crystalline phase, the higher content samples x=0.1 and 0.2 crystallize the spinel hausmanite Mn_3O_4 as residual manganese crystalline phase. Therefore the limit of solid solution is between 0.05 and

0.1. The pink powder darkens (L* decreases) with x and both red a* and yellow b* parameters of 5wt.% enameled in a double firing glaze (1000ºC) show a maximum at x=0.05 that is considered the optimal composition coinciding with the limit of solid solution.

Figure 4.12 shows the evolution with temperature of color (5wt.% enameled in a double firing glaze (1000ºC)), XRD and optical UV-Vis-NIR absorbance spectrum of sample $Mn_{0.05}Al_{1.95}O_3$ obtained from eskolaite (Cr_2O_3) and Mn_2O_3 as starting precursors.

At 900ºC/2h the XRD shows corundum as the main crystalline phase and residual peaks of eskolaite that gives a light pink color (L*a*b*=78/8.5/7) when 5 wt.% enameled in a double fire glaze (1000ºC). At 1100ºC/2h only corundum is detected by XRD and the enameled sample shows a pink color (L*a*b*=74.4/10.5/7.9). Finally at 1200ºC only corundum is detected by XRD and the glazed sample shows the best pink shade (L*a*b*=68.7/15.2/10.2).

UV-Vis-NIR absorption spectra on Figure 4.12 show that the unreacted sample at 900ºC presents a feature associated with Mn^{3+} (d^4) in octahedral coordination in eskolaite lattice: a cation-anion charge transfer band at 250 nm and a band centered at around 600 nm of Mn^{3+} (d^4) in octahedral coordination (low spin) which consists of an intensive, broad, often two-humped band due to $^5E_g \rightarrow {}^5T_{2g}$ transition with Jahn–Teller splitting of the $^5T_{2g}$ level or of two dichroic bands in the lower symmetry fields [31].

The $E_g \rightarrow {}^5T_{2g}$ transition band shifts from 485 nm (900ºC) to 495 nm (1200ºC) and the inflection point also shifts from 580 nm (2.14 eV) at 900ºC to 590 nm (2.10 eV) at 1100ºC and 600 nm (2.07 eV) at 1200ºC. To the contrary that in the case of Cr-corundum, when the solid solution is obtained (1100 and 1200ºC) the band shift to higher wavelengths. This shift cannot be explained by the entrance of the bigger ion (Mn^{3+}, Shannon-Prewit radius 0.72 Å in octahedral coordination low spin) replacing Al^{3+} (Shannon-Prewit radius 0.675 Å in octahedral coordination) which enhance the crystal field over Mn^{3+} (d^4) ion, therefore some other Mn valences (e.g. Mn^{4+}, 0.675 Å) associated to oxygen vacancies which relax the crystal field over Mn^{3+} should be considered. Therefore the pink color of Cr-corundum pink solid solution can be explained by the crystal field theory with a cooperative transition of Cr^{3+} influenced for Al^{3+} but in this case a component of the intervalence charge transfer mechanism Mn^{3+}- Mn^{4+} (IVCT) cannot be discarded.

c) Chromium-$LnAl_2O_3$ perovskites (Ln=lanthanide).

In the perovskite red pigments $Ln_x(Al_{2-x-y}Cr_y)O_3$ [32,33], the color is due to the introduction of Cr^{3+} ions into the octahedron AlO_6 of $LnAlO_3$ perovskite lattice, substituting Al^{3+} ions. As discussed above, chromium (III) ion has an ionic radius bigger than aluminum (III) ion (Shannon-Prewit ionic radius for six-folded coordination of 0.755 Å and 0.675 Å respectively). Submitted to an octahedral crystal field of oxygen ions, like in Cr_2O_3 oxide, Cr (III) presents two absorption bands in visible range due to ${}^4A_{2g}\rightarrow{}^4T_{1g}$ transition (blue) and ${}^4A_{2g}\rightarrow{}^4T_{2g}$ transition (yellow).

In the $LnAlO_3$ perovskite lattice, the crystalline field over Cr^{3+} ion can be controlled by ionic radius of Ln^{3+} ion located in the centre of the FCC lattice described by Al^{3+} (edge positions) and O^{2-} ions: if Ln^{3+} size increases (La, Sm, Gd, Y), Cr-O distance decrease and the high tetragonal distortion of d orbitals of chromium (III) enhances the crystalline field over de ion. Consequently absorption bands shift to higher frequencies and the ${}^4A_{2g}\rightarrow{}^4T_{2g}$ transition yellow band shift to green wavelength resulting in a reflection in red. In effect $Y_x(Al_{2-x-y}Cr_y)O_3$ solid solutions become the best red color of the $Ln_x(Al_{2-x-y}Cr_y)O_3$ series.

The color of these perovskites stems from intense absorption of the yellow-to-violet wavelengths by a band occurring at low Cr concentration when the pigment is synthesized with mineralizers. As the origin of such a band is still under debate, the effect of increasing chromium doping on the relaxation of the perovskite structure was investigated through the substitution of isovalent cations with different ionic radii (Cr^{3+} after Al^{3+} in yttrium orthoaluminates) as well as with almost equivalent ionic radii (Cr^{3+} after Ga^{3+} in La–Nd orthogallates). Although perovskites exhibit the weakest structural relaxation around Cr^{3+} ion, meaning that the chromophore is highly "compressed" even in small concentration (e.g., particularly in orthoaluminates), the incorporation of Cr ions occurs with increasing covalence and polarizability of the Cr-O bond. This mechanism is not replicated in orthogallates, where the tiny difference of ionic radii between octahedrally coordinated Cr^{3+} (0.615 Å) and Ga^{3+} (0.62 Å) induces subtle changes in crystal structural and optical properties. Thus, the response to Cr incorporation in terms of both cell and octahedral site distortion is different in Y, La, and Nd perovskites.

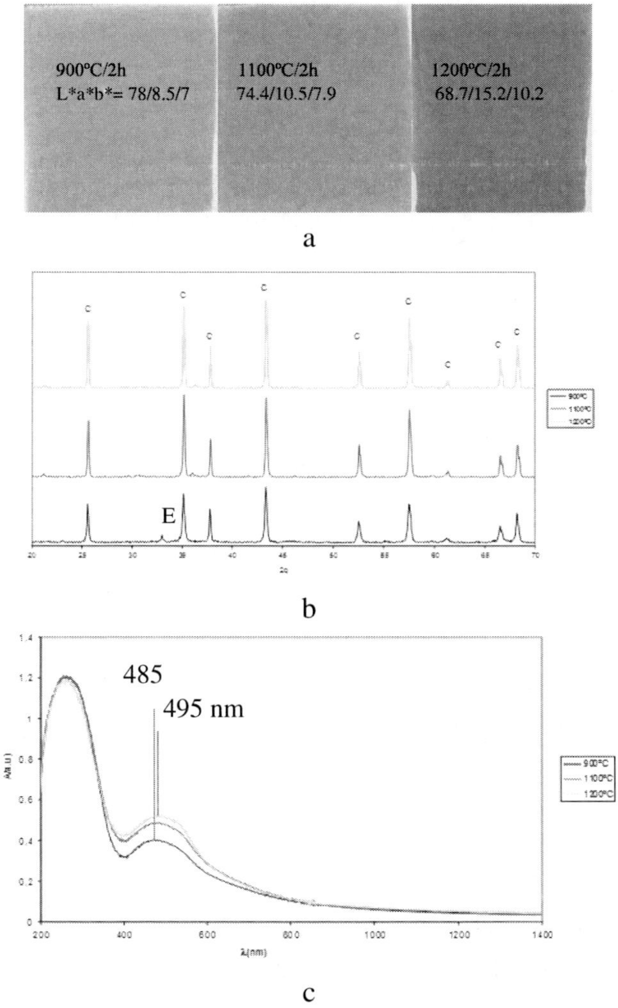

Figure 4.12. Color evolution with temperature of sample $Mn_{0.05}Al_{1.95}O_3$ obtained from eskolaite (Cr_2O_3) and Mn_2O_3 (5wt.% enameled in a double firing glaze (1000ºC)), XRD and optical UV-Vis-NIR absorbance spectra of sample $Mn_{0.05}Al_{1.95}O_3$. CRYSTALLINE PHASES: E (eskolaite), C (corundum).

Final remarks highlight that the red color of $Y(Al,Cr)O_3$ is developed only in the presence of mineralizers, with Ca salts being the most effective. A little amount of calcium has to be incorporated at the cubic site and the charge mismatch between Y^{3+} and Ca^{2+} is likely compensated by oxygen vacancies. The intense absorption band responsible for the red color is thought to be originated by the occurrence of Cr^{3+}–Ca^{2+} pairs with a possible involvement of intervalence Cr^{3+}–Cr^{4+} charge transfer [34].

d) Chromium-$Y_3Al_5O_{12}$ (YAG) garnet.

There is only one color cataloged by the DCMA based on the structure of garnet (cubic, space group *Ia3d*); it is Victoria Green based on the structural modification of the Cr^{3+} ion in the uvarovite garnet $Ca_3Cr_2Si_3O_{12}$.

In the colorless yttrium-aluminum garnet YAG ($Y_3Al_5O_{12}$), the Al^{3+} occupies both the octahedral and tetrahedral sites in the garnet structure, and the Y^{3+} occupies the dodecahedral position. These polyhedra are slightly distorted and connected by vertex sharing, and then each octahedron is connected with six tetrahedra and each tetrahedron with four octahedra. The yttrium ions are placed in the dodecahedral holes left vacant by the anterior skeleton, crystallizing with a cubic habit (a = 12.01 Å) and assigned to space group *Ia3d*.

The possibility of doping the YAG structure with other lanthanide cations is well known: Nd^{3+} is a common dopant (up to 6% molar) used in order to increase photoluminescence and cathodoluminescence by creating a metastable level of energy at which the states excited can decay non-radioactively. Likewise, Ce^{3+} can also substitute for Y^{3+} in the dodecahedral vacancy and the solid solution is used as a fast-decaying phosphorescent material for detectors that require low background noise in illumination. Cerium reaches 6 mol % doping levels in YAG.

The possibility of doping YAG with transition metals has also been investigated. YAG can be doped with Fe and Ga, but now the substitution takes place in the octahedral and tetrahedral positions, as the existence of YIG ($Y_3Fe_5O_{12}$) and YGG ($Y_3Ga_5O_{12}$) garnets was predictably known. Chromium is also capable of entering both positions depending on its oxidation state: Cr^{3+} is incorporated into octahedral positions and Cr^{4+} to tetrahedral positions, the load balance is compensated with codopants or by the formation of vacancies, in this sense, vanadium behaves in a similar way to chromium, locating itself based on its valence [35, 36].

Shirpour et al. [37] study the $YAl_{1-x}Cr_xO_3$ solid solutions mineralized by addition of a 6 wt.% of ternary mineralizer system (weight ratio $NaF/MgF_2/Li_2CO_3$ = 3/2/1) based on binary NaF–MgF_2 which shows low melting point around 835°C and evaporates slowly. But it was not effective mineralizer due to its high viscosity and in order to decrease the viscosity Li_2CO_3 was added. At 1500°C/6h only $YAlO_3$ is detected and the best red shade (the highest value of coordinates a*) is reached for x=0.04. But at 1300°C/6h crystallizes a mixture of cubic and tetragonal YAG garnet ($Y_3Al_5O_{12}$) that showed red color, but the highest red component (a*) was reached at 1500°C, due to higher diffusion rate of chromium in $YAlO_3$ structure. Therefore the red color is not only associated to $YAlO_3$ perovskite; chromium doped YAG also gives red color and at low temperature.

In order to analyze the Cr-YAG red pigment in the CaO-Al_2O_3- Y_2O_3 system, Figure 4.13 shown the molar composition of prepared samples (see table inserted in Figure 4.13) and its localization on the phase diagram (PD) [38] and the 5 wt.% enameled samples in a double firing frit (1000°C).

The powders were prepared from Cr_2O_3, $CaCO_3$, AlO(OH) and Y_2O_3 as starting precursors and 0.15 moles of NaF were added as mineralizer. The mixtures homogenized in a ball mill in acetone medium, were fired at 1250°C/6h. When powders were 5 wt.% enameled in a double fire frit (1000°C) (see Figure 4.13.c), the only composition that produce red color is 3 with L*a*b*=44/22/19. Compositions 4 and 5 produce a light red color and the others samples are light green or colorless (see Figure 4.13.c). Without calcium addition, color is not achieved.

XRD of samples 3 and 5 showed in Figure 4.14 indicate that in the efficient red pigment 3, the tetragonal-$Y_3Al_5O_{12}$ garnet crystallize as single crystalline phase (in the powder 5 the cubic-$Y_3Al_5O_{12}$ garnet is the phase predominant but some peaks associated to $YAlO_3$ are detected by XRD).

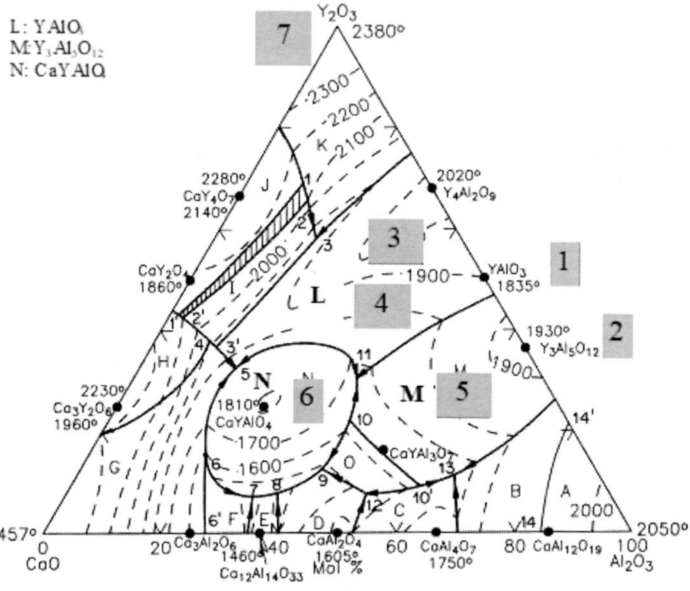

a

Sample (phase on PD)	CaO	NaF	Cr_2O_3	Al_2O_3	Y_2O_3	L*a*b* (enameled)
1($YAlO_3$)	-	0.13	0.1	0.50	0.50	-
2($Y_3Al_5O_{12}$)	-	"	"	0.60	0.36	-
3($YAlO_3$)	0.15	"	"	0.30	0.55	44/22/19
4($YAlO_3$)	0.20	"	"	0.35	0.45	70/16/12
5($Y_3Al_5O_{12}$)	0.15	"	"	0.60	0.25	-
6($CaYAlO_4$)	0.40	"	"	0.35	0.25	-
7(Y_2O_3)	-	"	"	-	1.00	-

b

Figure 4.13. (Continued).

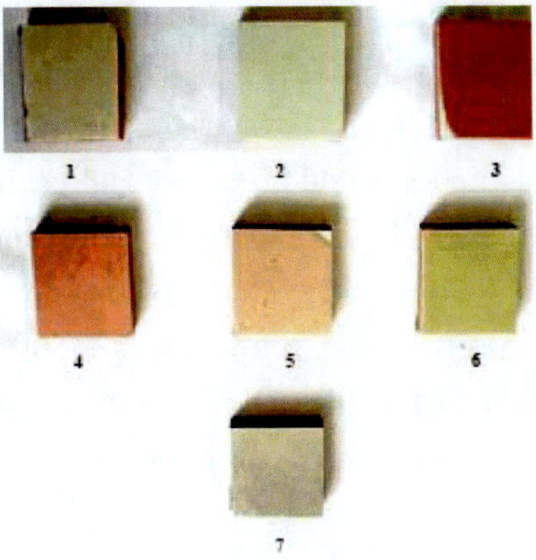

c

Figure 4.13. Molar composition of prepared samples (Table b) and its localization on the phase diagram (PD) of system CaO-Al_2O_3-Y_2O_3 (a) and the 5 wt.% enameled samples in a double firing frit (1000°C) (c).

UV-Vis-NIR spectrum of powder of this red sample 3 is compared with the spectrum of eskolaite (Cr_2O_3) with clear differences (Figure 4.15). As above discussed, in the literature [34,37] the color of Cr-$YAlO_3$ perovskite ($YAl_{1-x}Cr_xO_3$) is associated to Cr^{3+} chromophore ion replacing Al^{3+} but its UV-Vis-NIR spectrum is very similar (the same) to the spectrum of sample 3 in Figure 4.13 (b): in effect, Shirpour et al. [37] report that at the wavelength values smaller than 580 nm there is a region of nearly constant absorption, probably owing to matrix effect, and the distinction of the bands assigned to the allowed transition of the Cr^{3+} is impeded. It is also necessary to indicate that no absorption band appears in the visible spectrum between 580 and 700 nm (in the red region), contrary to the spectrum of green Cr_2O_3 where a wide absorption band can be appreciated in the same region. In effect, the bands associated to the allowed transition of the Cr^{3+} cannot be assigned and the spectrum is very different to the green eskolaite spectrum. Instead, the feature of the spectrum in Figure 4.13 for red sample is very similar to the Cr-malayaite (see Figure 4.8) compatible with Cr^{4+} (d^2) in octahedral coordination, with the spectrum dominated by its main Laporte-forbidden and spin-allowed transition $^3T_{1g}(3F)$ to $^3T_{2g}(3F)$ at 520 nm and the transition $^3T_{1g}(3F)$ to $^3T_{1g}(3P)$ at 410 nm.

In order to study the limit of solid solution of chromium in the tetragonal-YAG, were prepared samples with composition $xCr_2O_3.0.15CaO.Al_2O_3.1.1Y_2O_3$ x=0.02 to 0.2 mineralized by the addition of 5 wt.% of Na_2SiF_6 using the same precursors above described and fired at 1250°C/6h. Although NaF has low melting point (990°C), it easily evaporates and escapes from the system, Na_2SiF_6 slowly decomposes at 500°C in NaF and gaseous SiF_4. Figure 4.16.a shows 5wt.% enameled samples in a double firing frit (1000°C), XRD of powders and the UV-Vis-NIR absorbance spectra of enameled samples.

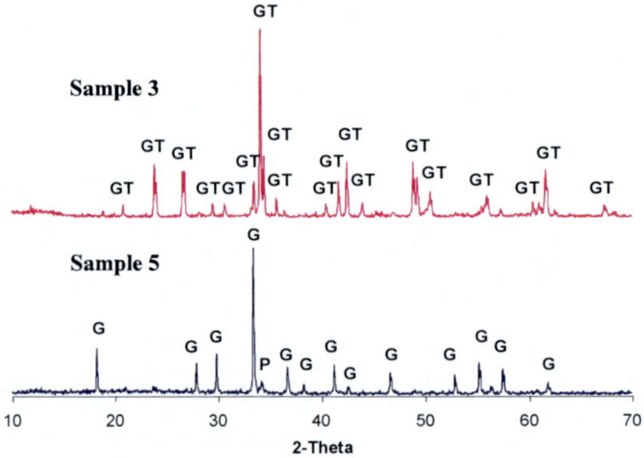

Figure 4.14. XRD pattern of samples 3 and 5 of Figure 4.12: CRYSTALLINE PHASES: G (YAG, $Y_3Al_5O_{12}$, cubic), P ($YAlO_3$, perovskite), GT (YAG, $Y_3Al_5O_{12}$, tetragonal).

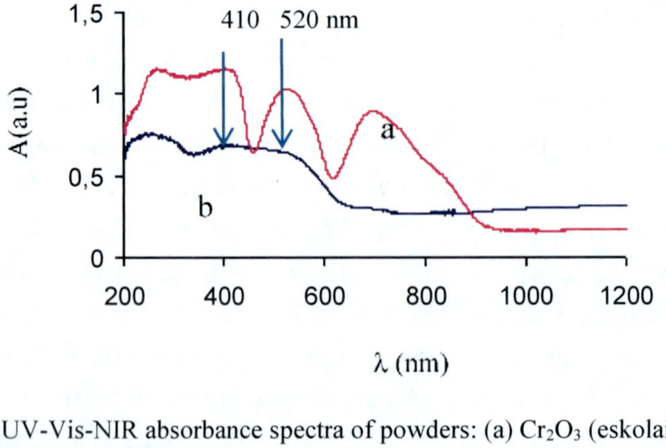

Figure 4.15. UV-Vis-NIR absorbance spectra of powders: (a) Cr_2O_3 (eskolaite), (b) red Cr-YAG (sample 3 on Figure 4.13).

XRD on Figure 4.16.b shows that the x=0.02 sample shows tetragonal-YAG as single phase detected, but peaks of undesired phases $Y(Y_2O_3)$ and $A(Al_2Y_4O_9)$ are detected from x=0.05; therefore the limit of solid solution of chromium is close to 5 mol%. However, developed color in the glaze increases the intensity (lower L*) as x increases, the red shade also increases up to the x=0.1 sample which shows the maximum red tone (a*=15.9) as the optimal performance. UV-Vis-NIR spectra on Figure 4.16.c shows a feature compatible, as discussed above, with Cr^{4+} (d^2) in octahedral coordination with the spectrum dominated by its main Laporte-forbidden and spin-allowed transition $^3T_{1g}(3F)$ to $^3T_{2g}(3F)$ at 520 nm and the transition $^3T_{1g}(3F)$ to $^3T_{1g}(3P)$ at 410 nm. From x=0.05 weak bands growth at 1200 and 700 nm respectively; the band at 1200 nm that can be associated to the overlapping contributions of spin-forbidden transitions $^1E_g(1D)$ and $^1T_{2g}(1D)$ in Cr^{4+} eight-folded and the band at 700 nm to the forbidden transition $^4A_{2g}(4F)$ to $^2T_{1g}(2G)$, respectively of Cr^{3+} (d^3) in octahedral coordination. Therefore some entrance of Cr^{3+} is evidenced at high chromium content in the tetragonal-lattice.

The detected bands increase in intensity as x increases but practically show the same position for all compositions (with a inflection point around 630 nm or 1.97 eV); due to the higher size of Cr^{4+} (Shannon-Prewit radius 0.69 Å in octahedral coordination) than Al^{3+} (Shannon-Prewit radius 0.675 Å in octahedral field) the crystal field over Cr^{4+} increases and bands should shift to lower wavelengths. The codoping with Ca^{2+} (Shannon-Prewit radius 1.14 Å in octahedral coordination) replacing Al^{3+} or Y^{3+} (Shannon-Prewit radius 1.04 Å in octahedral coordination) allow the lattice electroneutrality but it point in the same direction and bands should shift to lower wavelengths. Probably the more open lattice of tetragonal-YAG observed for red samples (not the normal cubic lattice) explain that the entrance of codopants do not modifies the crystal field. Therefore the reddish color of Cr-YAG solid solution show a mechanism in which both Cr^{4+} and Ca^{2+} are involved in a cooperative transition and the chromophore agent is Cr^{4+}eigth-folded.

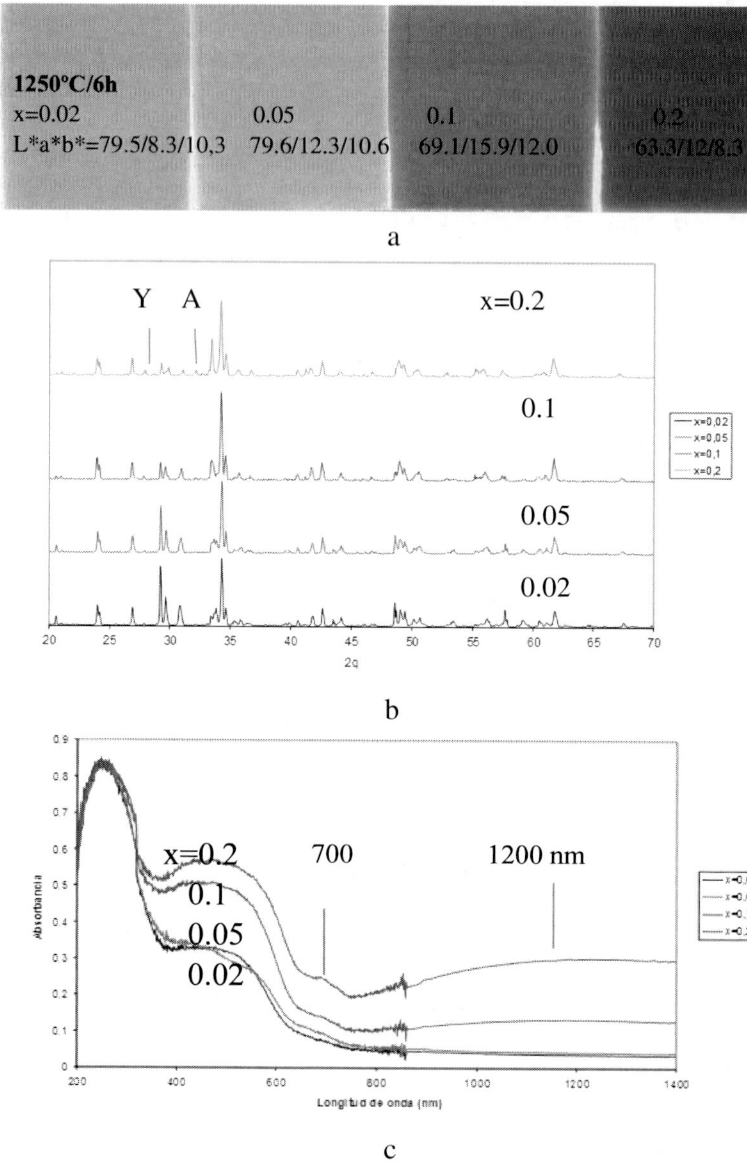

Figure 4.16. (a) 5wt.% enameled samples ($xCr_2O_3.0.15CaO.Al_2O_3.1.1Y_2O_3$ with 5 wt.% Na_2SiF_6 fired at 1250ºC/6h) in a double firing frit (1000ºC), (b) XRD of powders and (c) UV-Vis-NIR absorbance spectra of enameled samples: CRYSTALLINE PHASES: $Y(Y_2O_3)$, $A(Al_2Y_4O_9)$.

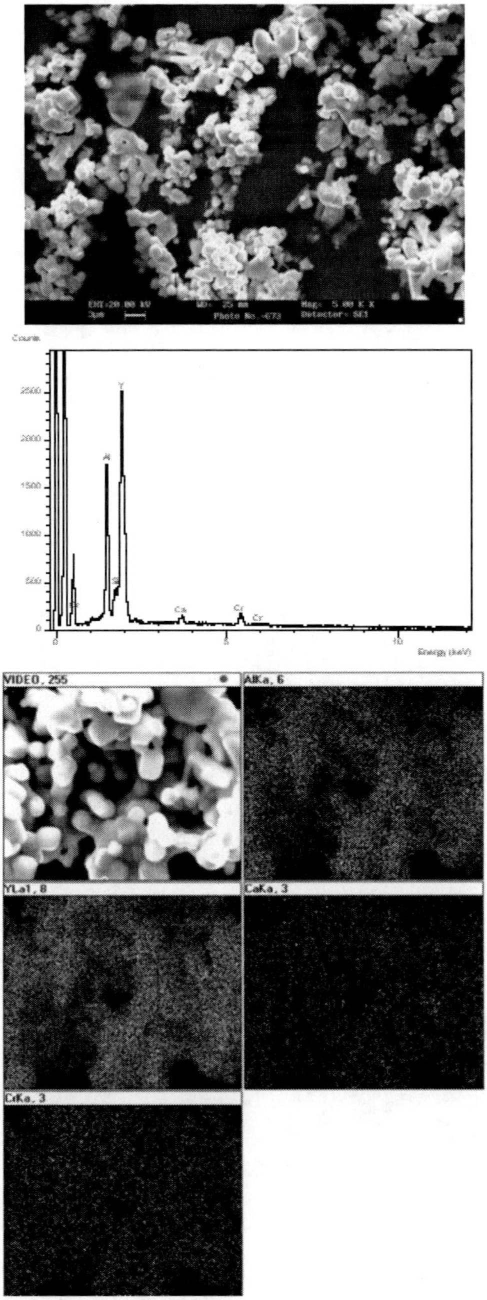

Figure 4.17. SEM micrograph of $0.1Cr_2O_3.0.15CaO.Al_2O_3.1.1Y_2O_3$ with 5wt.% Na_2SiF_6 fired at 1250ºC/6h with its EDX analysis (global and mapping).

Figure 4.17 shows the SEM micrograph of $0.1Cr_2O_3.0.15CaO.Al_2O_3.1.1Y_2O_3$ optimal reddish sample with 5wt.% Na_2SiF_6 as mineralizer fired at 1250ºC/6h and its EDX analysis (global and mapping). Regular tetragonal particles between 05-2 μm can be observed, the EDX spectrum detects the presence from Si of the mineralizer and the corresponding mapping shows a homogeneous atomic distribution of all the ions in sample.

4.6. Localized Surface Plasmonic Resonance (SPR)

A classical physics picture of plasmons can be the oscillation of electron density with respect to the fixed positive ions in a metal. When a metallic crystal is submitted to an external electric field, electrons will move to the positive potential producing positive ions on the negative potential: if the electric field is removed, the electrons move to positive ions direction, repelled by each other and attracted to the positive ions. They oscillate back and forth at the plasma frequency until the energy is lost in some kind of resistance or damping (Figure 4.18). Plasmons are a quantization of this kind of oscillation following the proposal of David Pines and David Bohm in 1952 [39].

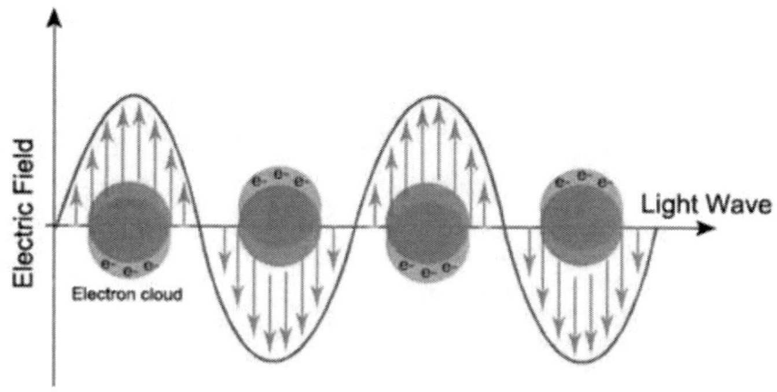

Figure 4.18. Basics of localized surface plasmon resonance (LSPR) of gold nanoparticles due to collective oscillation of surface electrons with incident light at a specific wavelength (from https://www.cytodiagnostics.com/pages/gold-nanoparticle-properties).

Therefore SPR (Surface Plasmonic Resonance) is a quantum electromagnetic phenomenon arising from the interaction of light with free electrons at a metal–dielectric interface emerging as a longitudinal electromagnetic wave in a two-dimensional gas of charged particles such as free electrons in metals. Under certain conditions the energy carried by the photons is transferred to collective excitations of free electrons, called surface plasmons. This transfer of energy occurs only at a specific resonance wavelength of light when the momentum of the photon matches that of the plasmon. The surface plasmons excited are strongly localized across the interface, and may be considered classically as electromagnetic surface waves that propagate along the interface and decay exponentially with distance normal to the interface. SPR is responsible for a dip in reflectance at the specific wavelength, the dip resulting from the absorption of optical energy in the metal. Since SP waves are tightly bound to metal–dielectric interfaces penetrating around 10 nm into the metal (the so-called "skin depth") and typically more than 100 nm into the dielectric (depending on the wavelength), they concentrate electromagnetic waves in a region that is considerably smaller than their wavelength, a feature that suggests the possibility of using surface plasmons for the fabrication of nanoscale photonic circuits operating at optical frequencies. This constitutes an important area of research, since surface-plasmon-based circuits are known to merge the fields of photonics and electronics at the nanoscale, thereby enabling them to overcome the existing difficulties related to the large size mismatch between the micrometer-scale bulky components of photonics and the nanometer-scale electronic chips [40].

Depositing plasmonic metallic nanoparticles on the surface of a semiconductor is also an effective surface modification for improving light harvesting. Plasmonic-enhanced water splitting was first reported in 2004. TiO_2 thin films were immersed into electrolyte to deposit gold and silver nanoparticles, and the plasmonic effect is attributed to the transfer of photo-induced electrons from the CB (Conductive band) of the metal to the CB of the semiconductor. Gold and silver particles have also been deposited on the surfaces of Fe_2O_3 and N-doped TiO_2 electrodes, resulting in enhancement that is attributed to plasmon resonance energy transfer from the metal nanoparticles to the semiconductor, which is due to the spectral overlap between the plasmonic metal and the semiconductor. Silver and gold are considered the best plasmonic metals so far, and have been deposited on the surfaces of CeO_2, CdS, and WO_3 [41].

Therefore in the localized surface plasmonic resonance the conducting electrons at the surface of metals are collectively oscillated when their frequency matches that of the incident electromagnetic radiation. As a consequence of this phenomenon, plasmonic metal nanoparticles (e.g., gold nanoparticles or silver nanoparticles) show strong absorption within certain wavelength located in visible region and therefore have corresponding colors. The LSPR is intimately related to particle size, composition, morphology, interparticle distances, and refractive index of the surrounding media [42].

Surface plasmons can be used as a mechanism for pigments by controlling the shape and size of particles which determines the types of surface plasmons that can be coupled into and propagate across it. This, in turn, controls the interaction of light with the surface. The particular wavelength of light where this occurs is strongly dependent on the nanoparticle size, shape, surface, and agglomeration state. For instance, anisotropic nanostructures such as gold nanorods, nanocages, and nanocubes the plasmon resonance can be tuned from 600 to 1400 nm by varying the aspect ratio. Au nanorods display two absorption peaks, one correlated with the short transversal axis at around 520 nm and the other one with the longer longitudinal axis situated between 640 and 1000 nm depending on the aspect rate (length/width) of the nanostructures (see Figure 4.19). These nanorods can be used in the "photothermal therapy of cancer": the nanoparticles are chemically modified with oligonucleotides that are localized on the tumor cells after injection into a vein. Local irradiation of the tumor site with a laser at the plasmonic resonance wavelength (e.g., 900 nm) causes nanoparticle heating and cell death followed by tumor eradication. Spherical particles with sizes above 100 nm have broader peaks spanning into the 600 nm range due to the presence of both transversal and longitudinal surface plasmon resonances, spherical 100 nm nanoparticles shows an absorption band at 570 nm (absorbs yellow and shows green color) and which shifts to

520 nm in the case of 20 nm nanoparticles (absorbs blue and shows a red color), gold nanoparticles with diameters below 2 nm do not exhibit surface plasmon resonance.

These effect are illustrated by the historic Lycurgus Cup, a 4th-century Roman glass cage cup made of a dichroic glass exposed in the British Museum, which shows the mythical King Lycurgus, who tried to kill Ambrosia, a follower of the god Dionysus; the cup shows a different color depending on whether or not light is passing through it: red when lit from behind and green when lit from in front (Figure 4.20). It is estimated that to a conventionally composed Roman glass flux 330 ppm of silver and 40 ppm of gold were added:

> "These particles were precipitated as colloids and form a silver-gold alloy. When viewed in reflected light the minute metallic particles are just coarse enough to reflect enough of the light without eliminating the transmission. In transmitted light the fine particles scatter the blue end of the spectrum more effectively than the red end, resulting in red transmission, and this is the color observed. Since it is impossible that the Roman artisans managed to add these incredibly low levels of silver and gold to the volume of the glass used to make the vessel deliberately, the levels were probably added at higher levels to a larger volume of glass-melt, and increasingly diluted by adding more glass" [43].

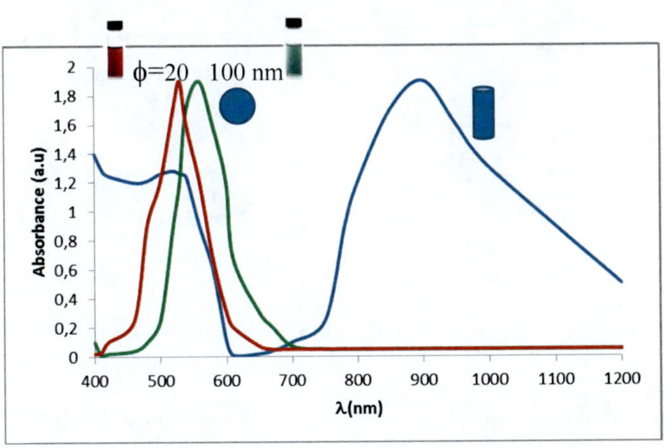

Figure 4.19. Optical spectra of spherical Au nanoparticles (●) and nanorods (▮).

As above discussed, the effect is associated to the interaction of light with electrons on the gold nanoparticle surface; at a specific wavelength of light, collective oscillation of electrons on the gold nanoparticle surface cause a surface plasmon resonance (SPR) resulting in strong extinction of light (absorption and scattering). In the case of Lycurgus cup the dichroism (two colors) is due to the presence of nanoparticles, silver 66.2%, 31.2% gold, and 2.6% copper, up to 100 nm in size, dispersed in a glass matrix. The red color observed is a result of absorption of light (~520 nm) by the gold particles. The purple color results due to the absorption by the larger particles while the green color is attributed to the light scattering by colloidal dispersions of silver particles with size >40 nm. The Lycurgus cup is recognized as one of the oldest synthetic nanocomposites [44].

Besides the Romans, medieval artisans also explored the effect of the addition of metal particles in glass to create stained glass windows. Stained glass was also produced in Chinese ancient culture and porcelain was also decorated using gold nanoparticles with sizes in the range of 20–60 nm. The use of gold had been extended even to areas in medieval medicine.

a b

Figure 4.20. Lycurgus Cup: (a) reflected light, (b) transmitted light (image from British Museum).

The use of plasmonic ceramic pigments based on metallic colloids has been known since ancient times. Thus, the red "sang de boeuf" colors were obtained by reacting in the furnace a colloidal dispersion of metallic copper with the glazes by reduction of copper (I) salts. This reduction was preceded by an oxidizing stage, the purpose of which was to control the size of the colloidal particles, since if they were very small, yellow instead of red colorations were obtained.

The discovery of the "*Cassius Purple*" stain by Joham Rudolph Glauber, dates back to 1659. In 1719, only 10 years after the discovery of porcelain at Meissen, this stain was used according to the note book of a color chemist of the Austrian factory named Samuel Stölzen who wrote [45]:

> "The finest purple is prepared from gold:
>
> 1. Take 3 batches of the best aqua regia and dissolve in it half a gold duchy in it, the liquid will turn yellow. Take 3 loth of best aqua regia and dissolve in it 1/2 a gold ducat, and the liquid will turn yellow.
> 2. Take then 4 batches of English tin and dissolve it in the necessary aqua regia. A fine tincture will be produced. Then take 4 loth of English tin and dissolve this in as much aqua regia as is necessary. It will likewise produce a fine tincture.
> 3. When both solutions are ready, add a few drops of each simultaneously to a quart glass of pure water — if it is bigger it is even better — and continue until all is gone it, when a red tinge will appear and the purple will fall to the bottom."

However, the production of the purple is not as simple as it seems, in 1757 the chemist from Sèvres, Jean Hellot, introduced a beautiful and fresh pink glaze based on the purple of Cassius that has probably been the most valued of all that have been developed. This enamel was known in England by merchants and known as "*Rose Pompadour*" in reference to Madame de Pompadour, the favorite of Louis XV who promoted the Parisian factory of Sèvres. Hellot's recipe remained as long as he lived. After his death in 1766, his successor Pierre Joseph Macquer was unable to produce it, admitting in his "Dictionnaire de chimie" the uncertainty in its manufacture:

> "Purple Powder of Cassius:
>
> This precipitate of gold may be made by several methods, but which of these is the best has not yet been decided. The cause of this uncertainty

is that the preparation requires delicate management and is uncertain in its result, the beauty of the color seeming to depend on some small circumstances not yet discovered."

The most important chemists of the nineteenth century were dedicated to unraveling the nature of the Cassius powder, Jeremias Benjamin Richter, discoverer of the law of multiple proportions of the chemical reactions and chemist of colors in Berlin porcelain factories, considers that the powder it is "*an intimate mixture of an extremely divided gold in tin lime,*" on the other hand Proust in 1806 in the royal laboratory of Madrid considers that the precipitate is a compound of metallic gold and tin oxide. Gay-Lussac in 1832 considers it a real compound formed between gold and tin oxide. Michael Faraday in 1857 in his entrance lecture in the Royal Society "*Experimental Relations of Gold (and other Metals) to Light*" considers the precipitate formed "*essentially by finely divided gold associated with more or less oxide of tin.*"

However, it was the Viennese chemist Richard Zsigmondy, founder of the Schott factory, who in 1897 intensively investigated gold dyes and colloidal golds. Together with Jena's physicist Heinrich Siedentopf, he developed the projection ultramicroscope that allowed them to observe the nanometric particles of gold colloids. Zsigmondy was able to synthesize the Cassius precipitate using red colloidal gold, concluding that it was therefore a dispersion of colloidal gold on $Sn(OH)_4$ stannic acid and that such a mixture was capable of behaving as a chemical compound.

For this research, Richard Zsigmondy was awarded the Nobel Prize in Chemistry in 1925. Figure 4.21 shows a TEM micrograph dated in 1949 [46] of a sample of the Cassius precipitate showing clusters of dark nanometric particles of colloidal gold on the light surface of tin oxide.

The uncertainty in the production of Cassius purple is linked to the fine colloidal chemistry involved in its synthesis and which presents two distinct steps: (a) the reduction to metal gold by stannous ion, (b) the stabilization of gold in colloidal particles.

(a) The reduction to gold metal. The simple reaction could be:

$Sn^{4+} + 2e^{-} \rightarrow Sn^{2+}$ E°= -0.15 V

$Au^{3+} + 3e^{-} \rightarrow Au$ E°= 1.5 V

$3Sn^{2+} + 2Au^{3+} \rightarrow 2Au + 3Sn^{4+}$ E°= 1.65 V

Actually the reaction is more complex and involves the chlorinated complexes of the species in a very acidic medium:

$SnCl_6^{2-} + 2e^{-} \rightarrow SnCl_3^{-} + 3Cl^{-}$ E° = 0.0 V (HCl 1M).

$AuCl_4^{-} + 3e^{-} \rightarrow Au + 4Cl^{-}$ E° <1.5 V

$3\ SnCl_3^{-} + 2\ AuCl_4^{-} \rightarrow 2Au + 3\ SnCl_6^{2-}$ E°< 1.5 V

Figure 4.21. TEM micrograph of purple of Cassius dated on 1949 [44].

Therefore, the process is slower in chlorinated media which is absolutely necessary in the synthesis, and allows a gold precipitation more in line with the purpose of obtaining a fine dispersion of colloidal gold. In historical methods, such as Stolze's method, aqua regia (3HCl: HNO_3) was used to dissolve both metallic gold and tin to their chlorinated forms [47].

(b) Stabilization of colloidal gold.

TEM studies such as those shown in Figure 4.21 indicate that to obtain a good Cassius precipitate the mean size of the gold particles must be between 10 and 15 nm. To stabilize these sizes, the gold sol must be stabilized over another larger colloidal particle. In Cassius purple, the stabilizing colloid is the stannic oxide $Sn(OH)_4$, and two hydrated forms of the oxide SnO_2 can precipitate: the so-called-ortho-stannic oxide which is produced by hydrolysis of the salts of Sn^{4+} soluble in acids and bases, and β-metastannic which is very inert and is obtained by aging the former or, by the attack of tin metal with nitric acid. Both hydroxides have a rutile structure, observed by X-ray diffraction, only differences in the size of the particles and the degree of hydration are detected. The formation of one type of oxide or another, as well as the conditions for obtaining are very important since it is the support that isolates and stabilizes the gold nanoparticles on its surface. In the classical method, the attack with "aqua regia" ruled over the tin had to leave in solution a mixture Sn^{2+}-Sn^{4+}, in an approximate 2:1 ratio, which allowed the adequate presence of ortho-stannic colloids. At present, the corresponding chlorides are used directly and the above molar ratio can be optimized in a stricter and controlled way, making the obtaining of the pigment less uncertain. If the powder is dried once precipitated, it loses its pigmenting properties, due to the fact that the gold particles coalesce and grow as a consequence of the contraction of the stannic support. To avoid the problem that arises in the necessary decantation process, and the washing process if desired (to eliminate chlorides), a paste is obtained that is mixed with the frit by wet grinding. In this stage the paste obtained can be filtered and dried, sometimes it is even calcined at 750°C to stabilize the material, grinding it wet until the required particle size is reached and finally the coloring powder is dried.

For inkjet applications, in addition to purple or rose gold, tin chrome rose (Cr-malayaite), considered a solid solution of Cr^{4+} in $CaSnSiO_5$ sphene (see section 4.4) can be used. In Figure 4.22 both colors are compared when Cassius purple (with composition Au:3SnO_2) and commercial tin chrome pink (obtained by solid state reaction between cassiterite, calcium carbonate and quartz to obtain $CaSnSiO_5$ composition to which 0.1 molar lead chromate is added, firing the mixture at 1200°C) were mixed with a polyglycol serigraphic vehicle and deposited on a conventional single firing glaze (1080°C). The colloid of gold shows greater intensity, even though the colloid of Cassius is in a concentration 29 times lower than the commercial pink.

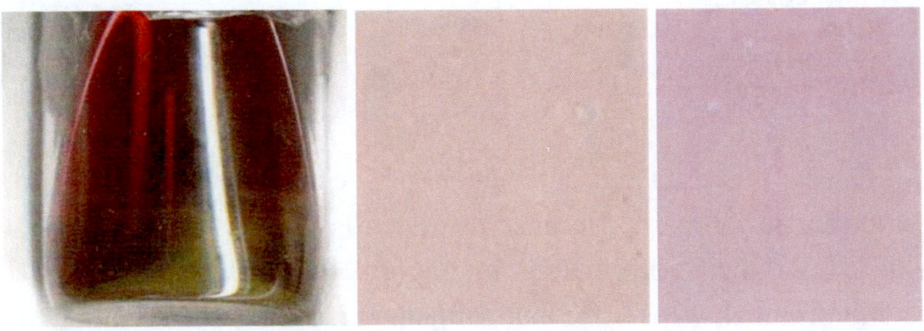

a b c

Figure 4.22. Silkscreen deposition comparison: (a) appearance of the wine red solution with the greenish black Cassius purple precipitate, (b) $0.1Cr_2O_3$-$CaSnSiO_5$ at 20 wt.%, (b) Au:$3SnO_2$ at 0.7 wt.%.

In the same way that Cassius purple is prepared, colloidal depositions of other precipitable metals by Sn(II) such as Ag or Cu can be developed. In highly chlorinated media of Cassius technology, silver produces Ag^0 colloid but also AgCl precipitate and $AgCl_2^-$ complex depending on the chloride concentration and pH. If the precipitated silver is added to Cassius purple in the grinding phase with the frit, allowing the modification of the red coloration towards redder tones if the frit incorporates lead, and when boron is avoided in the frit composition the coloration tends to blue. Copper colloids can also be prepared by Cassius processing, in this case, the formation of the green $CuCl_4^{2-}$ complex and the precipitation of white CuCl (which redissolves in highly chlorinated media in the form of $CuCl^{2-}$ and $CuCl^{3-}$ complexes), are observed. Cassius copper gives rise to turquoise colorations when deposited or mixed with conventional glazes [47].

4.7. Paired Spin Exchange Transitions (PET)

Two paramagnetic transition metal ions which are close typical distance of a few angstroms show magnetic interaction. The Heisenberg Hamiltonian, describing the spin interaction between two transition metal ions A and B, is shown in eq. 4.1.

$$H_{AB}= -J(S_A S_B) + j(S_A S_B)^2 \quad (4.1)$$

In theory, this exchange interaction can occur for both the 3d and 4f transition metal ion pairs. Because the electrons of the 4f lanthanides are shielded from their surrounding by the filled 5s and 5p orbitals, the interaction is much weaker for pairs of lanthanide ions. The second term of eq. 4.1 usually is very small and when this term is neglected, the energy values associated with the Heisenberg Hamiltonian can be determined, which is shown in eq. 4.2.

$$E(S)= - J[(S(S+ 1) - S_A(S_A + 1) - S_B(S_B + 1)] \quad (4.2)$$

From eq. 4.2 it can be observed that the energy is dependent on the total spin, S, and on the exchange coupling strength parameter J. The S can take all integer values between the sum and difference of the values of S_A and S_B. States with different values for S are at different energies [48]. As an example, the pair splitting of Mn^{2+} is discussed here. The influence of the crystal field on the position of the energy levels for the $3d^5$ transition metal ion Mn^{2+} is given in the Tanabe-Sugano diagram for $3d^5$ [49]. The ground state of $Mn^{2+}(3d^5)$ is $^6A_1$ ($S= 5/2$) and first excited state $^4T_1$ ($S = 3/2$). When both ions are in the ground state, both spins have the same value: 5/2. Therefore, the values of S for the exchange coupled pair can be: (S_A+S_B), ..., (S_A-S_B), so S=5, 4, 3, 2, 1, 0. When one of the Mn^{2+} or Fe^{3+} ions is in the first excited state, $^4T_1$, and the other ion is in the ground state, $^6A_1$, the total S value can be S=4, 3, 2, and 1. The sign of J in eq. 4.2 is determined by the type of magnetic interaction, $J > 0$ for a ferromagnetic and $J < 0$ for an antiferromagnetic interaction [50]. As an example for the splitting pattern for a Mn^{2+} pair with both ions in the ground state and one ion in the excited state, the interaction has been taken as an antiferromagnetic interaction for both the $^6A_1{}^6A_1$ ground state and the $^6A_1{}^4T_1$ excited state (Figure 4.23).

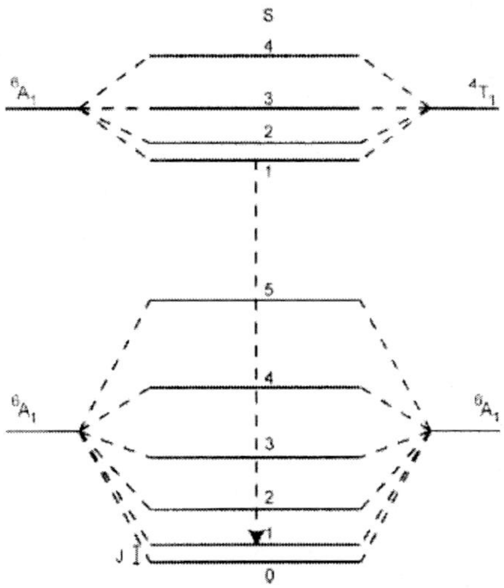

Figure 4.23. Splitting for energy levels in $^6A_1$-$^6A_1$ ground state and $^6A_1$-$^4T_1$ state resulting from exchange coupling for antiferromagnetic exchange-coupled $3d^5$ transition metal pairs. The vertical dotted line shows the spin allowed 1→1 transition (ΔS=0).

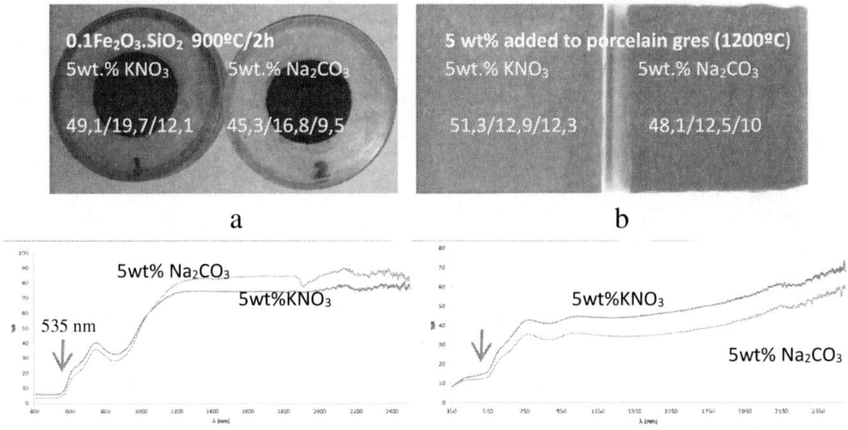

Figure 4.24. Hematite-silica pigments $0.1Fe_2O_3.SiO_2$ with 5 wt.% flux agent addition: (a) powders, (b) added to porcelain stoneware and (c) its diffuse reflectance spectra.

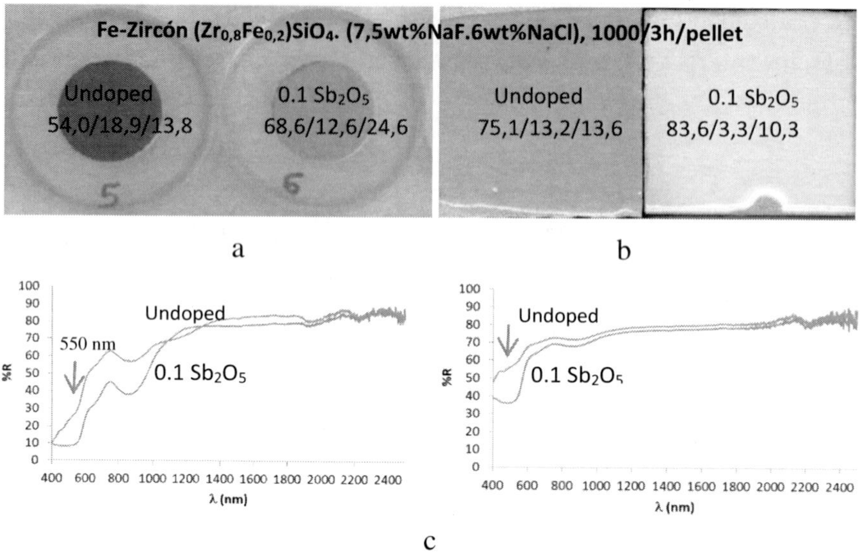

Figure 4.25. Samples of Fe-zircon $(Zr_{0,8}Fe_{0,2})SiO_4$. (7.5 wt.%NaF.6wt.%NaCl), 1000/3h/pellet starting with $FeSO_4.xH_2O$ and doped with 0.1 Sb_2O_5 $((Zr_{0,6}Sb_{0,2}Fe_{0,2})SiO_4$: powders (a), 5wt.% glazed in single firing frit (1080ºC) (b) and its diffuse reflectance spectra (c).

The value of *J* is dependent on the distance between ions in the pairs. Emissions originating from the pairs can be ascribed to transitions from the lowest excited spin state to different spin components of the ground state. Transitions where $\Delta S=0$ are spin allowed. From Figure 4.23, it can be observed that there are always transitions present, where $\Delta S=0$.

The absorption originating from the $^6A_1$-$^6A_1 \rightarrow {}^4T_1$-$^4T_1$ state excitation of the Fe^{3+}-Fe^{3+} pair are recognized in the literature as the base of several iron pigments such as hematite, Fe-silica [51], Fe-zircon $ZrSiO_4$ [52] and Fe-pseudobrookite Fe_2TiO_5 [53] ceramic pigments.

Figure 4.24 shows the results with hematite-silica pigments ($0.1Fe_2O_3.SiO_2$ with 5 wt.% flux agent addition) and in the Figure 4.25 with Fe-zircon $(Zr_{0,8}Fe_{0,2})SiO_4$.(7.5wt.%NaF.6wt.%NaCl flux agent addition), 1000/3h/pellet starting with $FeSO_4.xH_2O$ and doped with 0.1 Sb_2O_5 $((Zr_{0,6}Sb_{0,2}Fe_{0,2})SiO_4)$.

The Fe-silica pigment is not stable (solubilize) in glazes but it gives good yield of color in porcelain stoneware bulk additions, on the other hand the

Fe-zircon show good pink coral colors in glazes but unstabilizes in bulk ceramic matrices. Both pigments are considered obtained by inclusion of hematite crystals into silica (quartz or cristobalite) and zircon particles respectively. In effect, although in the case of Fe-silica always is detected hematite by XRD in the powders and rarely is observed in Fe-zircon (that usually show zircon as predominant crystalline phase and usually monoclinic zirconia as residual phase), UV-Vis-NIR diffuse reflectance spectra (Figure 4.24.c and Figure 4.25.c) is the same in both cases, with:

a) Intense charge transfer band at 290-310 nm (not shown),
b) Fe^{3+} spin forbidden transitions from the ground state $^6A_1$(6S) to ^4E and ^4A (4G) at 400 nm.
c) Fe^{3+} spin forbidden transitions from the ground state $^6A_1$ to the excited states $^4T_2$ and $^4T_1$, whose configuration is $(t_{2g})^4(e_g)^1$ at 650 and 900 nm respectively.
d) The band (marked on Figures 4.24.b and 4.25.b) at 535 nm (2.33 eV) for Fe-hematite spectra and 550 nm (2.25 eV) for Fe-zircon spectra, originated from the $^6A_1$-$^6A_1 \rightarrow {}^4T_1$-$^4T_1$ state absorption of the Fe^{3+}-Fe^{3+} pair, which is related to the hematite optical band gap that shifts slightly to higher values in the case of hematite occluded in zircon.

The addition of Sb_2O_5 unstabilizes the inclusion of hematite in zircon, giving orange colored powders that are colorless on glaze, indicating the delicacy of inclusion pigments.

REFERENCES

[1] Calbo, J., Sorlí, S., Llusar, M., Tena, M. A., Monrós, G., "Minimisation of toxicity in nickel ferrite black pigment," *British Ceramic Transactions* 103.1(2003) 3-9, https://doi.org/10.1179/096797804225012729.

[2] Levinson, R., Berdahl, P., Akbari, H., "Solar spectral optical properties of pigments–Part II: Survey of common colorants," *Sol Energy Mater Sol Cells*, 89(2000):3 51-389.

[3] Llusar M., Forés A., Badenes J. A., Calbo J., Tena M. A., Monrós G. "Colour analysis of some cobalt-based pigments." *J Eur Ceram Soc* 30(2001:1121–30.

[4] Forés A., Llusar M., Badenes, J. A., Calbo J., Tena M. A., Monrós G. "Cobalt minimisation in willemite ($Co_xZn_{2-x}SiO_4$) ceramic pigments." *Green Chemistry* 2(2000):93–100.

[5] Monrós G., Garcia A., Sorli S., "Heteromorphic Pigment Synthesis Mechanisms." *Ceramic Acta*, 2002, 14(2): 12-27.

[6] Jianfeng, W., Kun, L, Xiaohong, X., Yaxiang, Z., XiaoyanG, X., Xinbin, L., "Research and Development of Cadmium Sulphoselenide Red Pigment," *Journal of Wuhan University of Technology-Mater. Sci. Ed.* www.jwutms.net, 12(2015): 1247-1254.

[7] Robin, M. B., "The Color and Electronic Configurations of Prussian Blue," *Inorg. Chem.* 1, 2 (1962): 337-342.

[8] Verger, L., Dargaud, O., Chassé, M., Trcera, N., Rousse, G., Synthesis, properties and uses of chromium-based pigments from the Manufacture de Sèvres. *Journal of Cultural Heritage*, 30(2018): 26 - 33.

[9] Lopez-Navarrete, E., Caballero, A., Orera, V. M., Lazaro, F. J., Ocaña, M., "Oxidation state and localization of chromium ions in Cr doped cassiterite and Cr-doped malayaite." *Acta Materialia*, 51(8) (2003): 2371–2381.

[10] Cruciani, G., Dondi, M., Ardit, A., Lyubenova T. S., Carda, J. B., Matteucci, F., Costa, A. L., "Malayaite ceramic pigments: A combined optical spectroscopy and neutron/X-ray diffraction study," *Materials Research Bulletin* 44 (2009)1778–1785.

[11] Faurel, X., Vanderperre, A., Colomban, P., Pink pigment optimization by resonance, Raman spectroscopy. *Journal of Raman Spectroscopy*, 34(4) (2003): 290–294.

[12] Harisanov, V. et al., "Influence of crystallinity on chromatic parameters of enamels coloured with malayaite pink pigments." *Journal of the European Ceramic Society*, 23(3)(2003): 429–435.

[13] Lever, A. B. P., *Inorganic Electronic Spectroscopy,* 2nd ed. (Elsevier, Amsterdam, 1984).

[14] Wildner, M., Andrut, M., Rudowicz, C. Z., in: *A. Beran, E. Libowitzky (Eds.), Optical Absorption Spectroscopy in Geosciences. Part I. Basic Concepts of Crystal Field Theory, EMU Notes in Mineralogy, vol. 6*, (Eotous University Press, 2004 , p. 93).

[15] Burns, R. G., *Mineralogical Applications of Crystal Field Theory*, 2nd ed., (Cambridge University Press, Cambridge, 1993).

[16] Hochu, F., Lenglet, M., "Co(II) optical absorption in spinels: infrared and ligand-field spectroscopic study of the ionicity of the bond. Magnetic structure and Co^{2+}-Fe^{3+} MMCT in ferrites. Correlation with the magneto-optical properties," *Active and Passive Elec. Comp.,* 20 (1998): 169-187.

[17] Figgis, B. N., *Introduction to ligand fields* (Interscience publishers, 1966).

[18] Prim, S. R., García, A., Galindo, R., Cerro, S., Llusar, M., Folgueras, M. V., Monrós, G., "Pink ceramic pigment based on chromium doped $M(Al_{2-x}Cr_x)O_4$, M=Mg, Zn, normal spinel," *Ceram. Int.*, 39 (2013) 6981-6989.

[19] García, A., Llusar, M., Badenes, J., Tena, M. A., Monrós, G., "Encapsulation of hematite in zircon by microemulsion and Sol-Gel methods," *J. of Sol-Gel and Tech.,* 27(2003)267-276.

[20] Matteucci, F., Lepri Neto, C., Dondi, M., Cruciani, G., Baldi, G., Boschi, A. O., "Colour development of red perovskite pigment $Y(Al,Cr)O_3$ in various ceramic applications," *Advances in Applied Ceramics,* 105(2006)99-106.

[21] Monrós, G., Pinto, H., Badenes, J., Llusar, M., Tena, M. A., "Chromium (IV) stabilisation in new ceramic matrices by coprecipitation method: application as ceramic pigments," *Z. Anorg. Allg. Chem.* 631(2005)2131-2135.

[22] Serment, B., Gaudon, M., Toulemonde, O., Duttine, M., Brochon, C., Demourgues, A., "Tuning the CrIV/CrIII valence states in purplish Cr-doped SnO_2 nanopowders: the key role of CrIV centres and defects," *Inorg. Chem.* 2020, 59, 1, 678–686.

[23] Molinari, C., Conte, S., Zanelli, C., Ardit, M., Cruciani, G., Dondi, M., "Ceramic Pigments and Dyes beyond the Inkjet Revolution; from Technological Requirements to Constraints in Colorant Design," *Ceramics International*, doi.org/10.1016/j.ceramint.2020.05.302.

[24] Gargori, C., Cerro, S., Galindo, R., García, A., Llusar, M., Monrós, G., "Iron and chromium doped perovskite ($CaMO_3$ M = Ti, Zr) ceramic pigments, effect of mineralizer," *Ceramics International*, 38, 6(2012):4453-4460.

[25] Stoyanova Lyubenova, T., Ocaña, M., Carda, J., "Brown ceramic pigments based on chromium (III)-doped titanite obtained by spray pyrolysis," *Dyes and Pigments* 79 (2008) 265–269.

[26] O'Donoghue, M., *Gems: Their Sources, Descriptions and Identification*, (Elsevier; sixth edition, 2006).

[27] (a)Bondioli, F., Ferrari, A. M., Leonelli, C., Manfredini, T., Linati, L., Mustarelli, P., "Reaction mechanism in alumina/chromia (Al_2O_3–Cr_2O_3) solid solutions obtained by coprecipitation." *Journal of the American Ceramic Society*, 83, 8(2000):2036-2040. (b) Nguyen, D.K., Lee H., In-Tae Kim I. T., "Synthesis and Thermochromic Properties of Cr-Doped Al_2O_3 for a Reversible Thermochromic Sensor," *Materials (Basel),* 10(5)(2017):476.

[28] López-Navarrete, E., Caballero, A., González-Elipe, A. R., Ocaña, M., "Chemical state and distribution of Mn ions in Mn-doped α-Al_2O_3 solid solutions prepared in the absence and the presence of fluxes." *Journal of the European Ceramic Society*, 24,10-11(2004):3057-3062.

[29] Lopez-Navarrete, E., Ocana, M., "Aerosol-derived Mn-doped Al_2O_3 pink pigments prepared in the absence of fluxes." *Dyes and Pigments*, 61(3) (2004): 279-286.

[30] Ricceri, R., Ardizzone, S., Baldi, G., Matteazzi, P. (2002). "Ceramic pigments obtained by sol-gel techniques and by mechanochemical

insertion of color centers in Al_2O_3 host matrix." *Journal of the European Ceramic Society*, 22, 5(2002): 629-637.

[31] Czaja, M., Lisiecki, R., Chrobak, A., Sitko, R., Mazurak, Z., "The absorption- and luminescence spectra of Mn^{3+} in beryl and vesuvianite," *Physics and Chemistry of Minerals* 45(2018):475–488.

[32] Olazcuaga, R., et al., "Ceramiques Coloreés de Structure K_2NiF_4 [K_2NiF_4 Colored Structural Ceramics]," *Rev. Chim. Minér.*, 23(1986):55-60.

[33] Olazcuaga, R. et al., "Sur de Nouveaux Pigments Rouges de Type Perovskite stable à Haute Temperature [On New Red Pigments of the Perovskite type stable at High Temperature]," *Rev. Chim. Minér.*, 21(1984):221.

[34] Ardit, M., Cruciani, G., Dondi, M., Zanelli, C., *Pigments Based on Perovskite, in Perovskites and Related Mixed Oxides, Chapter 12*, (Wiley Online Library, 2015), https://doi.org/10.1002/9783527686605.ch12.

[35] Brown, K. R., Bonell, D. A., "Segregation in Yttrium Aluminium Garnet: I, experimental determination." *J. Am. Ceram. Soc.* 82(9) (1999):2423-30.

[36] Ren, F., Ishida, S., Takeuchi, N., Fujiyoshi K., "Chromium-Based Ceramic Colors." *Ceram. Bull* 71, 5(1992):759-764.

[37] Shirpour, A., Faghihi Sani, M. A., Mirhabibi, A., "Synthesis and study of a new class of red pigments based on perovskite $YAlO_3$ structure," *Ceramics International* 33 (2007):1427–1433.

[38] Richter, A., Gobbels, M., "Phase Equilibria and Crystal Chemistry in the System CaO-Al_2O_3-Y_2O_3," *Journal of Phase Equilibria and Diffusion* 31, 2 (2010)153-157.

[39] Pines, D., Bohm, D., "A Collective Description of Electron Interactions: II. Collective vs Individual Particle Aspects of the Interactions." *Physical Review.* 85, 2(1952) 338–353.

[40] Zalevsky, Z., Abdulhalim, I., *Plasmonics, in Integrated Nanophotonic Devices (Second Edition), Chapter 1*, (Elsevier, 2014).

[41] Wang, J., Lian Gao, L., *Metal oxide semiconductors for solar water splitting, in Metal Oxides in Energy Technologies,* Chapter 8, (Elsevier, 2018).

[42] Zeng, J., Xue, Y., Plasmonic Metallic Nanostructures as Colorimetric Probes for Environmental Pollutants, in *Novel Nanomaterials for Biomedical, Environmental and Energy Applications,* Chapter 11, (Elsevier, 2019).

[43] Barber, D. J., Freestone, I. C., "An investigation of the colour of the Lycurgus Cup by analytical transmission electron microscopy," *Archeometry*, 32(1990):33–45.

[44] Loos, M., *Nanoscience and Nanotechnology, in Carbon Nanotube Reinforced Composites,* Chapter 1 (Elsevier, 2015).

[45] Carbert, J., "Gold-Based Enamel Colours," *Gold Bulletin*, 8 (1997)144-150.

[46] Turkeyich, J., Hillier, J., "Electron Microscopy of Colloidal Systems," *Analytical Chemistry*, 21, 4, (1949):475-485.

[47] Monrós, G., García, A., Badenes J. A., Tena, M. A., *El Color de la cerámica: nuevos mecanismos en pigmentos para los nuevos procesados de la industria cerámica* [*The Color of ceramics: new mechanisms in pigments for the new processes of the ceramic industry*], (Athenea, Jaume I University, 2003) ISBN-13: 978-8480214490.

[48] Zhang, G. Y., Xu, Y. Y., Gao, D. Z., Sun, Y. Q., "α-Fe2O3 nanoplates: PEG-600 assisted hydrothermal synthesis and formation mechanism," *Journal of Alloys and Compounds,* 509 (2011):85–890.

[49] Tanabe, Y., Sugano, S. (1956), "On the absorption spectra of complex ions III," *Journal of the Physical Society of Japan*, 11 (1956) 864-877. doi: 10.1143/JPSJ.11.864.

[50] Vink, A. P., de Bruin, M. A., Roke, S., Peijzel, P. S., Meijerinka, A., "Luminescence of Exchange Coupled Pairs of Transition Metal Ions," *Journal of The Electrochemical Society*, 148, 7(2001): E313-E320.

[51] Vicent, J. B., Llusar, M., Badenes, J., Tena, M. A., Vicente, M., Monrós, G., "Occlusion of chromophore oxides by Sol-Gel methods:

Application to the synthesis of hematite-silica red pigments," *Bol. Soc. Esp. Cerám. Vidrio*, 39, 1, (2000):83-93.

[52] Llusar, M., Calbo, J., Badenes, J., Tena, M. A., Monrós, G., "Synthesis of iron zircon coral by coprecipitation routes," *Journal of Materials Science*, 36 (2001):153–163.

[53] Dondi, M., Matteucci, F., Cruciani, G., Gasparotto, G., Tobaldi, D.M., "Pseudobrookite ceramic pigments: Crystal structural, optical and technological properties," *Solid State Sciences*, 9 (2007):362-369.

Chapter 5

CERAMIC PIGMENTS, HISTORICAL CONTEXT

In the development of ceramic pigments for glazed ceramics, four main stages can be considered: (1) before the discovery of European porcelain in Meissen (1710), (2) after this discovery until 1800 (with the introduction of the Sèvres green as the main relevant novelty in 1802), (3) the 19th century with the explosion of color of Sèvres, (4) the 20th and 21th centuries with the discovery of the rutile and zircon families of stains, which will be described below, and the development of submicronic pigments for inkjet applications that drastically restricts the pigment palette in industrial ceramics, not always for technical reasons.

5.1. CERAMIC PIGMENTS BEFORE THE DISCOVERY OF WHITE PORCELAIN AT MEISSEN

It is well known that the Egyptians were producing unglazed Terra Cotta vases red and black colored. In effect, from 3000 to 1700 B.C. pictures on the walls of tombs illustrated Egyptian potters modelling vases and firing them in kilns. There were a great demand of colorants which was met by iron and copper compounds able to produce red to brown and turquoise to green colors respectively, at the 900-1200ºC temperature range. Egyptian blue, a more or less crystalline copper-lime silicate ($CaCuSi_4O_{10}$), can be considered the first synthetic pigment.

Since the early Chinese periods up to about 1200 the colors based on iron and copper were the mainstay of Chinese pottery. Later a new coloring palette was developed based on the systems shown below.

5.1.1. Blue

Blue cobalt silicate Co_2SiO_4 olivine shows orthorhombic crystal structure with the space group *Pnma* (Figure 4.1.b). The general chemical formula for olivine structure is A_2BX_4. The relatively large cations A^{2+} occupy the octahedral sites whereas relatively small cations B^{4+} are located in the tetrahedral sites. M_2SiO_4 (M=transition metal) compounds crystallizing with the olivine structure are characterized by a slightly distorted hexagonal close-packed HCP array of oxygen anions in which one-eighth of the tetrahedral interstices are filled by silicon and one-half of the octahedral interstices are occupied by the divalent metal cation M.

When cobalt is added to glazes, the called "*cobalt leaching*" effect appears; XAFS studies [1] confirm that cobalt plays the dual role of chromophore and network-former in the blue glaze of ancient Chinese porcelains, the tetrahedral Co^{2+} ions being responsible for an intense blue coloring. Conversely, non-coloring pseudo-octahedral Co^{2+} ions occupy available coordination sites of the tetrahedral silica-rich glassy matrix. Cobalt has been used to color glass since the Bronze Age. The excavation of the Uluburun shipwreck yielded an ingot of blue glass, cast during the 14th century BC. Blue glass from Egypt was either colored with copper, iron, or cobalt. The oldest cobalt-colored glass is from the eighteenth dynasty of Egypt (1550–1292 BC). The source for the cobalt the Egyptians used is not known. The word *cobalt* is derived from the German "*kobalt,*" from "*kobold*" meaning "goblin," a term used for the ore of cobalt by miners due to its high toxicity and because its presence degrade the ores of copper or nickel. It turns out that the workers tried unsuccessfully to capture an evil spirit named "*Kobold,*" which took refuge in caves and forests; a goblin whose main mischief was to replace or transform the silver that, supposedly, they obtained from the bowels of the earth, into a material that they feared and rejected because it left a grit in their body that caused them diseases. In effect, the first attempts to smelt those ores for copper or nickel failed, yielding simply powder (cobalt (II) oxide) instead. Because the primary ores of cobalt always contain arsenic, smelting the ore oxidized the arsenic into the highly toxic and volatile arsenic oxide, adding to the toxicity of the ore.

Following Mimoso [2] the ancient ceramics use the known "*saffer*" also known as "*saffre*" and "*zaffre,*" this processed blue mineral, which is roasted cobalt ore, or impure cobalt arsenate. The sources of cobalt are a fascinating subject. Cobalt was found in Early Egyptian, Persian, and Roman glass, but was rather seldom. In medieval times one, two or even three sources of cobalt were known in East Asia (one or two in China and, possibly, one in Sumatra) but that pigment, used in Chinese porcelain, did not reach Europe until around 1500, pigment from at least one known source in Persia could reach Europe. There might be a second source of cobalt pigment in Iraq, possibly near Samarra, where the earliest tin-glazes (9th-10th century AD) were found in archaeological contexts. At least two cobalt pigments with different associations were used in Europe before 1500. Both had little or no arsenic in their compositions but one of them had high nickel content while the other had only traces of Ni. The nickel content is that which is more readily visible in the color because it darkens the blue in a quite undesirable way. The blue cobalt used in Europe during 12th-14th medieval centuries do not content arsenic, and the detection by XRF (X-Ray Fluorescence) of this element allow to date the paints because the paints with cobalt blue based pigments that contain As should be dated on 15th and following centuries [3].

In his excellent paper Josef Horschik [4], with substantial sources, points out that bismuth was discovered in the Erzgebirge (a mountainous region that follows the present border of Germany's Saxony with the Carlsbad District of the Czech Republic) during the 15th century, probably before 1450. It was mined, refined and used for several purposes including the lead alloys used to cast printing types. It occurs associated with cobalt, but the presence of a substance that could produce a blue pigment was unknown at the time. Bismuth was extracted by roasting the ores. When the metal separated, cobalt, with a higher point of fusion, remained as part of the slag and was discarded as waste without value. The name "*kobold*" (a mountain and mine devil) was associated with this slag possibly because of the nefarious fumes rich in arsenic caused by its roasting. Cobalt ores and cobalt occurring under different forms and compounds were known to the miners of the Erzgebirge Mountains at least since the 15th century. Before the invention of the blue color, the simple mountain men called all arsenic and sulphur containing rocks and ores "*kobold.*" When they did not contain bismuth, they were cast away as unusable rubble. Usable bismuth ores often contained cobalt, nickel, sulphur and arsenic. Therefore, the miners also spoke of "*bismuth-kobold.*" The bismuth low melting point facilitated its extraction. It was "roasted" outdoors over an open wood fire. This resulted in the bismuth giving away toxic smoke. The slag-like residues were named "*bismuth barley*" and only at a higher temperature did cobalt melt from them. About the meaning of the word "*barley*" the "Oeconom Encyclopedia" of JG Krünitz states [2]: "*In the mine mills, the coarsest pieces of pounded ore are called grains or barley.*" The discovery that a blue pigment could be obtained from this slag is not well documented, partially because the local authorities later aimed to keep the details of its production secret to try and ensure a monopoly.

Therefore although the product itself was long known, the method of manufacture there was only explained in the late seventeenth century, it involved heating the ore in a reverbaratory furnace until the arsenic came off in a white smoke that settled and was collected in a long wooden tunnel. The zaffre was cooled, mixed with pulverized stones, moistened and packed into barrels. Here it set so hard that it had to be broken up with sledge hammers [6, 7]. The roasting of zaffre is described in 1679 by Johannes Kunkel [8]:

> "This cobalt is initially thrown in a furnace. When it starts to glow it smokes a white smoke that comes out of it. Such is collected in a wooden construction and this is the arsenic. This fumed and then roasted cobalt is

> then thrown into a crushing mill. When it goes through it several times the cobalt is the smallest fraction that passes through a narrow sieve and taken into a storage. From this cobalt is taken one portion and mixed well with two or more parts of the finest and smallest crushed gravel pieces which the miners call quartz, mixed and fired together, put in barrels until it is hard and strong like stone.

Then following Johannes Kunkel the ancient roasting blue cobalt is a fired mixture of cobalt oxide and quartz, that is to say cobalt silicate Co_2SiO_4 or cobalt-olivine above described.

In 19th century a significant part of the world's production of cobalt blue (a pigment made with cobalt compounds and alumina that produce after firing spinel $CoAl_2O_4$) and smalt (cobalt glass powdered for use for pigment purposes in ceramics and painting) was carried out at the Norwegian Blaafarveværket. The first mines for the production of smalt in the 16th century were located in Norway, Sweden, Saxony and Hungary. Smalt is a potassium glass (SiO_2 + K_20 + Al_2O_3+CoO),containing cobalt and other chemical elements (Ni, As, Bi), has an intense blue color when first applied, but changes color over time and can become totally transparent or turn greenish-brownish [3]. With the discovery of cobalt ore in New Caledonia in 1864, the mining of cobalt in Europe declined. With the discovery of ore deposits in Ontario, Canada in 1904 and the discovery of even larger deposits in the Katanga Province in the Congo in 1914, the mining operations shifted again [9].

Today cobalt is a basic element for technology: a) in metallurgy the cobalt-based super alloys (such as stellite alloys, containing chromium with varying quantities of tungsten and carbon), with high temperature stability makes them suitable for turbine blades for gas turbines and aircraft jet engines, although nickel-based single-crystal alloys surpass them in performance. Cobalt-based alloys are also corrosion and wear resistant, making them, like titanium, useful for making orthopedic implants that don't wear down over time, b) in lithium-ion battery where lithium cobalt oxide ($LiCoO_2$) is widely used in the battery cathodes, c) as catalyst in polymer's industry; the cobalt acetate is used to convert xylene to terephthalic acid, the precursor of the bulk polymer polyethylene terephthalate or cobalt based compounds are also a catalyst in the Fischer–Tropsch process for the hydrogenation of carbon monoxide CO into liquid fuels. And obviously in ceramic and glass industry as blue pigment, but also as green pigment in $Zn_{1-x}Co_xO$ wurtzite solid solution up to x=0.7 and also in Co-M_2TiO_4 (M=Mg,Zn) spinels, which is known as Rimans's green because was reported in 1780 by Sven Rinman before in 1802 Louis Jacques Thénard reported the cobalt blue ($CoAl_2O_4$) [10], also a cobalt based yellow pigment called Aureolin was first described in 1848 by Nikolaus Wolfgang Fischer, the potassium cobaltinitrite also known as Indian Yellow or potassium hexanitritocobaltate (III) $K_3[Co(NO_2)_6]$ [11].

The Mukondo Mountain project, operated by the Central African Mining and Exploration Company (CAMEC) in Katanga Province, may be the richest cobalt reserve in the world. It produced an estimated one-third of the total global cobalt production in 2008. In February 2018, global asset management firm Alliance Bernstein defined the Democratic Republic of Congo as economically "the Saudi Arabia of the electric vehicle age," due to its cobalt resources, as essential to the lithium-ion batteries that drive electric vehicles. But there are two main problems associated to cobalt production; human rights activists have alleged, and investigative journalism reported confirmation, that child labor is used in mining cobalt from African artisanal mines, this revelation prompted cell phone maker Apple Inc., on 2017, to stop buying ore from suppliers such as Zhejiang Huayou Cobalt who source from artisanal mines in the Congo, and begin using only suppliers that are verified to meet its workplace standards. On the other hand, although cobalt is essential in the human diet in the form of vitamin B12, in some other contexts it is hazardous which is why five cobalt compounds (as substances and in mixtures) have been proposed for restriction by the REACH (Registration, Evaluation, Authorization and restriction of Chemicals) regulation (sulphate, dichloride, dinitrate, carbonate and diacetate) and this is currently under review in EU. These five substances are already classified as REACH Substances of Very High Concern (SVHC) as they are category 1B carcinogens. The proposed restriction does not appear to be a full ban as these can be used under certain specifically controlled use conditions, according to the draft proposed restriction. This introduces a new concept, the "Reference Exposure Level" where manufacturers must demonstrate that workers are exposed to cobalt levels of less than 0.01μg/m^3. Cobalt already has an occupational exposure limit in the EU of 0.1μg/m^3, so this new limit is 10 times lower than the current limit. There is an issue, however, in that measurement of cobalt in the air at 0.01μg/m^3 is technically very challenging and may not be possible such that this requirement may not practically be monitored by employers or enforced by market surveillance authorities. With the new CLP regulation (Classification, Labeling and Packaging of hazardous substances), the European REACH classifies Co_3O_4 as a category carcinogen 2 (C), skin sensitizer (Ss) and respiratory sensitizer (Sr). The European Parliament and the Council have two months to object to the ruling. If no objection is filed, the classification undergoes a transition period of 18 months before taking full effect; in this case, it would start to apply in the summer of 2021.

5.1.2. Red

Among ancient red pigments realgar (α-As_4S_4) was, along with orpiment (As_2S_3), a significant item of trade in the ancient Roman Empire and was used as a red paint pigment. Early occurrences of realgar as a red painting pigment are known for works of art from China, India, Central Asia, and Egypt. It was used in European fine-art painting during the Renaissance era, a use which died out by the 18th century. It is commonly held that after a long period of exposure to light (λ<610 nm) realgar changes form to a yellow powder known as pararealgar (β-As_4S_4). It was once thought that this powder was the yellow sulfide orpiment, but has been recently shown to be a distinct chemical compound [12].

The hematite (corundum or ilmenite lattice, rhombohedral, space group *R-3c*) is the oldest known red-brown pigment as discussed in section 4.7 (Paired spin exchange transitions (PET)). The name hematite is derived from the Greek word for blood αἷμα *(haima)*, due to the red coloration found in some varieties of hematite and was imported from Latin *lapis haematites* originated from Ancient Greek αἱματίτης λίθος (*haimatitēs lithos*, "blood-red stone"). On the other hand Ochre is clay colored by varying amounts of hematite, varying between 20% and 70%. Red ochre contains anhydrous hematite, whereas yellow ochre contains hydrated hematite or limonite ($Fe_2O_3 \cdot H_2O$). The principal use of ochre is for tinting with a permanent color. Hematite present different varieties of red-brown pigments that receive different names according to the color shade and composition: sanguine or red chalk is chalk of a reddish-brown color, so called because it resembles the color of dried blood, that in natural form is a mixture of hematite and clay, yellow ochre ($FeO(OH) \cdot nH_2O$) is a hydrated iron hydroxide (limonite) also called gold ochre, red ochre takes its reddish color from the mineral hematite, which is an anhydrous iron oxide, purple ochre is identical to red ochre chemically but of a different hue caused by different light diffraction properties associated with a greater average particle size, brown ochre, also FeO(OH) goethite, is a partly hydrated iron oxide, sienna contains both limonite and a small amount of manganese oxide (less than 5%), which makes it darker than ochre and finally umber pigments contain a larger proportion of manganese (5-20%), which makes them a dark brown. The use of hematite or ochre is very wide since prehistory as describe Leroi-Gourhan [13]:

> "the use of ochre is particularly intensive: it is not unusual to find a layer of the cave floor impregnated with a purplish red to a depth of eight inches. The size of these ochre deposits raises a problem not yet solved. The coloring is so intense that practically all the loose ground seems to consist of ochre. One can imagine that the Aurignacians regularly painted their bodies red, dyed their animal skins, coated their weapons, and sprinkled the ground of their dwellings, and that a paste of ochre was used for decorative purposes in every phase of their domestic life. We must assume no less, if we are to account for the veritable mines of ochre on which some of them lived."

But the pigment based on hematite was used in all times, yellow and red ochre pigments were widely used in painting with greatly variation of colors depending upon whether the local clay was richer in yellowish limonite or reddish hematite: the red earth from Pozzuoli near Naples was a salmon pink, while the pigment from Tuscany contained manganese, making it a darker reddish brown called "*terra di siena,*" or sienna earth (Figure 5.1) [14].

Figure 5.1. Red ochre (with anhydrous hematite), yellow ochre (with limonite) and sienna ochre (anhydrous hematite with MnO_2).

A historical and very interesting use of hematite pigment is the "*Socarrat*" a Valencian term to designate the large unglazed fired clay slabs, decorated with paintings on one side, which were placed in the 15th and 16th centuries between the beams of the ceilings. It is a ceramic with Arab reminiscences. The first record of its existence dates back to 1604, when D. Feliciano de Figueroa, Bishop of Segorbe refers in a work containing considerations on Islamic civilization, that in a house adjoining the Church of San Pedro were found "*the bricks of the cover and walls scripts of many sentences of the Koran in Arabic with large letters of the Koran colored and preserved to this day.*" They were manufactured in Paterna, and perhaps also in Manises and other towns in Valencian Community (Spain) [15]. The decorative motifs were made with oxides, in black and red. The dry brick, covered by a layer of uncooked lime, was later painted with hematite (red ochre) and manganese oxide. Finally it was fired at around 900ºC, and the result was a matte decoration. Figure 5.2 shows a socarrat with pigment analysis carried out using a EDXRF (Energy Dispersion of X-Ray

Fluorescence) portable spectrometer consisting of a tube of X-rays with W anode operating with a potential of excitation of 30 kV and a current of 0.1 mA and a detector of Si-PIN semiconductor with 180 eV resolution (FWHM 5.9 keV) [16]. The black zone shows the presence of manganese associated to the band at 6 keV (but also Cu), the pigmented red area shows iron at 6.3 keV which is also detected in the ceramic body with Ca and Pb.

Figure 5.2. “Socarrat” with XRF analysis of black (Mn) and red (Fe) areas [16].

An ancient red color is based in red cuprous oxide Cu_2O obtained in reducing atmosphere; it crystallizes in a cubic structure where Cu^+ ions arrange in a FCC sublattice, the O atoms in a BCC sublattice. One sublattice is shifted by a quarter of the body diagonal and belongs to space group *Pn-3m*. It is the base of red medieval stained glass, where trapped copper colloidal particles in the glass matrix generate the ruby red color in the windows. If oxidant atmosphere is used, the obtained color could be blue due to dissolved Cu^{2+} in octahedral coordination highly distorted (in basic alkali glass environment resulting in a square-planar geometry due to Jhan Teller effect) or green when distortion is low (in acid Ca,Pb,Zn,Ti glass environments) (Figure 5.3). Medieval stained glass also trapped silver nanoparticles which gave it a deep yellow color. Possibly stained glass artisans were the first nanotechnologists. The called Lycurgus Cup (see section 4.6) is a 4th century Roman glass cup made of a dichroic glass which shows a very nice dichroism, from red by transmitted light, changing to deep green by reflected light, due to dispersion of silver and gold nanoparticles into the glass (see Figure 4.20).

Figure 5.3. Blue copper dye from Cu_2O 5 wt.% added to a alkali glaze (1080°C) and green from similar addition to a Ca-Zn matte glaze (1200°C).

5.1.3. Yellow

The oldest classic yellow pigments known in Europe since at least the thirteenth century are the "lead-tin-yellow type I" (orthorhombic Pb_2SnO_4), and its Si-doped modification called "lead-tin-yellow type II" (cubic pyrochlore $PbSn_xSi_{1-x}O_3$) [17]. Lead-tin yellow type I was obtained by heating a powder mixture of lead and tin oxides, and Type II by combining this mixture with quartz. The pigment is opaque and resistant to light, but it is also poisonous due to the presence of lead. Initially used in ceramics and glass, this pigment was also used in fresco painting in the Type II variety. The lead stannate Type I was the standard yellow used in oil painting until the eighteenth century, for instance by painters such as Titian (Bacchus and Ariadne, 1520-23), or Velázquez (Apollo in the Forge of Vulcan, 1630); it is a saturated yellow that was later replaced by the yellow of Naples, still more intense (Figure 5.4). In the 19th century its existence was forgotten, being rediscovered in 1941 by the German chemist Richard Jakobi [18].

Figure 5.4. a) “lead-tin-yellow type I” in Titian (Bacchus and Ariadne, 1520-23), or Velázquez (Apollo in the Forge of Vulcan, 1630), b) “yellow of Naples” in Angelica Kauffmann (The royal family of Naples and Sicily, 1784).

The successor of the “lead-tin-yellow " in the European ceramic practice during the eighteenth century was the yellow of lead (II) antimonate $Pb_2Sb_2O_7$ cubic pyrochlore, space group *Fd3m*. The called Naples’ Yellow arrives to Venice in Italy from the Byzantine Empire. The history of lead antimonate goes back at least 3500 year. As early as the 18th century Egyptian dynasty (ca.1500 BC) lead and antimony compounds were used to give glass objects a yellow color. Lead antimonate was also known in the Mesopotamia, Babylonian and Assyrian cultures. It has been found on an Islamic 14th century enameled bottle from Syria. Dyk [19] study the recipes for the synthesis of Naples yellow that describes Picolpasso [20] from the compiled by the miniaturist Valerio Mariani da Pesaro (1568-1625?). Mariani was a professional miniaturist who worked the larger part of his productive years for the Duke of Urbino, Francesco Maria II Della Rovere. He describes three types of Naples Yellow. (a) “*one which appears to be more subtle and works well for coloring drawings.., five ounces of burnt lead in powder and sieved, and three ounces of antimony and two ounces of plain salt*” which gives, fired above 650ºC, lead (II) antimonate $Pb_2Sb_2O_7$ (95%)

and other phases as residual NaCl used as mineralizer and Pb_3O_4 that gives a red surface color of the powder surface, (b) "*another which shows more body*" which used wine residue (potassium tartrate) as mineralizer and gives the $Pb_2Sb_2O_7$ cubic pyrochlore in a 75wt.% in the 650-900ºC range together $K_3Sb_30_9$, and in a 95wt.% degree in the 1000-1100ºC range with minor crystalline unidentified phase with a $NaSbCl_6$ structure, (c) "*a fuller color and more beautiful by taking 6 ounces of burnt lead and 4 ounces of antimony and one ounce of Alexandrine Tutty and I ounce of salt,*" which gives when fired above 700ºC $Pb_2Sb_{2.x}Sn_x0_{6.5}$ solid solution indicating that the inclusion in the recipe of "*Alexandrine Tutty*" (which results to be SnO_2, although the solid solution with ZnO gives a similar product) produces a better yellow color based on the solid solution of Sn in the pyrochlore structure. At present, Naples Yellow cannot be industrially prepared and commercialized due to its high toxicity related to the antimonite and lead.

5.1.4. Black

Pyrolusite (β-MnO_2) tetragonal *P42/mnm space group* crystallizes in the rutile crystal structure, with three-coordinate oxide and octahedral metal centers. It is nonstoichiometric, being deficient in oxygen. Some of the most famous early black cave paintings in Europe (Lascaux, Altamira…) were executed by means of manganese dioxide. Manganese compounds were used by Egyptian and Roman glassmakers, to either remove color from glass (glassmaker's soap) or add color to it [21]. It was widely used in Islamic ceramics as discussed in the above dichroic ceramics of "socarrats" with hematite red ochre. Manganese impart purple color to leaded glazes and blue-purple to alkaline glazes, but as above indicated, acts as glass soap when added to glass that has been colored green due to iron impurities. The decoloring process is due to the presence of violet Mn(III) (absorbs yellow-green wavelengths) complementary to green due to Fe(II) (absorbs red-violet lights), then the transmitted light becomes colorless. The black characteristic of Attic Greek pottery was obtained by a quasi-Sol-Gel methodology: the vessels are enameled with a fine suspension (colloidal fraction) of illitic clay with very low calcium oxide content and rich in iron and hydroxides, the body of the vase, in opposition to slip, show relative high calcium content. The stability of the Attic black slip suggests the use of added deffloculants such as wood or other plant ashes, urea, tannins… as in our current glazes. The enamel was applied on the areas intended to become black after firing. In the kiln the black color was obtained by a three-phase firing: (a) one oxidizing phase at 800°C with open air windows in the kiln which produces the red hematite Fe_2O_3 crystallization from clay, (b) a reducing phase at 950°C closing windows and firing fresh wood which gives to reduction of hematite to black magnetite Fe_3O_4 by released CO, and finally (c) a new oxidizing firing phase at 900°C which deoxidizes magnetite (accelerated by CaO presence [22]) in the bulk of vase but not in the enameled, without CaO, which remains black.

Painters of the Quattrocento and Renaissance used enamels and pigments developed by potters and the first “technological books” became largely available at that period [23].

5.2. Ceramic Pigments after the European Discovery of White Porcelain to 1800

Following Queiroz [24], in the context of the movement born at 18th century which receives the name of "Enlightenment" based in a critical spirit where philosophy and science were the most valued sources of knowledge, a real “task force,” as a group of people who are brought together to do this particular job, was the origin of the European discovery of white porcelain: on the one hand an entrepreneur (the financial, political and executive leader), also a scientific leader, a technological partner and finally an innovation communicator-coordinator.

a) The financial, political and executive leader (the entrepreneur); August the Strong (1670-1733), Elector of Saxony and King of Poland, following the Enlightenment spirit gave Tschirnhaus (1651-1708) the assignment of searching the whole Saxon territory for deposits of precious stones, while developing methods of polishing and lapping these valuable resources. At the same time, he pressed Johann Friedrich Böttger (1682-1719) to make gold. These two projects merged in the hard-porcelain industrial project.
b) The scientific leader; Ehrenfried Walther von Tschirnhaus (1651-1708), the German mathematician, physicist, physician, and philosopher who through the 1680's applied his scientific knowledge to the development of industrial processes, focusing at the manufacture of glass and porcelain. He also established the first Saxon glass manufactures and developed polishing technology [25]. By 1682, he studied theoretically the envelope of light beams emitted from a point source after reflection on a parabolic surface [26], as a preliminary step towards the development of large burning lens and mirrors. The last ones were an improved version of the mirrors developed earlier by François Villette, in France. The use of such equipment allowed him to reach temperatures within 1500-2000°C, higher than what he could achieve in contemporary combustion furnaces. The method was welcome in laboratory research because the samples could be easily observed and many trials could run in a short time. From 1693-1694 onwards, Tschirnhaus found that porcelain could melt, and studied the refractory properties of several possible raw minerals. Understanding eutectics, he discovered that porcelain could be obtained from a mixture of clay and fusible minerals (flux). Tschirnhaus conducted most of this research guided by his prior knowledge of glass technology, whereby he seemingly believed that porcelain was a glass. By 1699, he actually obtained "porcelain" by melting and cooling mixtures of raw materials [25]. Tschirnhaus started the porcelain research long before Böttger joined the quest, as his assistant, in 1705 [1].

After the preliminary laboratory trials provided promising porcelain samples, a scale up procedure began. It started in 1706, when August ordered metallurgist experts to be hired in the Freiberg region, since they held valuable knowledge on thermal processes and kilns.

c) The technological partner: the Freiberg Mining Counsellor, Gottfried Pabst von Ohain (1656-1729), expert in mineralogy and metallurgy, was chosen to direct the iron-metallurgist workers hired in January 1706 [27]. Tschirnhaus was godfather of Ohain's son; therefore, he supported Ohain's cooperation. Ohain, contributed with his vast practical knowledge of ores and minerals, as an expert furnace constructor, and developed formulas for overglaze enamels. Similarly, to Tschirnhaus, his contribution has been largely disregarded. Ohain provided Böttger with selected mineral samples, which were otherwise inaccessible to him. Also in Tschirnhaus's period (before 1708), Samuel Stöltzel specialist in past and kilns (see Chapter 4) provided his knowledge on this subject, and after Dr. Bartholomäei (1670-1742) and Dr. Hienrich W. Nehmitz (to whom in 1714, in the middle of the porcelain commercialization period, out of prudence, the composition of the paste was revealed to Bartholomäei and to Nehmitz that of the glaze). Since it doesn't matter which of the manufacturing workers can discover the secret, the work is rigorously divided. Böttger is the only one, for a long time, who knows the whole process [28].

d) The innovation-communicator coordinator: Johann Friedrich Böttger born in 1682 at Schleiz. After he started working as a pharmacist apprentice in Berlin, he devoted himself to alchemy, around 1700, as an 18-year-old apprentice chemist with the pharmacist Zorn in Berlin, also an alchemist in pursuit of the "philosopher's stone," locked himself away to discover in private the "*Alltinktur or Goldmachertinktur*" (gold/maker/tincture), an alchemist's secret substance with which supposedly any disease could be cured and base metals converted into gold. In 1701, he made his first demonstration in the presence of trustworthy witnesses. He pretended to transmute two silver coins into gold. After his "success," the Prussian King Frederick I, tried to lock him as a treasure, but he managed to escape to Saxony. However, his fame preceded him, and August the Strong kept him watched and confined in Albrechtsburg Castle (Meissen), trying to exploit his "gold maker" skills. Later on, as no gold was forthcoming, August directed him to porcelain research. At first Böttger had no interest in von Tschirnhaus' own experiments, but with no results of his own and by then fearing for his life, by September 1707 he began to slowly cooperate. He did not wish to be involved with porcelain, which he thought was von Tschirnhaus' business, but ordered by the monarch and probably thinking the deciphering of porcelain's secrets his only option left (as then, second to gold) to both satisfy the monarch's agreement and save his own neck, he began cooperating in earnest. At stake was not just Augustus's displeasure. As Hoffmann refers [27], Figure 5.5 is a reproduction of a broadside, a contemporary account of what happened to a Neapolitan alchemist, Count Domenico Emanuele Cajetano, who was found cheating. He was hanged. Presumably by involving Böttger in his experiments, von Tschirnhaus spared him the fate that overtook many former alchemist adventurers [29].

In September 1707, three cellar vaults were dedicated to porcelain production at the Dresden fortress, providing an intermediate level between laboratory and factory's production scales. The production of red stoneware was soon started, from iron-rich refractory clay mixed with a fusible flux fired to very high temperature. A "red porcelain" is obtained using red earth from Nüremberg, which contains iron oxide, and a mixture of clay from Plauenscher Grund, a place near Dresden. This stoneware was however extremely hard. Thus, it was cut, polished and wheel-engraved like a semi-precious stone, with the aid of equipment developed by Tschirnhaus.

Figure 5.5. Hanging of a Neapolitan alchemist: Count Domenico Emanuele Cajetano, who was found cheating. (https://de.wikipedia.org/wiki/Domenico_Manuel_Caetano).

But this "red porcelain" is not an original discovery in Europe as similar stoneware had already been obtained in Delft by Ary de Milde between 1677 and 1678, as well as in England by the Elers brothers and by John Dwight [30]. They became part of the extensive European quest to discover porcelain, which had already resulted in Rouen soft-paste porcelain in France [31], and also, almost a hundred years before in France, Duke Francesco de Medici had already opened a soft-paste porcelain factory in Florence [32].

The key point of the search of Tschirnhaus and Böttger is their interest in finding the correct natural materials that, when fused, can produce that white vitrified substance whose translucency does not depend on the presence of glass in the paste. Those two natural ingredients, in Chinese porcelain, are kaolin and petuntse. Kaolin is non-fusible clay obtained from the degradation of granite rocks; it is the purest of natural clays, and it was obtained from the "high hill" or Kao-ling that gave it its name, while in England it will be known as china clay. Petuntse (or "white bricks," the form in which this material reached Chinese factories) is a partially kaolinized feldspathic rock, is a less degraded fusible material, a variety of the same rock, which has by nature the binding properties that are obtained artificially with a frit (mixture of cold materials that are vitrified and then pulverized to be added to the paste as a binder for the non-fusible clay particles). In England the petuntse will be known as "China stone" or "Cornish stone." Once the porcelain body is obtained by mixing kaolin and petuntse, it is covered with a glaze composed of the same feldspathic rock, which obtains greater fusible property by the addition of a flux such as potassium feldspar with calcium carbonate. Then a white porcelain similar to china for its brightness, hardness and transparency, is obtained at the Dresden fortress after twelve hours of firing at 1300/1400ºC; it consists of a non-fusible component that turns white upon firing (Kolditz kaolin) and another fuse (Northausen calcined alabaster), both feldspathic in nature.

This period was characterized by the decline of Tschirnhaus's influence, and by the abandon of the former Tchirnhaus's program, based on the use of sun light kilns, which were appropriate for laboratory trials but unsuitable to be scaled-up to the factory's level. This decay was compensated by the rising influence of the two new members who joined the porcelain "*contubernium*" (the task force): Pabst von Ohain and Dr. Bartholomäei. The latter was a medicine physician and naturalist who, working from 1707 onwards carried out numerous trials of paste preparation, intended to find the most favorable clay raw materials among those found in Saxony region. He mostly established the relative shares of the chosen clays. A protocol dated from 15.01.1708 certifies the invention of the European hard paste porcelain, establishing the transition from the laboratory trial stage to the production scale maturity. This document has been attributed either to Böttger [33] or to Bartholomäei [25]. On October 11, 1708 the count Tschirnhaus dies in Dresde.

In 1708-09, the first Saxon white clay (kaolin) was found in Erzgebirge. This discovery is often attributed to Böttger, but the research was actually conducted by Ohain [25] and Bartholomäei [27]. This finding was crucial for the success of the manufacture of white porcelain. New deposits were found later at Aue, near Schneeberg, in Saxony, and a mixture of the two clays with calcified alabaster and white silica was found particularly successful. After Tschirnhaus' death, Böttger wrote a report to the Elector (28.03.1709), claiming that he had prepared "good white porcelain" equivalent or even better than the Chinese one. The official startup date of the porcelain manufacture (at Dresden) dates from 23.01.1710, when the Saxon Court Chancellery issued a "Supreme Decree" in Latin, French, German and Dutch, announcing the discovery to the World. For security reasons, a Decree from 7.03.1710 transferred the factory to Albrechtsburg (Meissen), being officially opened in August 6, the same year [33].

The inspector of Meissen manufactory, Johann Melchior Steinbrück (1673-1723), was largely responsible for the initial promotion of Böttger as a genius, to whom he claimed that all credit of the discovery should be given. He stated, in a laudatory text [34], that his "*inventoribus rerum*" would hardly be repeated in a hundred years. Steinbrück was a close relative of Böttger (his brother in law), and under Steinbrück's suggestion (1715), Augustus chose Böttger as director of porcelain manufacturing and attributed him the title of Baron. The secrecy policy followed by Augustus also contributed for the relative oblivion of Tschirnhaus's role, since early documents were secret. Earlier historians, following Steinbrück, emphasized Böttger's contribution instead of Tschirnhaus's, starting a chain that still echoes.

Hoffmann refers [29] that the story of the rediscovery of porcelain, is told beautifully in Janet Gleeson's 1998 book Arcanum [35]. She cites an emotional poem that Böttger wrote to Augustus late in 1709:

> "The King will yearn for golden fruit
> Which the weak hand yet cannot present.
> It puts but crystals of porphyry and borax
> Before the King's throne in place of such sacrifices.
> Yes, the hand extends even the heart, in vessels of porcelain."

Hoffmann writes [29]:

> "This rings true. Böttger was caught. I sense from the poem that he would have liked to get out of the alchemist's bind and rest on his great practical synthesis, the making of the translucent "white gold." But he was not allowed to do so. Although the porcelain was of immense value and eventually contributed to the King's coffers, it took time to establish a market for it. Meanwhile Augustus did not give up his dream for gold, and Böttger remained a prisoner. Time and again he was pressed to make good on his promises of making gold, and he finally staged a "successful" transmutation in 1713. Böttger was freed in the end, even ennobled. But he was ill and exhausted. He died in 1719, the year porcelain was first made outside of Meissen. It's not easy to keep an industrial secret."

but, as in general, the story is more complex and there are many more people involved, the story in Arcanum of a Böttger as "the teenage alchemist who is imprisoned and forced, given the required materials, to find a reliable way to produce gold and who tries but ends changing course towards the search for a commercially viable formula for porcelain, which in this period was more valuable than gold, by its innovative approach, to bake clay at higher temperatures than previously achieved in European kilns, produced the breakthrough that European potters had pursued for a century"… is a heroic story but also too simple and prone to the exaltation of the individual over the team (the task force) with Tschirnhaus, Ohain, Samuel Stöltzel, Bartholomäei. Nehmitz…and many more people.

Although initially August and Böttger have no preferential interest in obtaining the hard paste and instead find a reliable way to produce gold; ceramics and preferentially white porcelain was a substitute of gold and silver in tableware applications. This is demonstrated for example by the "Pragmatic against luxury," promulgated in 1601 by the Spanish court of Felipe III, by means of which the Duke of Lerma promoted the gradual replacement of gold and silver tableware by ceramic pieces, contributing to the proliferation of ceramics tableware such as the Talavera's manufactures, in order to obtain materials with which to make money due to the scarcity of noble metals. Talavera tableware was highly appreciated for its brilliance and whiteness and stimulating properties were even attributed to it:

> "they awaken the taste in delicacies (...) enhancing the tasty with the clean and pleasant cleanliness of the glasses in which they serve ..." [36].

Although the porcelain manufactory at Meissen started the production of porcelain in 1710, the decoration was poor because the ceramic pigments and dyes used for low temperature glazes are not stable in the thermal range of the new porcelain stoneware. It was necessary to investigate new ceramic pigments and pigmenting systems (e.g. a second firing for decoration). From 1710 the research on ceramic pigments progress and chemistry replaces the alchemy. The manufactures of ceramic produces their own colorants using the raw chromophore oxides stabilized with the admixture of developing and fluxing agents. The "secret formulae book" was part of the know-how of any ceramic factory and problems arose when color maker died: the secret of their blue or green colorant died with him [37], but since Meissen these color recipes were transcribed and kept secret, but secrets always transcend and colors are copied, and they probably reach the higher quality in Sèvres, where the true color revolution in ceramic will take place, as will be discussed in the next section.

Following Colomban [37] Chinese Eastern Han to Sui potters (2th century B.C.) initiated the development of very high temperature firing technology: Yue wares were fired above 1300°C. White porcelains appeared during Sui and Tang Dynasties (7th–8th c.). At the same period the Islamic potters found how opacification by cassiterite tin oxide combined with low temperature firing could compete with high temperature fired porcelain to produce artifacts with complex depictions [38]. This innovation resulted from the imitation of the three-color Chinese porcelain wares with lead-rich blue and green glaze dots deposited on a faience/terra cotta body covered with an alkaline glaze. The chemical reaction between the two glazes led to precipitation of white cassiterite [39]. This innovation was transmitted to Europe with the extension of the Islamic World (Ifriqaya (7th c.), al Andalous (8th c.), Sicilia (9th c.)) and then Italy (majolica, 13th c.) and

France (15th–16th c.). Attempts to imitate the Blue-and-white Chinese porcelain were first made in Anatolia with Ottoman “Iznik firewares” from ~1450 to ~1620 [40] and then in Italy with Medici porcelain from ~1575 to 1587 [41], as well as in the whole of Europe one century after (18th century porcelain Manufactures). The availability of kaolin permitted the preparation of hard-paste alumina-rich porcelain, though the soft-paste porcelain route, which could be considered as an evolution of the Iznik fritware technology.

A revolution took place after 1750 with the development of faience and porcelain factories (Figure 5.6). A second one occurred during the 20th century with the development of advanced ceramics for electronic devices, porcelain stoneware and the inkjet decoration for tiles. But kaolin, abundant throughout the East, was very scarce in Europe, and this led to the proliferation of "secretists"; expert ceramists and chemists who traveled Europe offering supposed recipes for the manufacture of real porcelain to the best bidders, that different types of clays were mixed, sometimes with elements as strange as calcined sheep bones or ground rock crystal. With all this, obviously, true porcelain was not achieved, but pastes that were generally more fragile than usual, whose inability to withstand sudden thermal shocks made it impossible to use them in tableware, making them more suitable for ornamental objects, which explains the true legion of “figurines” that arise from European manufactures, in the 18th century.

No country escaped this contagious obsession to produce porcelain, and he hired a good number of foreign technicians and “secrets.” None of them achieved their goal, but they did achieve excellent quality pastes especially suitable for sculpture. This obsession produces an interesting technical competition between countries but also within the same country.

For example, in England the Leeds Manufacture, also known as Hartley Greens & Co., founded around 1756 in Hunslet, just south of Leeds, England. It is best known for its creamware, which is often called Leedsware and was the "most important rival" of the highly popular ware of Wedgwood, who had invented the improved version used from the 1760s on. Wedgwood manufacture was founded on May 1, 1759 by Josiah Wedgwood and was first incorporated in 1895 as “Josiah Wedgwood and

Sons Ltd." It was rapidly successful and was soon one of the largest manufacturers of Staffordshire, exporting across Europe as far as Russia, and to the Americas. It was especially successful at producing fine earthenware and stoneware that were accepted as equivalent in quality to porcelain (which Wedgwood only made later) but were considerably cheaper [42].

Also in Spain there is a pottery competition between the classic tableware of Talavera de la Reina or the manufacturers of Manises and Valencia from the Gothic-Mudejar tradition with the novel producers of the new porcelain fever; the early in Alcora in 1727 near Valencia, promoted by the IX Count of Aranda, D. Buenaventura Jiménez de Urrea (1699-1742), also "lord of tenure and honor of Alcalaten" to which it belonged Alcora. It represents an authentic revolution in the dormant panorama of Spanish ceramics at the beginning of the 18th century. Coinciding with the access to the Bourbon throne, just in 1700 Felipe V, the first of the dynasty will imbue his Madrid court with French taste and refinement. The Count of Aranda will be the first to found in Spain one of these aristocratic manufactures and luxurious production. The count obtained the tutelage of King Felipe V, as well as rights, such as exemption from paying taxes, which favored the factory and the employees.

But in 1759 Carlos III of Spain succeeded to the Spanish throne, prior to becoming King of Spain, Carlos reigned as King of Naples and Sicily. A factory, the "Porcellana di Capodimonte," had been established by the king and his wife, Maria Amalia of Saxony in Naples. When the couple move part of the factory (53 specialized Italian craftsmen from Naples came to Madrid, including Giuseppe Gricci (1700–1770), the main modeler, along with three shipments containing the necessary equipment and nearly five tons of porcelain paste, and establish a factory to produce royal porcelain in Madrid, the "Real Fabrica de Porcelana del Buen Retiro." The porcelain's quality was internationally recognized, and its manufacturing techniques were kept a state secret. Maria Amalia died in 1760, a year after moving to Madrid with her husband, but she was an important influence on both the Naples and the Madrid factories.

In 1742, died the 9th Count, Alcora factory is inherited by his son the 10th Count of Aranda, D. Pedro Pablo Abarca de Bolea y Ximenez de Urrea, who was ambassador in Lisbon (1755-1756) and Warsaw, then obtained the rank of captain general, with which he led the Spanish army that invaded Portugal in 1762, with disastrous results, because during the so-called "Fantastic War," the army led by Aranda suffered 20,000 casualties, failing in his attempt to capture Lisbon. Then he was appointed Governor of Valencia, a position he had to resign to preside over the Council of Castile in 1765 and to be "Captain General of Castilla la Nueva" (April 11, 1766). In search of "real porcelain," the Count successively hire French and Saxon technicians such as François Haly, Johan Christian Knipfer, François Martin, Pierre Cloosterman and Christoph Boull and the Swiss Petro Reinhart in contrast to the Italian line of Buen Retiro factory promoted by the king. The product achieved, after numerous unsuccessful trials, was similar to French porcelain, but it was not the long-awaited Meissen-style "hard porcelain" but soft-paste porcelain because the count never wanted to use imported paste. But when the Count of Aranda died in 1798 without direct descendants (our only son died in 1755 at 14 years old). The new management, D. Pedro Alcántara duke of Híjar, given the few benefits obtained, suppressed the production of porcelain to focus more on the elaboration of sculpture in earthenware that mimics Wedgwood frit porcelain (called in Spain as "tierra de pipa" or "pipe's clay" from which the famous Wedgwood anti-slavery tobacco pipe was made [42]). But due the disruptions of the Napoleonic Wars, the Buen Retiro factory was destroyed during the battles in Madrid; the Italian potters Poggeti and Palmerani moved to Alcora, a circumstance that has caused confusion to distinguish ceramics between the two factories. In 1817 the factory moved a short distance across Madrid to become the Royal Factory of La Moncloa, again taking such molds and equipment as survived, and the employees. This operated until 1849 [43].

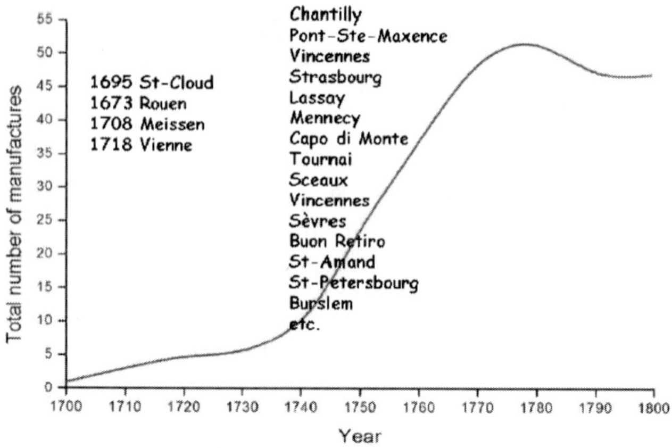

Figure 5.6. Evolution of the number of porcelain manufactures in Europe during the 18th century, after P. Ricciardi et al. [44].

5.3. Ceramic Pigments in 19th Century: The Color Explosion of Sèvres

If the white porcelain rediscovered in Meissen is the disruptive innovation for ceramic pastes in 18th century, Sèvres is the disruptive innovation for ceramic pigments based on chromium. Following the report of Verger et al. [45] there is, as in the Meissen case, a "task force" with the entrepreneur), the scientific leader, the technological partner and the innovation-communicator coordinator.

a) The entrepreneur (the patron); Jeanne Antoinette Poisson, Marquise of Pompadour, known as Madame de Pompadour, the official chief mistress of Louis XV from 1745 to 1751, remained influential as court favorite until her death. She found in 1740 the soft porcelain manufacture of Vincennes in order to compete with the productions of Chantilly and Meissen. In 1756, the manufacture was transferred to Sèvres in a building built on the initiative of Madame de Pompadour, near her castle of Bellevue: 130 meters long and four storeys high, it was built between 1753 and 1756 by the architect Lindet on the site of the so-called "Guyarde" farm. On the ground floor, the building contained the land reserves, the pyre and the raw materials depots. The first floor housed the molding, plastering, sculpture and engraving workshops as well as the ovens. On the second floor were the sculptors, turners, repairers and upholsterers. Finally, the attic floor housed painters, gilders, animal artists and sculptors in earthenware. The manufacture was attached to the Crown in 1759.

As Rebecca Leonard [46] refers, some has recognized Madame de Pompadour's heavy financial support for Sèvres, but scholars frequently debunk her contributions as being conclusively significant in regard to eighteenth-century French visual culture. While some do not seek to downplay her influence, they still do not consider the broader significance of the aspects of Sèvres' production on which she made an impact. Rebecca Leonard [46] situates the manufacture within the growing material culture of porcelain, considering how European porcelain looked before Sèvres began and the influences on that earlier porcelain. The study of Rebecca Leonard [46] shows how Sèvres was able to create a product that aligned with already-existing trends in the market while drawing on new influences that came from French culture itself. Madame de Pompadour helped push the factory toward that French influence. She also cultivated the particular clientele that allowed Sèvres to receive continuous support. She helped the factory develop its new artistic style in the porcelain medium and its target market through her patronage of French artists at Sèvres, her influence on factory decisions and her direct promotion of Sèvres toward her fellow noble company domestically and abroad.

In the early 1740s in France, two brothers left their jobs at the Chantilly porcelain factory to start their own porcelain company. Robert and Gilles Dubois moved into a chateau in the town of Vincennes, France in 1741. They started producing and selling porcelain there by 1748. In 1753 the factory of Vincennes became Royal Manufactory and its products identified with a monogram (Figure 5.7). Two-crossed 'L' forms the monogram with a rhomboidal space in between where the manufactory used to write the year of the production. The letter 'A' indicated the 1753, each of the other letters marked the following years until 1776. Subsequently double letters, 'AA', 'BB' and so on until 1793, when they began to use other methods, identified each year. This factory relocated to Sèvres eight years later (1756), soon becoming one of the most well-known porcelain manufacturers in the world. Sèvres is located between Versailles and Paris, the two cities from which the factory's clientele was mostly likely to be coming. Clients were either French royals or nobles, or they were wealthy foreigners who purchased through the Parisian art dealers; in any of these cases, the proximity to both Versailles and Paris wonderfully facilitated factory visits. The King could have considered the benefit of this location on his own, but it is important to note Sèvres' proximity to Madame de Pompadour's chateau at Bellevue, at present-day Meudon. She built Bellevue in 1750, six years before the factory moved to Sèvres. It is likely she brought about the move, convincing Louis XV to move the factory so that it could be nearer to her. Furthermore, the marquise gave to the manufactory the rights she owned to a glass-bottle factory in the town of Sèvres so that its workers could live there. She not only urged the king to move the factory to Sèvres, but she also helped to settle into the new location.

b) The scientific leaders: Louis-Nicolas Vauquelin in 1797 discovered and isolated chromium from a crocoite ($PbCrO_4$) sample from Siberia [47].). The name chromium, as proposed by Haüy a prominent mineralogist who previously described the minerals emerald and beryl, originates from the word *χρῶμα* (*khrôma*) meaning color in Greek referring to its ability to impart various coloration to minerals, such as green for eskolaite Cr_2O_3 and red for lead chromate $PbCrO_4$. Soon after this discovery, by analyzing ruby, spinel and emerald, Vauquelin has evidenced that a low content of chromium was responsible for the red and green colors of these three minerals [47, 48, 49] associated to its various oxidation states and chemical environments. Based on these discoveries, Vauquelin has already foreseen in his report to the "Journal de la Société des Pharmaciens de Paris" in 1797 [47], that the use of chromium oxides will provide "solid and nice colors to Painting and Enameller's art." The association of pigments with crystals and chemistry in order to avoid the "*try and error*" system in its preparation is due to mineralogist and crystallographers. Following Kubbinga [50] the term 'crystallography' was introduced, in 1723, by Moritz Anton Cappeller (1685–1769). From about 1800 crystallography definitely branched off from mineralogy. It is noteworthy that René-Just Haüy (1743-1822) first manual, published in 1801, was entitled "Traite´ de Minéralogie." It was followed, in 1822, by a "Traité de Cristallographie." Ever since, the science of crystals (often considered as a form of applied mathematics), has been called crystallography. It not only assessed the crystalline part of Nature's minerals, but also the crystals grown in the chemical laboratory. In the French framework, the molecular theory reigned supreme. Those molecules were extremely small, that much was certain; their exact dimensions defied the imagination. Their particular 'symmetry', however, had to be related to that of a crystal. We will see how Haüy's pilings of polyhedra metamorphosed into Bravais' point lattices. Those lattices, then, came to be analyzed from the viewpoint of the new group theory. Laue, Friedrich and Knipping, in a way, put a final end to that development [50].

Figure 5.7. Cobalt blue monograms of identification of Meissen and Vincennes porcelain respectively. (http://gothphotographer.blogspot.com/2011/06/all-meissen-logos.html).

c) technological partners; the chemist Jean Hellot (1685–1766) [51], who is considered the inventor of the most pigments of the manufactory's in the first step in 18th century, these pigments are really a compilation of the pigments already common in Europe described in section 3.1, but perfected and adjusted to the Sèvres glazes and pastes, reaching probably the highest quality in Europe. Each color is described by its historic name, often based on references found in the factory sales archives (Figure 5.8):

i. Le *BLEU LAPIS*, (Co_2SiO_4 zaffre, see section 4.1). The earliest ground color used at Vincennes was blue, known as *bleu lapis*, a reference to the stone, lapis lazuli. It was introduced in 1751, probably to imitate Chinese powdered blue grounds. It is uneven and mottled as it was applied directly onto the unglazed body, with powdered cobalt added as a colorant, before glazing and firing. In the 1770s, it was known as *'gros bleu'* (dark blue). In Figure 5.8 is shown a detail from a Vincennes cup (*Gobelet litron*), 1754. (Private Collection).

ii. Le *FOND JAUNE* (Naple's yellow, lead (II) antimonate $Pb_2Sb_2O_7$ see section 5.1.3). As early as 1751, the Vincennes factory records note *'fond jaune'* to describe a variety of yellow grounds based on Naples' yellow. A rich and even yellow ground was difficult to produce and earlier experimental attempts were very pale. The inspiration was the yellow ground produced at the Saxon royal porcelain manufactory at Meissen. In Figure 5.8, is shown a Vincennes mustard pot and stand (*moutardier ordinaire*), 1753. Metropolitan Museum of Art, Gift of R. Thornton Wilson, in memory of Florence Ellsworth Wilson, 1954).

iii. Le *BLEU CÉLESTE* (copper (II) solved in a alkaly glaze modified by Co_2SiO_4 addition, see section 5.1.2, Figure 5.3). The much-admired turquoise-blue ground color was introduced in 1753. The colorant is copper, rich in soda and potash, and cobalt. It was applied as a powder, sieved onto a tacky undercoat on the glaze, before firing. It was famously used on the

Vincennes dinner service for Louis XV in 1753, where it was described as *'bleu ancien'*. It was one of the most expensive colors ever used at the Sèvres manufactory and again imitates a Chinese glaze color. In Figure 5.8, is shown a Vincennes covered sugar bowl and stand (*sucrier et presentoir*) from the Louis XV service, 1753. Boughton House, Northamptonshire).

iv. Le *FOND VERT* (copper (II) solved in an acidic glaze, see section 5.1.2, Figure 5.4). Solid green grounds from copper oxide were not common until 1756, introduced shortly after the factory was moved from Vincennes to Sèvres. The earliest are slightly bluish but by 1757, greener. The records sometimes described it as '*fond vert de Saxe*', paying homage to the green grounds used earlier at Meissen. In Figure 5.8, is shown a Vincennes sugar bowl or *pot à sucre Hébert*, part of a tray and tea service, 1755-6. (Royal Collection Trust/ © Her Majesty Queen Elizabeth II 2017).

v. Le *FOND ROSE* **(**Cassius Purple, colloidal gold, see section 4.6 and Figure 4.22). The celebrated pink ground was first recorded as '*fond rose*' in 1758. It was based on the colloidal suspension of very small solid particles of gold in the enamel glaze. The term 'Rose Pompadour' was coined in England in the 1760s but this was a crimson or claret color not the French *rose* or pink with which Pompadour's name later became linked. In Figure 5.8, is shown a Sèvres pot pourri vase (in the form of masted ship) or *pot-pourri "à vaisseau,"* 1758. (Metropolitan Museum of Art, Gift of Samuel H. Kress Foundation, 1958).

vi. Le *BEAU BLEU*, **(**$CoAl_2O_4$ spinel, see section 4.1). The popular solid dark-blue ground, applied over the glazed surface, was described as *beau bleu* from 1768; earlier, between 1763 and 1768, it was known as *bleu nouveau*. Again, the source of the blue was cobalt. In Figure 5.8, Detail, Sèvres *Vase Royal*, c. 1768-70. (Royal Collection Trust/© Her Majesty Queen Elizabeth II 2017).

vii. Le *FOND NOIR*, ($CoFe_2O_4$ spinel). The elegant *fond noir* is a blue-black pigment produced from iron and cobalt. It is applied in thick layers in a technique developed at Sèvres for the new hard-paste body in the early 1780s. The color imitated rare Chinese black-grounds (*wujin*) on porcelain, identified as 'mirror black' in the letters of Père François-Xavier d'Entrecolles (1664–1741), a French Jesuit who lived in China, and studied its ceramic production. In Figure 5.8, Sèvres wine bottle cooler (*seau à demi-bouteilles ordinaires*), one of a pair, 1791. (The J. Paul Getty Museum, Los Angeles).

But is in the 19th century when under the vigorous impulsion of Alexandre Brongniart, chemist and director of the manufacture, chromium was used as pigment for deep green paints and background glazes. Brongniart might have known chromium possibilities thanks to his closeness with Haüy in Paris and his work at the *Mines de Paris* at the very same moment, as shown by its publication a few months earlier in the same journal as Vauquelin [52].

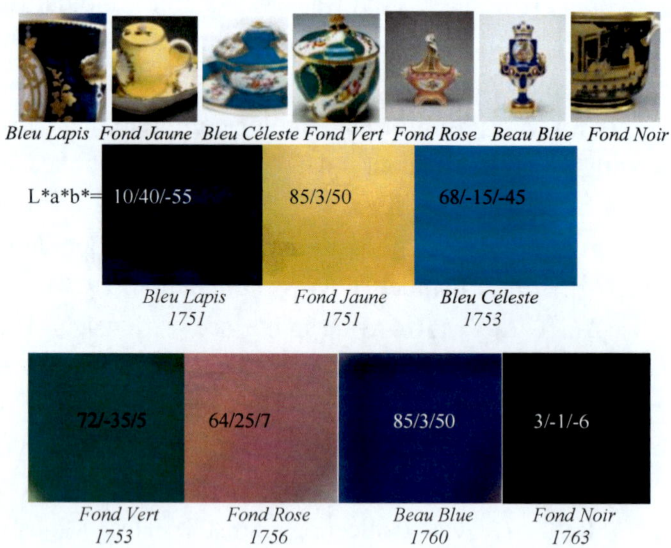

Figure 5.8. "Les coleurs de Sèvres" in the first step Vincennes (https://www.thefrenchporcelainsociety.com/about-us/ground-colours/).

The development of chromium green colors appeared relatively simple since the color is obtained with the direct addition of the precipitated oxide in a lead flux to produce a glaze [53]. The first trials of chromium based glazes at the manufacture date back to 1802 [54]. In 1806, the new green glazes obtained with chromium were used in production and subsequently replaced copper based green glazes [43]. During the first half of the 19th century, many chromates were also identified as potentially useful pigments: lead chromate (red), barium chromate (yellow-green) [55], potassium chromate (yellow) [56], mercury chromate (red/violet and green) [57], and iron chromate [55]. In 1845, chromium green under-glaze is eventually introduced as a genuine marker of Sèvres porcelains with a mark stamped for each product (Figure 5.9).

Figure 5.9. Chromium green under-glaze introduced as a genuine marker of Sèvres porcelains since 1845 [45].

When Brongniart died in 1847, Ebelmen succeeded Brongniart as the head of the factory. Besides his work to improve porcelain production, Ebelmen conducted research on the synthesis of minerals. Drawing its inspiration from solution chemistry, Ebelmen developed the crystalline growth method by the use of fluxes: precursors are mixed with a flux, such as boric acid or borax, and brought at high temperature in platinum foils. The flux plays the role of solvent at high temperature. Using this method, Ebelmen succeeded to synthesize some minerals for the first time, such as spinels [58]. After the sudden death of Ebelmen (1852), Alphonse Louis Salvétat, in charge of chemical works at the *Manufacture de Sèvres* until 1880, benefited from Ebelmen's experience and continued his work. The size of crystals obtained by Ebelmen or Salvétat was small; they were only observable under an optical microscope. However, since pigments are ground to be used in porcelain decoration, this new synthesis method was perfectly suited to obtain new pigments composed of chromium. This crystalline growth method using fluxes initiated by Ebelmen is now commonly used to synthesize pigments or improve its preparation (addition of mineralizers or flux agents).

Brongniart and his successors are at the origin of series of new pigments composed of chromium oxides: research conducted at the laboratory leads to a palette of pigments which was finalized in the middle of the 20th century. Chromium is the most used coloring element in the palette at Sèvres: among the 138 pigments regularly prepared, 76 are composed of chromium oxides [53] and used in a wide range of temperature from 900°C to 1400°C (Figure 5.10). Pigments are regularly synthesized and mixed in different uncolored frits. The mixture is painted on porcelains and fired at high or low temperature. The colorless frit vitrifies during firing and its role is to fix the color on the porcelain. Glazes are obtained, i.e., glassy decorative layers on the porcelain Surface [45].

While the synthesis protocol of pigments of Sèvres is known [53] the crystalline phases obtained after the thermal treatment should be characterized in order to know the chromium speciation. Such knowledge is essential to address (a) environmental issues associated with the toxicity of Cr(VI), (b) technical issues in order to better understand the history of synthesis of the pigments and (c) scientific issues surrounding the origin of the colors of pigments. Furthermore, such analysis is mandatory to improve the coloring properties of the pigments from Sèvres and to update their synthesis. The crystalline phases associated to Sèvres pigments are the followings [45,59] (Figure 5.10):

i. Eskolaite (Green of eskolaite Cr_2O_3). In 1802, eskolaite Cr_2O_3 came into an intense and new green pigment (green of Sèvres) with high thermal and chemical stability in glazes. Some hue variation can be obtained controlling the particles size: at low particle size (0.3 μm) the hue is green and turns to green-blue when particle size increases (0.5 μm). A black pigment is based on chromium doped hematite (or iron doped eskolaite): firing a mixture of iron and chromium oxides the $Cr_xFe_{2-x}O_3$ solid solution is used as black pigment [60-62].

ii. Garnet (Green of uvarovite garnet $Ca_2Cr_3Si_2O_7$ (Green Victoria)). The first synthesis of uvarovite at Sèvres is proven by a mention in the laboratory notebooks dating back to 1885. Furthermore, this kind of pigment is only used on one porcelain type (new hard-paste porcelain), the use of which started in 1882.

iii. Corundum (Green (x>1.9) and pink (x<0.2) of Chromium doped corundum $Cr_xAl_{2-x}O_3$). Soon, firing eskolaite homogenized with alumina a slight pink was obtained (chrome doped corundum pink pigment) which was improved adding fluorspar (CaF_2) as mineralizer in England (English pink), introduced about 1820 [63]. The first synthesis of corundum doped with chromium were conducted by Ebelmen in the kilns of the Manufacture de Sèvres and presented to "the Academy" in 1847. Ebelmen obtained it by mixing alumina with borax, adding chromium oxide (1% of the alumina mass), and calcining it on a platinum foil in a porcelain kiln. By this process, he obtained small transparent red crystals, embedded in an amorphous part. He isolated the crystals by the addition of hydrochloric acid at 70-80°C. Ebelmen's works were continued by Salvétat to design new colored porcelain glazes. In the laboratory notebooks, numerous synthesis using alumina and potassium dichromate are reported [45, 59].

iv. Sphene (Pink of Chromium-doped sphene or malayaite $Ca(Cr_xSn_{1-x})SiO_5$). The name *Pink* was kept since the introduction of this pigment in Sèvres from a British sample. At the beginning of the 19th century, the British industry had developed a pink pigment to color under-glazes on earthenware. The synthesis being secret, Brongniart asked Malaguti, chemist at the *Manufacture de Sèvres* in 1838, to find the composition of this pigment. Malaguti succeeded to obtain the chemical composition and to identify the elements necessary to retrieve the pink color, as well as unneeded constituents which were added on purpose to mask the real pigment elements [64]. He identified the key role of calcium, tin, silica and chromium.

v. Cassiterite (Pink of Chromium-doped cassiterite $Cr_xSn_{1-x}O_2$). The researches of Malaguti on the pink of chromium malayaite also led him to synthesize a new lilac pigment, by simply calcining tin oxide and chromium oxide, leading to Cr-doped cassiterite (SnO_2).

vi. Spinel: (a) pink of chromium doped gahnite $Zn(Cr_xAl_{2-x})O_4$, (b) pink of chromium doped spinel $Mg(Cr_xAl_{2-x})O_4$, (c) greenish-blue to blue of chromium doped cobalt aluminate spinel $Co(Cr_xAl_{2-x})O_4$, (d) pink, greenish-blue to blue of multicomponent $(Mg,Al,Co,Zn)(Al,Cr,Co)_2O_4$ spinel, (e) dark green, brown to black of multicomponent $(Mg,Al,Mn,Fe,Co,Zn)(Al,Cr,Co,Fe)_2O_4$ spinel.

vii. Spinel $MgAl_2O_4$ is a mineral which was named in 1779 by Jean Demeste from Latin "*spinella*," that means "little thorn," alluding to its sharp octahedral crystals and confused with ruby in former times because the solid solution of chromium in spinel gives a ruby appearance. Spinel gives also brown, green and blue pigments as below discussed.

viii. Spinels were synthesized for the first time at the Manufacture de Sèvres by Ebelmen, around the middle of 19th century. In 1847, he reported the synthesis of Cr-doped red spinel ($Cr^{3+}:MgAl_2O_4$), chromite ($FeCr_2O_4$) and magnesiochromite ($MgCr_2O_4$) in the wood-kiln for porcelain at Sèvres. Ebelmen used also kilns from a manufacturer of buttons, Jean-Félix Bapterosses, which can maintain a constant temperature. With those kilns he succeeded to obtain colorless gahnite ($ZnAl_2O_4$)

and Cr-doped gahnite. In the laboratory notebooks from 1882, syntheses of spinels are frequent, representing more than 50% of the laboratory's production between 1882 and 1904. The use of iron and manganese leads to pink brown to black shades [45].

ix. Willemite: greenish-blue to blue of chromium doped chromium doped willemite $Zn(Cr_xCo_{1-x})SiO_5$). This pigment is actually used in Vincennes and Chantilly factories based on zaffre.

d) The innovation-communicator coordinator: as described above, Madame de Pompadour was included in the efforts of the crown and the factory to serve a specific clientele and make Sèvres a strong economic venture. There were still further ways in which she promoted Sèvres within its customer niche and spread her taste throughout France and beyond. It produced these pieces all during its first fifty years. But Sèvres' continuity is surprising, given inauspicious conditions during this time. The factory emerged when China and the Meissen factory in Germany were already dominating the porcelain scene. For the first thirty years, its chemists could not produce a substance as sturdy as Meissen's and China's porcelain. And the French Revolution at the end of the eighteenth-century overthrew the values of the French nobility, which was closely tied to Sèvres [46].

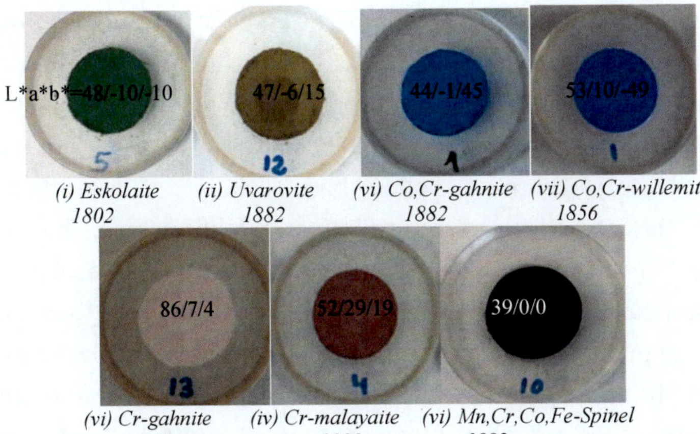

(i) Eskolaite 1802 *(ii) Uvarovite 1882* *(vi) Co,Cr-gahnite 1882* *(vii) Co,Cr-willemite 1856*

(vi) Cr-gahnite 1882 *(iv) Cr-malayaite 1838* *(vi) Mn,Cr,Co,Fe-Spinel 1882*

Figure 5.10. Crystalline phases associated to Sèvres pigments including chromium.

Yellow pigment is a lack on Sèvres palette which is based on ancient Naples Yellow, the discovery of a new yellow pigments based on cadmium is dated at begins of 19th century: the rise and fall of cadmium salts as pigments is a strange story, driven by misunderstandings and market forces [65]. Cadmium itself wasn't discovered until recently 1817, and then only by a chance observation of abnormal yellow coloration of a sample of what should have been zinc carbonate. The brilliant yellow color of its salt cadmium sulfide was noticed the following year, but it wasn't exploited as a pigment until the 1840s, when it became possible to manufacture it in quantity. CdS with a refractive index of 2.5 show two polymorphs: greenockite (hexagonal, space group $P6_3mc$) discovered as mineral in 1840 in Scotland and hawleyite (cubic, space group $F\text{-}43m$) discovered in 1955 in Canada. Nevertheless, it has been claimed that it was used as a pigment for oil paints as early as 1829, and by 1851 it was shown by Winsor & Newton at the Great Exhibition at Crystal Palace in London. But it remained extremely costly. By the end of the First World War, cadmium production had increased greatly, and the cost of Cadmium Yellow had fallen as a result. Another development which helped popularize the colors was the manufacture of lithopones by admixing barium sulphate, during the 1920s. These were half the cost of pure Cadmium Yellow, yet proved almost as lightfast and stable. The general commercial availability of cadmium sulfide from the 1840s led to its adoption by artists, notably by impressionist painters (style derived from the title of a Claude Monet work, "*Impression, soleil levant*" (*Impression, Sunrise*), 1872), such as Van Gogh, Monet and Matisse. The presence of cadmium in paints has been used to detect forgeries in paintings alleged to have been produced prior to the 19th century [66].

In the case of ceramics, yellow but specially a true bright red color available for ceramics is the cadmium-sulphoselenide, but they are unsuitable for underglaze applications because oxidize in air above 550ºC and sublimates up to 900ºC, besides they are incompatible with other inorganic pigments and many lead fluxes to produce the blackening of the color, finally react violently with glazes at high temperature [61]. The stabilization of the cadmium pigments (see section 4.3) were achieved by wrapping them in a protective envelope: the zircon crystals as will be discussed later.

In the twentieth century [67], cadmium pigments became widely used to make brightly-colored plastics, particularly those for domestic use, and the consumption of pigment for artists' paints was relatively tiny. It was then that cadmium fell from favor. Cadmium salts are notoriously toxic, something discovered soon after their introduction. However, when locked into pigment particles, their insolubility makes them of very low risk, unless they are dispersed in dust which is inhaled, as could be the case when incorporated into soft pastels. This lack of immediate threat lulled users into a false sense of security: it turned out that cadmium (as with other toxic pigments) posed major environmental risks. In the twenty-first century, evidence has accumulated that cadmium pigments can become concentrated in sewage sludge, which is often applied to fertilize agricultural land, resulting in increased dietary intake of cadmium, which is in turn associated with long-term medical problems such as osteoporosis and bone fracture, and some cancers. In opinion of some artists [43], none of this should have been any worry for artists, whose use of cadmium salts is tiny relative to industrial uses, including most notably rechargeable batteries. Except that most artists had almost no awareness of their impact on the environment, and have been happily discharging pigment-rich waste into the sewage system.

The red pigment is associated to the solid solution $Cd(S_xSe_{1-x})$ of cadmium sulfoselenide that varies from yellow for low x to red at high content of selenium (Figure 5.11). The cadmium red was available as a commercial product from 1919. When introduced, there were hardly any stable pigments in the yellow to red range, especially orange and bright red was very troublesome, when the cadmium pigments replaced mercury sulfide (the original vermilion), the light-fastness was greatly improved. About 2000 tons of cadmium pigments are produced annually as of 1982, representing about 25% of the cadmium processed commercially. Nowadays, the ECHA regulation in EU [68] indicates that shall not be used or placed on the market in paints with codes [3208] [3209] (hexagonal color code, both usually obtained from cadmium sulfoselenide) in a concentration (expressed as Cd metal) equal to or greater than 0.01 wt.%. For paints with codes [3208] [3209] with a zinc content exceeding 10 wt.% of the paint, the concentration of cadmium (expressed as Cd metal) shall not be equal to or greater than 0.1 wt.%. Painted articles shall not be placed on the market if the concentration of cadmium (expressed as Cd metal) is equal to or greater than 0.1 wt.% of the paint on the painted article. Therefore practically the uses of cadmium pigments are banned in EU. The cadmium pigments have been partially replaced by azo pigments in paints. These are similar in lightfastness to the cadmium colors and have the advantage of both being cheaper and non-toxic.

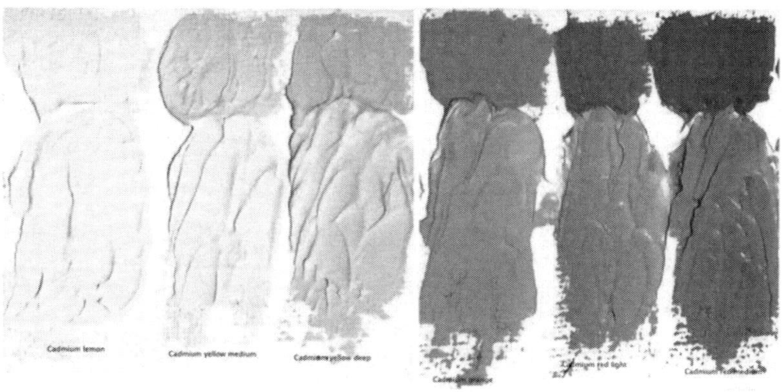

Figure 5.11. Cadmium colors (from Williamsburg Oil Paints: https://www.williamsburgoils.com/).

5.4. The Rutile Pigment Family

Towards the end of 19th century a specialization on ceramic pigments manufacture of some factories starts and this allow to improve the strength and stability of the ceramic pigments. At the middle of 20t century the in-house manufacturing of pigments had ceased.

Titanium dioxide TiO_2 shows different polymorphs: rutile and anatase (tetragonal), brookite (orthorhombic). Rutile is the only stable phase, while the two metastable phases, anatase and brookite, irreversibly transform to rutile on heating. In 1911 the first pigment "Rutile White/Anatase (PW6,77891)" was patented by A.J. Rossi (US Pat. 1,184,1311), but titanium dioxide has only been manufactured commercially since 1918 and has increased as the most widely used white pigment, with a world average production of more than 3 million tons each year, replacing the toxic traditional pigments based on lead, zinc and antimony. In the 1930s, anatase obtained by the "sulfate process" competed with traditional toxic targets based on Zn, Sb or Pb such as white lead $(PbCO_3)_2Pb.(OH)_2$ or lithopone white (an equimolecular coprecipitate $BaSO_4$ and ZnS).

In the sulphate process [69], finely ground ilmenite (or high-TiO_2 slag) is digested in concentrated sulfuric acid. After initial heating, an exothermic reaction starts between the titanium raw material and sulfuric acid. As the reaction proceeds, a solid porous cake is formed. This cake is dissolved in dilute acid and water to yield a solution of titanyl sulphate and iron sulphate according to the following reaction:

$$FeTiO_3 + 2H_2SO_4 \rightarrow TiOSO_4 + FeSO_4 + 2H_2O$$

In a separate reduction tank, any ferric iron present is reduced to the ferrous state by adding scrap iron. After reduction, the solution is clarified by settling and filtration to remove unreacted solids. Iron sulphate is then removed from the solution as crystallized ferrous sulphate heptahydrate ($FeSO_4.7H_2O$), called copperas. Following iron removal, titanyl sulphate is hydrolyzed to hydrated titanium dioxide, from which TiO_2 is precipitated as either rutile or anatase. Finally, the precipitate is chemically purified, filtered, calcined and milled to proper particle size. Contaminants that would affect the quality of the finished product are removed from the filtered precipitate by a multistage wash before calcination.

Titanium dioxide in the form of anatase (Tetragonal $I4_1/amd$) is a white powder with high opacity, brilliant whiteness, excellent covering power and chemical and physical resistance. These excellent optical features, coupled with the high refractive index, made it a valuable pigment and opacifier for a broad range of applications in paints, plastic goods, inks and paper [70].

In the 40s., rutile (Tetragonal *$P4_2/mnm$*, obtained by seeding in the crystallization of the sulfate process), began to replace anatase pigments in coatings and plastic applications due to its greater opacity and better stability of resistance to degradation due to its low photocatalytic activity compared to anatase. Titanium dioxide stands out for its high refractive index (2.73 for rutile and 2.55 for anatase (the cassiterite used in the previous stage shows n=2.01) and presents a very pure white. Anatase was used from 1918 as a bleach and opacifier in paints.

The opacifying power of the TiO_2 is progressively increased with the anatase to rutile transformation which can take place at very different temperatures between 400 and 1000°C, depending on several factors [71,72]. However, the use of rutile as white was only applied in exceptional cases during the 19th century, given that small impurities of iron or chromium stained yellowish the mass as previously discussed [73].

The high refractive index of rutile makes it very interesting as a base crystal structure for ceramic pigments:

i. Yellow of rutile.
In 1939 Ni-Sb Rutile Yellow (PY53.77788) was patented by H.H Schaumann, (EI Du Pont de Nemours and Comp. US Pat. 2,257,278) a solid solution of Ni (II) and Sb (VI) in rutile obtained by calcination from oxides and anatase.

ii. Ocher of rutile.
Also in 1939 the Cr-Sb Rutile Ocher-Brown was discovered and patented by CJ Harbert (Harshaw Ch Company patent US 2,251,829) with the same philosophy but now the substitution of two Ti (IV) for each Cr (III) and Sb (V) allows the electroneutrality of the lattice [74].

iii. Brown of rutile.
In 1974, rutile Mn-Nb brown was patented (Ferro, US Patent 3,832,205). These pigments soon became part of the ceramic palette as they were stable in ceramic glazes.

The mechanism of rutile pigments is based in the presence of a Ti^{4+} - oxygen charge transfer band occurring in the near UV; it makes the Ti-oxygen based compounds colorless and crystal field effects associated to optical absorbance of chromophores transition metals substituting Ti^{4+} in octahedral coordination give interesting colors [75]. This charge transfer band partially shifts to the visible region when the anatase-to-rutile transformation occurs, so that a typical cream color arises. In fact, the charge transfer band of anatase, centered at about 29800 cm^{-1} (336 nm) in the case of rutile is progressively shifted towards 26500 cm^{-1} (377 nm) approaching the visible region.

The contemporary presence of a chromophore ion entering the TiO_2 structure in solid solution determines the presence of a crystal field, whose effects give rise to different colorings patterns [74]. In the synthesis strategy of rutile pigments to allow the transition metals chromophores solid solutions in rutile, two facts are very important:

a) The entrance of transition metals chromophores with a valence other than four of titanium requires the codoping with so-called counter-ions in order to preserve the lattice electroneutrality. Thereby Ni^{2+} has as counter-ion W^{+6} and Cr^{3+} or Mn^{3+} has W^{+5} , Sb^{+5} or Nb^{+5} as counter-ions. The counterion (i.e., Mo, Sb, Nb or W) is always added in order to achieve the desired coloration and/or improve the technological properties (e.g., chemical-physical resistance in ceramic bodies and glazes). Though the absorbance bands are broad and frequently overlapped, the UV–Vis–NIR spectra suggest the occurrence of Cr^{3+}, Mn^{2+}, Mn^{3+}, Ni^{2+}, V^{3+}, and V^{4+} as chromophores. Rutile pigments are suitable for through-body (up to 1250°C) and glaze applications (up to 1100°C). The best coloration of porcelain stoneware bodies is achieved with Sb or W as counterions, though the higher stability is ensured by Sb. The best glaze colors are accomplished by W-bearing pigments, which however are less stable than Nb or Sb containing ones, except than for the V + W coupling. This latter represents a new and very interesting Co-free and Cr-free dark blue-black pigment for low temperature applications [76] (Figure 5.12).

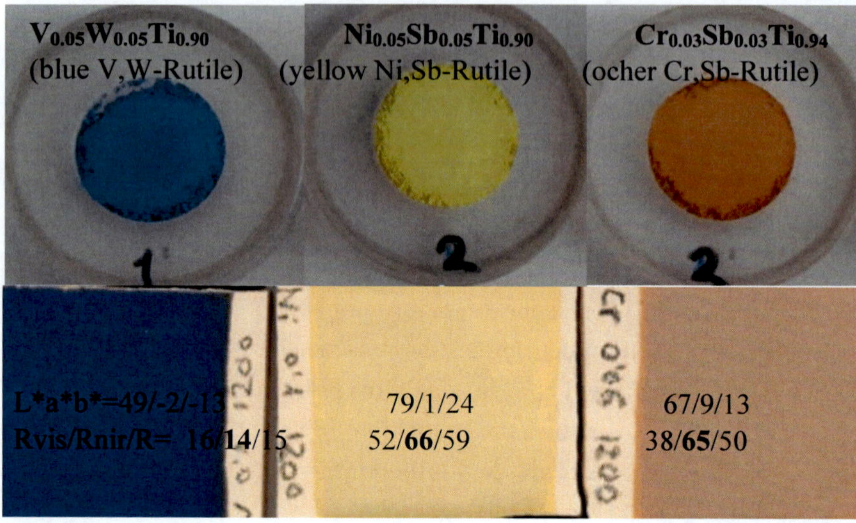

Figure 5.12. Rutile ceramic pigments: powders (1100°C/3h) and 5wt.% glazed in a double firing frit (1050°C).

b) On the other hand the direct doping of rutile is not successful, it is necessary to use anatase as precursor of titanium. The entrance of the chromophores transition metals occurs as the anatase-rutile transition develops [77]. Thereby, the use of non conventional methods such as coprecipitation or Sol-Gel procedures, are very interesting for the synthesis of these pigments since they start from a homogeneous amorphous gel, and first crystallizes anatase in the heat treatment and after transforms into rutile, in the presence of chromophores and counter ions that are easily introduced homogeneously into the rutile neo reticulum to form the solid solution of the pigment. [75, 78].

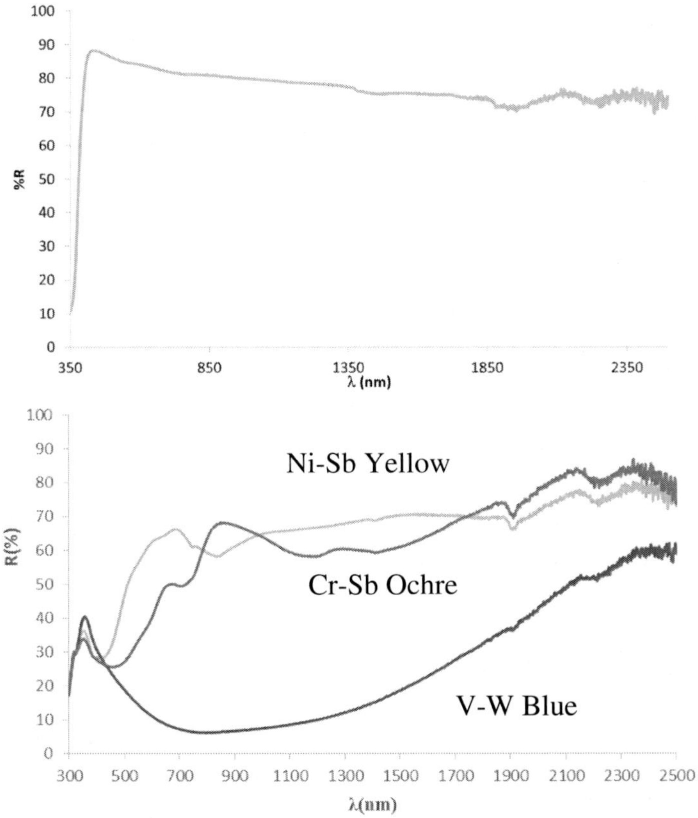

Figure 5.13. UV-Vis-NIR diffuse reflectance spectra of rutile powder and the rutile pigments 5wt.% glazed in a double firing frit (1050°C).

Figure 5.12 shows the powders of some rutile ceramic pigments obtained by ceramic method an fired at 1100°C/3h and 5wt.% glazed in a double firing frit (1050°C). The $V_{0.05}W_{0.05}Ti_{0.90}$ (V,W-Rutile) shows a dark blue color, $Ni_{0.05}Sb_{0.05}Ti_{0.90}$ (Ni,Sb-Rutile) is yellow and $Cr_{0.03}Sb_{0.03}Ti_{0.94}$ (Cr,Sb-Rutile) shows the classic ocher-orange color. The UV-Vis-NIR diffuse reflectance spectra of the rutile pigments 5wt.% glazed in a double firing frit (1050°C) showed in Figure 5.13 are compatible with V^{4+}, Ni^{2+} and Cr^{3+} in octahedral coordination [76]. The reflectance measurement Rvis/Rnir/R=80/80/80 for white anatase decreases to 52/66/59 for yellow, 38/65/50 for ochre and 16/14/15 in the V,W-rutile associated to absorbance of the above mentioned ions. Yellow and ocher show high NIR reflectance (there are actually cool pigments) but the NIR reflectance of V-W blue pigment is very low (only 14%) and it cannot be suitable for cool building coatings.

REFERENCES

[1] Figueiredo, M. O., Silva, T. P, Veiga, J. P., Prudêncio, M. I., Dias, M. I., Matos, M. A., Pais A. M., Blue Pigments in XVI-XVII Century Glazes: A Comparative Study between Portuguese Faiences and Chinese Porcelains, *2nd Latin-American Symposium on Physical and Chemical Methods in Archaeology, Art and Cultural Heritage Conservation & Archaeological and Arts Issues in Materials Science* – IMRC, Cancún, Mexico (2009).

[2] Mimoso, J. A., *Origin, early history and technology of the blue pigment in azulejos*, (Glaze Arts 2015, Lisbon, 2015).

[3] Kriznar, A., Muñoz, M. V., de La Paz, F., Respaldiza, M. A., Vega, M., A Diptich by Marcellus Coffermans Analysed by Portable XRF, *Proceedings of the 37th International Symposium on Archaeometry*, I. Turbanti-Memmi (ed.), Springer-Verlag Berlin Heidelberg 2011.

[4] Seccaroni, C., Moioli, P., *Fluorescenza X: Prontuario per l'analisi XRF portatile applicata a superfici policrome* [*Fluorescence X: Handbook for portable XRF analysis applied to polychrome surfaces*], (Nardini editore, Firenze, 2002).

[5] Horschik, J, Beiträge zur Geschichte der Kobaltfarbe und ihre Verwendung in der Keramik [Contributions to the history of the cobalt colour and its use in ceramics], *Keramos* 85(1979)119-142.

[6] BHO, *British history on line*, https://www.british-history.ac.uk/no-series/traded-goods-dictionary/,1550-1820/zaffre-zoobditty-match,

[7] Mackenzie, *Mackenzie's Five Thousand Receipts in All the Useful and Domestic Arts*, (Pottery: p 369, Philadelphia, 1845). https://archive.org/details/mackenziesfiveth00mack.

[8] Kunkel, J, *Ars vitraria experimentalis* [*Experimental technique vitraria*], (Berlin, 1679).

[9] Shijie, W., "Cobalt: Its recovery, recycling, and application," *Journal of the Minerals, Metals and Materials Society* 58, 10, (2006): 47-50.

[10] Wagenknecht, F., Juza, R., Rinmann's Green in *Handbook of Preparative Inorganic Chemistry*, 2nd Ed. Edited by G. Brauer, Academic Press, NY. Vol. 1., 1963, p. 1092.

[11] Gates, G., A Note on the Artists' Pigment Aureolin, *Studies in Conservation*. 40,3(1995): 201–206.

[12] Douglass, D. L., Shing, C., Wang, G., The light-induced alteration of realgar to pararealgar, *American Mineralogist*, 77 (1992):1266-1274.

[13] Leroi-Gourhan, A., *The Art of Prehistoric Man in Western Europe*. (Thames & Hudson, London, p. 40. 1968).

[14] Cennini, C., *Il Libro dell'Arte* [*The Book of Art*], (Lanciano, R.Carabba editore, 1913).

[15] Pinho da Silva, L., Los socarrats valencianos [The Valencian socarrats]. *Estudios de caracterización y técnicas de limpieza*, Tesis de Master en Conservación y Restauración, Lisboa, 2007.

[16] ICMV, *Unidad de Arqueometría del ICMUV* [*ICMUV Archeometry Unit*], https://www.uv.es/lablaser/archivo/Documents/Informe.pdf,.

[17] Cerro, S., Llusar, M., Gargori, C., Monrós, G., "Cool and photocatalytic yellow ceramic pigments; from lead-tin to Cr doped scheelite pigments," *Ceramics International* 45 (2019) 4613–4625.

[18] Eastaugh, N., Walsh, V., Chaplin, T., Siddall, R., *Pigment Compendium: A Dictionary and Optical Microscopy of Historic Pigments*, (Elsevier; 2005).

[19] Dik, J., *Scientific analysis of historical paint and the implications for art history and art conservation: the case studies of Naples yellow and discoloured smalt*, Thesis, Faculty of Sciences, University of Amsterdam, (2003)115–120.

[20] Picolpasso, C., *Tre Libri dell'Arte del Vasaio* [*Three Books of the Potter's Art*], (a.c.d. Giovanni Conti, Florence, (1976) for example: p.118: Colori a I'Urbinata, p.119: Colori della Marca, p.120: Colori Castellani).

[21] Uda, M., Sassa, S., Yoshimura, S., Kondo, J., Nakamra, M., Ban Y., Adachi, H. "Yellow, red and blue pigment from ancient Egyptian

palace painted walls. Nuclear Instruments and Methods," *Physics Research B* 161–163(2000): 758–761.

[22] Sayre, E. V., Smith, R. W., "Compositional Categories of Ancient Glass," *Science* 133(1961): 1824–1826.

[23] Wiberg E., Wiberg N. (2001) *Inorganic Chemistry*, (Academic Press, p. 1413, 2001).

[24] Queiroz, C. M., Agathopoulos, S., *The discovery of European porcelain technology,* in Prudêncio, M. I., Dias, M. I., Waerenborgh, J. C., eds. Understanding people through their pottery (Trabalhos de Arqueologia; N.º 42, ISBN 972-8662-25-4. Lisboa: Instituto Português de Arqueologia, p. 211-215, 2005).

[25] Bauke, D. (1990), *Ehrenfried Walther von Tschirnhaus: Der Weg zum Porzellan* [*Ehrenfried Walther von Tschirnhaus: The way to porcelain*]. 1990 http://www.tschirnhaus.de/porzellan.html.

[26] O'Connor, J. J.; Robertson, E. F., *Ehrenfried Walter von Tschirnhaus*. 1997, http://wwwhistory.mcs.standrews.ac.uk/history/ Mathematicians/ Tschirnhaus.html.

[27] Goder, W.; Walter, H., *Die Erfindung des europäischen Porzellans* [*The invention of European porcelain*], in Johann Friedrich Böttger zum 300 Geburtstag. (Meissen: Staatliche Kunstsammlungen VEB-SPM, p. 74-82, 1982).

[28] Giménez, M., *El Descubrimiento de la Fórmula de la Porcelana en Europa y la Posterior Historia de la Manufactura de Meissen (un intento de periodización)* [*The Discovery of the Porcelain Formula in Europe and the Later History of the Meissen Manufacture (an attempt at periodization)*], in Romero, Alicia (dir.) (De Artes y Pasiones. Buenos Aires: 2005). www.deartesypasiones.com.ar.

[29] Hoffmann, R., Meissen Chymistry, *American Scientist*, http://www.americanscientist.org/issues/id.3431,y.0,no.,content.true, page.1,css.print/issue.aspx.

[30] Divis, J., Ernould-Gandoulet, M., *L'Art de la Porcelaine en Europe, ("Beaux objets d'autrefois")* [*The Art of Porcelain in Europe, ("Fine objects from the past")*], (Paris, Gründ, 1984).

[31] Honey, W. B., *German Porcelain*, (London, Faber & Faber, 1947).

[32] Miller, M., Miller, J., *Miller's Antiques Checklists: Porcelain, ("Viking Studio Books"),* (New York, Viking Penguin, 1991).

[33] Hofmann, F. H., *Das Porzellan der europäischen Manufakturen* [*The porcelain of the European manufactories*], Frankfurt, Ullstein (Propyläen Kunstgeschichte; 1,1980).

[34] Steinbrück, J. M., Charakteristik Böttgers [Characteristics of Böttgers] (manuscript). In *Johann Friedrich Böttger zum 300 Geburtstag*. (Meissen: Staatliche, Kunstsammlungen VEB-SPM, p. 57-58, 1717).

[35] Gleeson, J, *The Arcanum: The Extraordinary True Story*, (Warner Books, 1998).

[36] Frothingam, A. W., *Talavera Pottery.* Hispanic Society, New York, 1944.

[37] Colomban, P., "The Destructive/Non-Destructive Identification of Enameled Pottery, Glass Artifacts and Associated Pigments-A Brief Overview," *Arts*, 2(2013): 77-110.

[38] Cooper, E., *Ten Thousand Years of Pottery*, (British Museum Press: London, UK, 2000).

[39] Colomban, P.; Truong, C. "A Non-destructive Raman Study of the Glazing Technique in Lustre Potteries and Faiences (9th-14th centuries): Silver ions, Nanoclusters, Microstructure and Processing." *J. Raman Spectrosc.*, 35 (2004): 195–207.

[40] Simsek, G., Geckinli, E., "An assessment study of tiles from Topkapı Palace Museum with energy-dispersive X-ray and Raman spectrometers." *J. Raman Spectrosc.* 43 (2012): 917-927.

[41] Colomban, P., Milande, V., Lucas, H. "On-site Raman analysis of medici porcelain." *J. Raman Spectrosc.* 35(2004): 68–72.

[42] Godden, G., *An Illustrated Encyclopaedia of British Pottery and Porcelain*, (Magna Books, 1992), ISBN 1 85422 333 X.

[43] Monrós, G., *El Color de la Cerámica: Nuevos Mecanismos en Pigmentos para Los Nuevos Procesados de la Industria Cerámica* [*The Color of Ceramic: New Mechanisms in Pigments for the New Processes of the Ceramic Industry*], (Universitat Jaume I, 2003), ISBN 9788480214490.

[44] Ricciardi, P., Colomban, P., Fabbri, B., Milande, V. "Towards the establishment of a Raman database of early European porcelain." *e-Preserv. Sci.*, 6 (2009): 22–26.

[45] Verger, L., Dargaud, O., Chassé, M., Trcera, N., Rousse, G., "Synthesis, properties and uses of chromium-based pigments from the Manufacture de Sèvres." *Journal of Cultural Heritage*, 30(2018): 26 - 33.

[46] Leonard, R., Madame de Pompadour and the Sèvres Porcelain Factory: Cultivating a French Artistic Identity and Clientele, *Proceedings of The National Conference On Undergraduate Research (NCUR)*, University of North Carolina Asheville Asheville, North Carolina April 7-9, 2016. https://www.ncurproceedings.org/ojs/index.php/ NCUR2016/article/view/2099/1024.

[47] Vauquelin, N. L., Du plomb rouge de Sibérie, et expériences sur le nouveau métal qu'il contient [Siberian red lead, and experiments on the new metal it contains], *Journal des Mines* 6(1797)737–761.

[48] Vauquelin, N. L., Analyse de l'Emeraude du Pérou [Peruvian Emerald Analysis], *Journal des Mines, 7(*1798):93–98.

[49] Vauquelin, N. L., Analyse du rubis spinelle [Spinel ruby analysis], *Journal des Mines,* 7 (1798): 81–92.

[50] Kubbinga, H., "Crystallography from Haüy to Laue: controversies on the molecular and atomistic nature of solids," *Z. Kristallogr.* 227 (2012): 1–26 /doi: 10.1524/zkri.2012.1459.

[51] Wisniak, J., Hellot. J., "A pioneer of chemical technology," *Revista CENIC Ciencias Químicas*, 40, 2 (2009): 111-121.

[52] Brongniart, A., "Note lithologique-Sur la colline de Champigny près de Paris [Lithological note-On the hill of Champigny near Paris]," *Journal des Mines 5*(1797)479–487.

[53] D'Albis, A., *Traité de la Porcelaine de Sèvres* [*Treaty of Sèvres Porcelain*], (Editions Faton, 2003).

[54] Brongniart, A., "Sur les couleurs obtenues des oxydes métalliques, et fixées par la fusión sur les différents corps vitreux [On the colors obtained from the metallic oxides, and fixed by the fusion on the different vitreous bodies]," *Journal des Mines 12*(1802):76–77.

[55] Brongniart, A., "Préparation chimique des couleurs vitrifiables et des métaux. In Traité des arts céramiques ou des poteries: considérées dans leur histoire, leur pratique et leur théorie [Chemical preparation of vitrifiable colors and metals. In Treatise on ceramic arts or pottery: considered in their history, practice and theory]," *Béchet jeune*, (1854) 514–582.

[56] Salvétat, A. L., *Leçons de Céramique - Tome 1 et 2* [*Ceramic Lessons - Volume 1 and 2*], (Mallet Bachelier, Paris, 1857).

[57] Godon-Saint-Memin, Sur la belle couleur verte que le chrôme peut fournir à la peinture [On the beautiful green color that the chrome can give to the painting]. In *Annales du Muséum national d'histoire naturelle*, (Paris, 1804:238). http://www.biodiversitylibrary.org/item/93709.

[58] Ebelmen, J. J., "Sur une nouvelle méthode pour obtenir des combinaisons cristallisées par la voie sèche, et sur ses applications à la reproduction des espèces minerales [On a new method for obtaining crystallized combinations by the dry route, and on its applications to the reproduction of mineral species]." *Chimie, Céramique, Géologie, Métallurgie*. 12(1861):125–225.

[59] Ren F., Ishida S., Takeuchi N., Feyiyoshi K., "Chromium Based Ceramic Colors," *American Ceramic Society Bulletin*, 71, 5(1992): 759-764.

[60] Fan X, Gan M., Jiang M, Yuan L., Chen X, "Influence of flux additives on iron ore oxidized pellets," *Journal of Central South University of Technology*, 17(2010):732-737.

[61] Airey, A. C., Roberts W., "Advances in Ceramic Colors," *Ceram. Eng. Sci. Proc.*, 8(1987):1168-75.

[62] Escardino, A., Mestre S., Barba A., Beltrán V., Blasco A., "Synthesis Mechanism of an Iron–Chromium Ceramic Pigment," *Journal of the American Ceramic Society* 83,1(2000):29-32.

[63] Bell B. T., "*The Development of Colorants for Ceramics, Review of Progress in Coloration and Related Topics*," 9(1987): 48–57.

[64] Gay-Lussac, Arago, *Annales de Chimie et de Physique, Tome LXI* [*Annals of Chemistry and Physics, Tome LXI*], (Paris, 1836).

[65] Fiedler, I., Bayard, M., *Artists' Pigments,* vol 1, ed Robert L Feller, Archetype, 1986, ISBN 978 1 904982 74 6.

[66] Taft, W. S., Mayer, J. W., Newman, R., Kuniholm, P., Stulik, D., *The Science of Paintings*, (Springer, 2000) ISBN 0-387-98722-3.

[67] Hoackley, *Pigment: Controversial Cadmiums, yellow to red*, june, 2018, https://eclecticlight.co/2018/06/15/pigment-controversial-cadmiums-yellow-to-red/.

[68] ECHA, *Restrictions adopted in the framework of REACH for Cadmium and its compounds*, https://echa.europa.eu/es/substances-restricted-under-reach/-/dislist/details/0b0236e1807e2518.

[69] Chernet, T., "Applied mineralogical studies on australian sand ilmenite concentrate with special reference to its behaviour in the sulphate process," *Mineral Engineering*, 12, 5 (1999):485-495.

[70] Loughbrough, R., "TiO_2 pigment, Overcapacity hits again," *Industrial Minerals*, 6 (1992):37-53.

[71] Ren, R., Yang, Z., Shaw, L. L., "Polymorphic transformation and powder characteristics of $Ti0_2$ during high energy milling." *J. Mat. Sci.* 35, 23 (2000):6015-6026.

[72] Shannon, R. D., Pask, J. A., "Kinetics of the anatase to rutile transformation," *J. Am. Ceram. Soc*, 48 (1965):391- 398.

[73] Maloney, J., *Titanate Pigments. Part I.6*, in High Perfomance Pigments, (Wiley-VCH, H.M Schmit ed., Weinheim, 2002).

[74] Gargori, C., Cerro, S., Galindo, R., Monrós, G., "In situ synthesis of orange rutile ceramic pigments by non-conventional methods," *Ceramics International*, 36 (2010):23–31.

[75] Matteucci, F., Cruciani, G., Dondi, M., Guarini, G., Raimondo, M., Colouring mechanisms in rutile-based ceramic pigments, *Qualicer* 2004, http://www.qualicer.org/recopilatorio/ponencias/pdfs/0413201e.pdf

[76] Dondi, M., Cruciani, G., Matteucci, F., Guarini, G., "The role of counterions (Mo, Nb, Sb, W) in Cr-, Mn-, Ni- and V-doped rutile ceramic pigments: Part 2. Colour and technological properties," *Ceramics International*, 32, 4, (2006):393-405.

[77] Rodrigues-Talavera, R., Vargas, S., Arroyo Murillo, R., Montiel-Campos, R., Haro-Poniatowski, E., "Modification of the phase transition temperatures in titania doped with various cations," *J. Mater. Res.*, 12 (1997):439-443.

[78] Sacerdoti, M. E., Carotta, L., Crema, G., Martinelli, G., "Sol-gel processed $Ti0_2$-based nano-sized powders for use in thick film gas sensors for atmospheric pollutant monitoring," *J. Sol-Gel Tech.* 22 (2001)167-79,

[79] Ishida, S., Hayashi, M., Fujimura, Y., "Spectroscopic study of the chemical state and coloration of chromium in rutile," *J. Am. Ceram. Soc.,* 73, 11 (1990): 3351-3355.

Chapter 6

THE ZIRCON PIGMENT FAMILY

The most important novelty in the field of ceramic pigments in the 20th century was the discovery of the family of zircon pigments by Seabright, consisting of the following pigments, namely: a) the turquoise blue V-zircon: a solid solution of both V^{4+} and F^- in zircon $ZrSiO_4$ discovered in 1948 [1], b) The yellow of praseodymium in zircon, Pr-zircon; a pigment of praseodymium oxide in zircon discovered in 1961 [2.a], c) The pink coral of iron in zircon, an inclusion of hematite in zircon crystals discovered in 1965 [3]. This zircon pigments show an excellent coloring yield in glazes due to its both high chemical and thermal stability.

As above discussed, if the discovery of European white porcelain was the fruit of the modern experimental method, the discovery of zircon stains can be considered the fruit of the application of the scientific knowledge (Research) to technology (Innovation). In effect, as discussed in Chapter 1, the introduction of zircon in ceramics was led for the high increase of the cassiterite (SnO_2) price about 1914. Tin oxide was, par excellence, the glaze's opacifier and withe stain for glazes since Rome, only antimony oxide and titanium oxide were sometimes used, but antimony oxide Sb_2O_3 release at relative low temperatures (m.p. 656ºC) and titanium oxide yellowish the glaze when contains traces of iron or chromium.

The high refraction index and high melting point of both zirconia and zircon (2.4 and 1.94 respectively), together their low solubility in glazes was emphasized by American and European researchers [4, 5, 6]. In 1930-1940 the new opacifier was introduced by several manufactures, such as Titanium Alloy Manufacturing Company [7], which patent the white and brilliant glazes obtained with zirconia and zircon.

Zirconia and zircon accomplish the three main characteristics of ceramic pigments: high thermal stability, compatibility with glaze and ceramics matrices (insoluble in these media), and high coloring capacity, but for give true ceramic pigments it was necessary they should produce solid solutions by doping with chromophore ions. Thus yellow of vanadium doped zirconia is developed but unstabilizes by reaction with silica from the glaze, precipitating zircon. On the other hand the direct doping of zircon is not successful.

In 1948 Clarence A. Seabright thought that probably it was necessary to change the synthesis strategy and, as it was led with corundum and rutile stains where anatase or aluminum hydroxides are used instead rutile or corundum respectively, use a mixture of silica and zirconia with chromophores. Thus the turquoise of vanadium doped zircon using NaF as mineralizer had a great success [1]. Later, in 1960, Seabright develops the yellow of praseodymium in zircon (although this pigment it had already been handled and cited in Japan) [2a, 2b]. and finally in 1965 the iron coral of hematite in zircon was patented by the same author [3].

The quality of zircon pigments depends on many factors, some of them limiting the synthesis [8]:

a) The particle size of the precursors in the case of ceramic synthesis must always be below 5 μm; especially the precursors of the chromophores, in fact the corresponding oxides are seldom used (NH_4VO_3 is preferably used over V_2O_5, praseodymium nitrate hexahydrate instead of Pr_6O_{11} or iron (II) sulfate heptahydrate instead of hematite).
b) It is absolutely necessary to introduce halides as mineralizer; its composition can include any one or a combination of alkali or

alkaline earth halide modifiers, a NaCl-NaF mixture is usually used depending on the calcination temperature.

c) The firing temperature and soaking time must also be optimized for each industrial facility.

d) The use of pelleting and addition of modifying cations such as Al_20_3, Fe_20_3, In_20_3, Mg_2O_3, SiO_2, SnO or TiO_2 can be used in order to optimize the pigment.

These conditions and the chemical nature of the each pigment are discussed and analyzed below taking as reference three pigments optimized for industrial application. Figure 6.1 shows the characterization of three optimized zircon powders and 5wt.% glazed in a double firing frit (1050°C) with its CIEL*a*b* color parameters, its diffuse reflectance values (in visible range Rvis, in near infrared range Rnir and the total reflectance R) and the respective XRD patterns of powders. Figure 6.2 shows the UV-Vis-NIR reflectance spectra of optimized yellow, cyan and pink zircon respectively, and Figure 6.3 shows representative SEM micrographs of these yellow, cyan and pink zircon pigments.

The CIEL*a*b*of powders show a good saturation of the color for the three pigments with high b* (58.8), negative b* (-17.9) and relative high a* (11.2) for yellow, blue and pink pigments that improves significantly in glazed samples to b*=54.2, b*=-23.8 and a*=25.3 respectively (data on Figure 6.1). The three pigments show zircon as the main crystalline phase detected by XRD but in these optimized pigments very weak peaks associated to baddeleyite (monoclinic zirconia ZrO_2, B in Figure 6.1) are detected in yellow and blue samples; instead in iron zircon sample weak peaks pf quartz (Q) and hematite (H) are detected.

The micronized and washed powders (with HNO_3 0.3M) of the yellow pigment show regular orthohedral particles 3-5 μm in length and 1-2 μm in width-height, the orthoedra are more irregular in the blue sample showing a particle size of 2-5 μm, finally iron zircon pink pigment show the more irregular particles with a particle size by 5 μm.

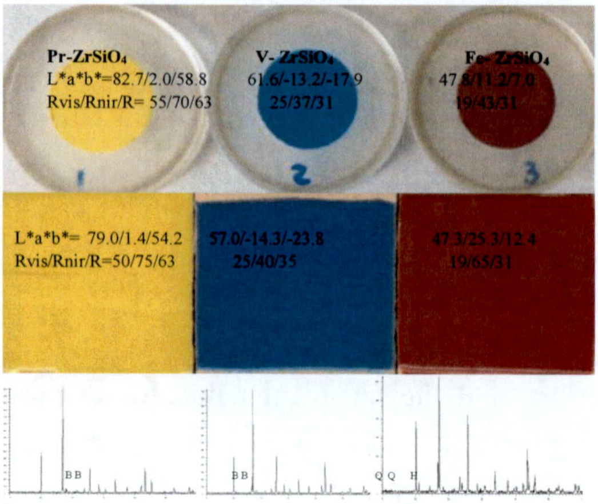

Figure 6.1. Powders and 5wt.% glazed in a double firing frit (1050°C) Yellow, cyan and magenta zircon. CRYSTALLYNE PHASES: B (Baddeleyite, monoclinic ZrO_2), Q (Quartz SiO_2), H (Hematite Fe_2O_3).

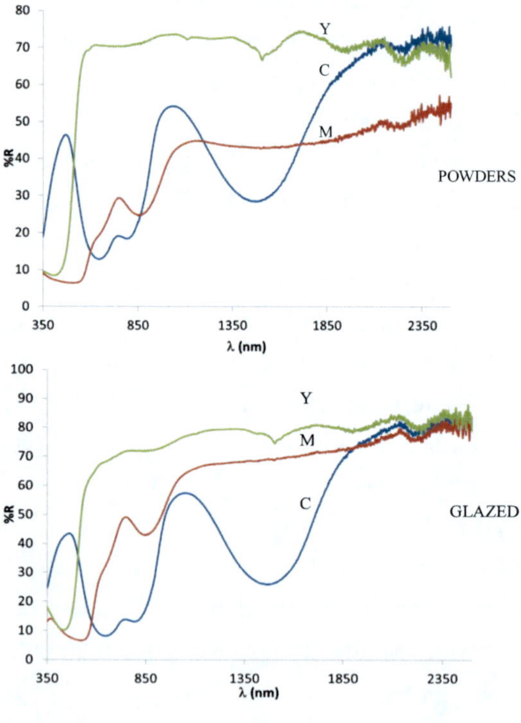

Figure 6.2. UV-Vis-NIR reflectance spectra of yellow (Y), cyan (C) and pink (M) zircon pigments.

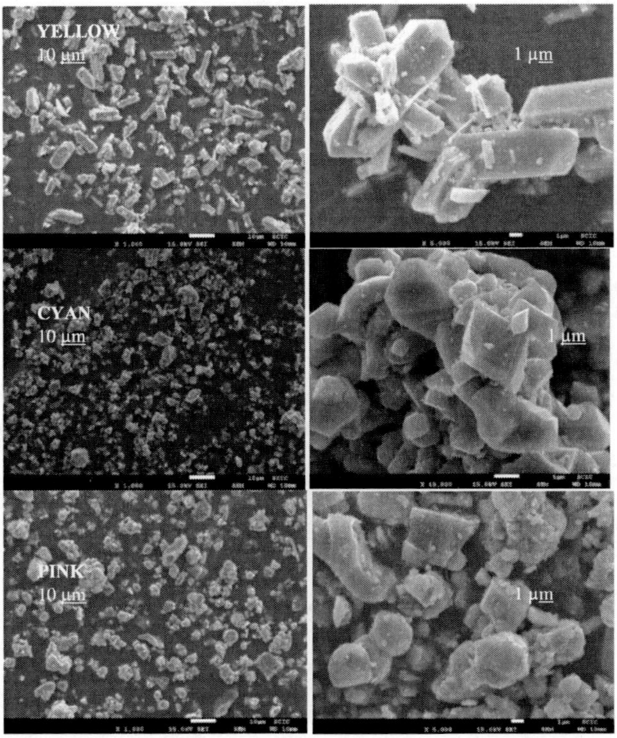

Figure 6.3. SEM micrographs of yellow, cyan and pink zircon pigments.

6.1. The Vanadium Turquoise Blue

This pigment is a solid solution of both V^{4+} and F^- in zircon $ZrSiO_4$ discovered in 1948 [1]. The sample shown on Figure 6.1 has been prepared from quartz (SiO_2 content=95.5%, D_{50}=3.9 μm particle size), zirconia (ZrO_2 content=95.0%, D_{50}=5.6 μm particle size), and NH_4VO_3, with the composition $(Zr_{0,81}V_{0,19})SiO_4$ and addition of a 5wt.%NaF+3wt.%NaCl as flux, the precursors were homogeneously mixed in a ball milling in acetone medium an fired at 800ºC/3h, the fired sample was washed and milled to optimized particle size around 5 μm (Figure 6.4).

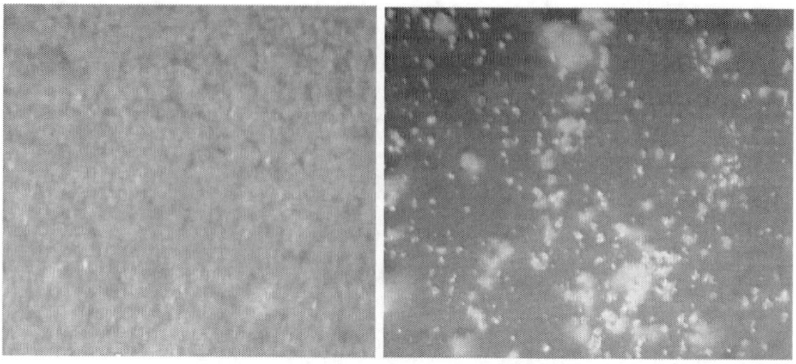

a b

Figure 6.4. Optical images of V-Zircon powders of Figure 6.1: a) x200, b) x500.

The blue vanadium zircon ceramic pigment is accepted to be a solid solution of tetravalent vanadium in the zircon lattice, V^{4+}-$ZrSiO_4$. In this system, vanadium acts not only as chromophore, but also as mineralizer for the synthesis of zircon allowing this phase to be formed at lower temperatures and within a narrower temperature range. Similarly, alkaline fluorides are widely used as mineralizers in the synthesis of the pigment, basically as flux agent (see Ebelmen method at Sèvres factory) and also because F^- ions can substitute for O^{2-} ions in the zircon lattice, easing the entrance of vanadium in the solid solution.

Structural studies indicate that V^{4+} enters at low temperatures in the tetrahedral sites substituting for Si^{4+}, but thermal activation promotes V^{4+} migration to dodecadeltahedral sites substituting for Zr^{4+} at higher temperatures [10]. Considering the studies reported by Matkovitch and Corbett [11,12] which were conducted in the absence of mineralizers; zirconium pyrovanadate (ZrV_2O_7) forms as reaction intermediate around 690ºC, decomposing around 760ºC prior to zircon formation.

The pigments obtained without mineralizers are usually green, but a blue pigmentation can also be obtained without fluorides depending on the vanadium content and on the firing temperature (the color is blue with vanadium amounts lower than 0.03 moles per formula weight, then turns greenish but give blue shade on glazes up to 0.3). Thereby, it is assumed that the relative amount of V^{4+}:V^{5+} incorporated in the zircon lattice determines the pigment color: it becomes greener as the V^{5+} content increases and a solid solution V^{4+},V^{5+}-$ZrSiO_4$ is allowed. Fluoride addition is believed to stabilize V^{4+} and to substitute for O^{2-} in the zircon lattice forming the turquoise blue F^-,V^{4+}-$ZrSiO_4$ solid solution. However, fluoride addition does not guarantee the blue color, since the firing temperature and the preparation method are also decisive [10].

Figure 6.5 shows the evolution of 5wt.% glazed samples (in both transparent and matte glazes with its CIEL*a*b* parameters), with firing temperature of unmineralized sample and optimized samples with NaF and NaCl mineralizers addition. It can be observed that unmineralized samples darken showing higher L* (around 70 in matte glaze) than the mineralized samples, but the best blue shades are obtained with NaF or NaCl additions (b*=-18.4 in matte glaze at 800ºC) [12, 13].

In the case of the non-mineralized composition when firing at temperatures between 650 and 750ºC, powders show brown tones, increasingly yellowish as the temperature increases. At 650ºC the major crystalline phases detected by XRD (not shown) are monoclinic zirconia (baddeleyite) and quartz, also observing peaks of very weak intensity corresponding to vanadium pentoxide. By increasing the calcination temperature to 700 and 750ºC, zirconia and quartz remain the majority phases and, as a noteworthy aspect at these temperatures, is the formation of zirconium pyrovanadate (ZrV_2O_7). The appearance of this phase of pyrovanadate would agree with the observations made by Matkovich [11]. From 800ºC the formation of the zircon phase is already appreciated, although the pigment obtained has a green color (see Figure 6.6).

In the mineralized composition with NaF, the evolution of the color of the powder with temperature is different: at 650ºC the sample with fluorides is white, at 700ºC shows slightly bluish-green hue, while from 750ºC the pigment shows a turquoise blue color, which is more intense at 800 and 850ºC. Whether analyzes the evolution of the crystalline phases by XRD (not shown) with the temperature in the sample with fluorides the following aspects can be highlighted:

a) Unlike the non-mineralized composition, the peaks corresponding to unreacted vanadium pentoxide are not detected, nor to zirconium pyrovanadate (ZrV_2O_7), which confirms the different mechanism of reaction in the presence of fluorides. On the other hand, peaks of very weak intensity are appreciated around 27.5°2θ over the entire temperature range, which may be due to the presence of mixed oxides of sodium and vanadium, probably $NaVO_3$. The samples with fluorides below 700°C are white as already been described in the bibliography, being attributed to the formation of a compound (not identified) resulting from the interaction between NaF and vanadium, silicon and zirconium oxides. This compound subsequently reacts with silica and zirconia to give the solid solution of V(IV) and F^- in the $ZrSiO_4$ network.

b) The mineralizing action of fluorides is demonstrated by achieving a good development of the zircon phase even at 750°C, reaching the highest development at 800°C. From 750°C the pigment shows turquoise blue color (see Figure 6.5).

Regarding the mineralized composition with NaCl, at low firing temperatures (650-700°C) the powders show a light brown color, unlike the mineralized composition with fluorides. Likewise, in this composition is also detected by XRD throughout the temperature range a mixed oxide of sodium and vanadium (probably $NaVO_3$). It should also be noted that the use of NaCl does not develop the zircon phase up to a temperature of 800°C, observing the maximum development of this phase at 850°C. The bluish coloration obtained from 800°C is indicates that the use of chlorides also stabilizes the solid solution of vanadium tetravalent in the zircon lattice. As in the mineralized composition with NaF, the presence of residual V_2O_5 is much less evident than in the composition without mineralizers, only very weak peaks are seen at 850 and 900°C. On the other hand, the use of vanadium and mineralizers by the industrial ceramic method brings environmental problems due to the emission of V_2O_5 vapor and HF, as well as the leaching of vanadium and fluorides when washing the pigments. These environmental problems must be treated by appropriate optimization of vanadium and mineralizer additions, as well as by appropriate thermal treatment analyzed by Llusar et al. [13]:

1) The formation of different reaction intermediates for different mineralizing agents (ZrV_2O_7 without mineralizers, $NaVO_3$ and/or sodium-vanadium bronzes ($5Na_2O.xV_2O_4.(12\text{-}x)V_2O_5$) with sodium halides, and a borosilicate vitreous phase with borax) proves to have an important effect on pigment synthesis, leading to different zircon yields and to a different color performance in the pigments.
2) The strong blue powder (unwashed powder L*a*b*=71.3/-13.2/-16.4) with composition $(ZrSiO_4)(V_2O_5)_{0.19}(NaF)_{0.05}(NaCl)_{0.10}$ has been found to be optimal environmentally (with loss in the firing 5.8%, vanadium leached=284 ppm and L*a*b*=71.3/-13.2/-16.4 glazed in double firing matte glaze (1000ºC)), though the green powder non-mineralized composition (with 0.19 mol of V_2O_5 per zircon formula weight) also presents good behavior in double matte firing glaze (1000ºC) giving adequately blue shades (L*a*b*=70.1/-13.1/-14.5) [13].

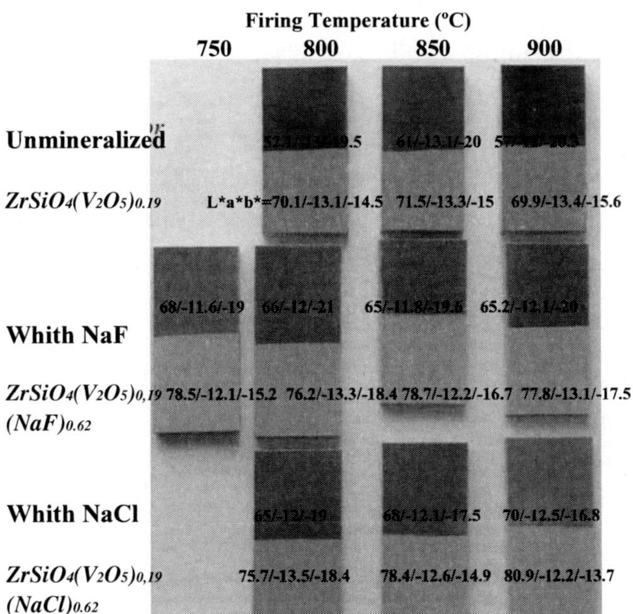

Figure 6.5. Compared evolution of 5wt.% glazed samples with firing temperature of unmineralized sample and optimized samples with NaF and NaCl mineralizers (up transparent glaze, down matte glaze (with CIEL*a*b* parameters)).

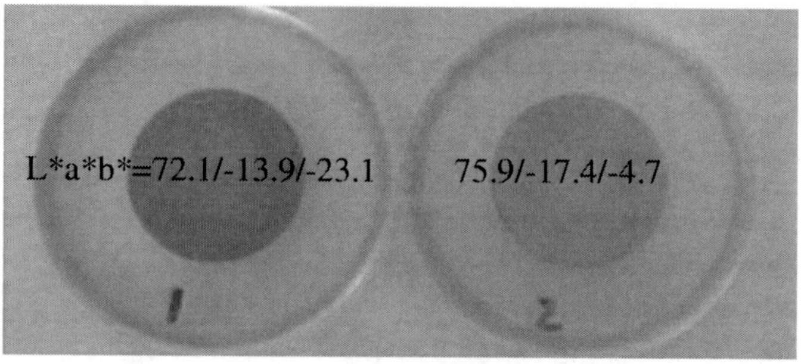

Figure 6.6. Blue NaF mineralized zircon powder $(ZrSiO_4)(V_2O_5)_{0,19}(NaF)_{0.62}$ and green unmineralized sample (both fired at 800ºC).

To explain the appearance and disappearance of the green color in the ceramic sample, in agreement with Monrós et al. [12] at low temperatures (around 700ºC), the samples show yellowish tones due to the existence of vanadium encapsulated in the zirconia lattice (baddeleyite); with increasing temperature (700-800ºC) this composite reacts with silica to form the zircon phase, in which process V^{+5} reduces to V^{+4} and joins both the dodecahedral and tetrahedral positions of the zircon lattice, obtaining the solid solution V^{+4}-$ZrSiO_4$ of blue color. Before stabilizing the blue color may have green colorations as a result of the mixture of both ions. Also if these samples are not stabilized by the addition of mineralizers (fluorides or chlorides) or in the case that the amount of V_2O_5 is greater than 0.03 moles per mole of zircon, increasing the temperature results in the destabilization of the blue color (from 900ºC) and samples show green color.

Figure 6.7 shows the UV-Vis-NIR diffuse reflectance spectra of the NaF mineralized zircon powder $(ZrSiO_4)(V_2O_5)_{0,19}(NaF)_{0.62}$ and green unmineralized sample fired at 800ºC. According to the model proposed by Hong Xiayou et al. [14] V^{+4} ($3d^1$ configuration) replaces both Zr^{+4} in the vanadium-zircon blue pigment in the dodecadeltahedral positions such as Si^{+4} in the tetrahedral positions. Analyzing optical spectroscopy data previously obtained by Demiray [15] and Di Gregorio [16] as well as using different models, Hong Xiayou et al. [14] proposed a distribution of energy levels for the tetrahedral and dodecahedral coordination environments of the V^{+4} such as the one presented in Figure 6.8. The assignments of the optical spectrum bands made by these authors are the following:

1) 190-212 nm, forbidden transition.
2) 290-297 nm, $^2B_1 \rightarrow {}^2A^1$ dodecadeltahedral coordination)
3) 640-644 nm, $^2B_1 \rightarrow {}^2E$ (both dodecadeltahedral and tetrahedral coordination)
4) 922-927 nm, $^2B_1 \rightarrow {}^2B_2$ (both dodecadeltahedral and tetrahedral coordination)
5) 1381-1405 nm, $^2B_1 \rightarrow {}^2A^1$ (tetrahedral coordination)
6) 750-770 nm, shoulder (forbidden transition)

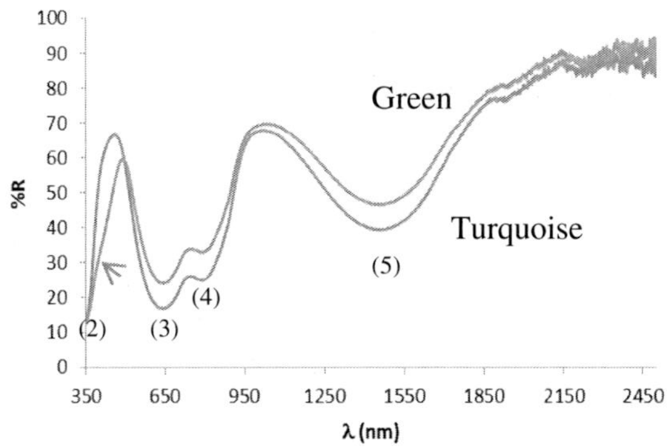

Figure 6.7. UV-Vis-NIR diffuse reflectance spectra of the NaF mineralized zircon powder $(ZrSiO_4)(V_2O_5)_{0,19}(NaF)_{0.62}$ and green unmineralized sample fired at 800ºC.

The above bands can be observed on the UV-Vis-NIR diffuse reflectance spectra of Figure 6.7. Based on the evolution of the optical spectra of the samples in Figure 6.7, it is concluded that the greening phenomenon is due to the entry of V^{+5} into the zircon network, associated with an increase in the charge transfer band absorption (reflectance decreases) associated with this ion (around 400 nm), indicated with an arrow in Figure 6.7. Similarly, the spectra when firing temperature increases (not shown) the disappearance of the absorption band around 1450 nm (5 in Figure 6.7) corresponding to V^{+4} in tetrahedral coordination, shows that with the increase in temperature, V^{+4} leaves the tetrahedral positions SiO_4 to occupy preferably the ZrO_8 dodecahedral positions. The decrease seen in the size of the unit cell of zircon as temperature increases seems to confirm these hypotheses, since that the change in the coordination environment of the V^{+4}

ions and the entry of the V^{+5} ions in the zircon lattice should cause a decrease in the cell size.

Vicent et al. [17] study the green V-zircon, synthesizing containing vanadium samples which gradually turn greenish by controlling the composition (fluoride addition) and the method of preparation (Sol-Gel, ceramic and solid impregnation) in order to investigate how to affect the shape and size of the vanadium particles, and their location, clustering and distribution. The authors prepared F^{-},V^{4+}-$ZrSiO_4$ solid solution samples in the commercial molar composition of $0.84ZrO_2.SiO_2.0.16NH_4VO_3.0.7NaF$ by three different methods: the ceramic method , the impregnation method, and a polymeric gel route (without NaF addition and with NaF addition). Samples were characterized by DTA-TG analysis, and measure of the amount of vanadium in the washing water (by H_2O_2 colorimetry in sulfuric acid medium and $FeSO_4.7H_2O$ addition for $V^{5+} \rightarrow V^{4+}$ reduction). The washed samples were enameled in a commercial borosilicate glaze at 5 wt.% in order to check its coloring properties. Samples have been studied by XRD, IR and optical UV-Vis-IR spectra, CIE-L*a*b* measurements of enameled samples, magnetic susceptibility measurements and SEM-EDX microstructural characterization.

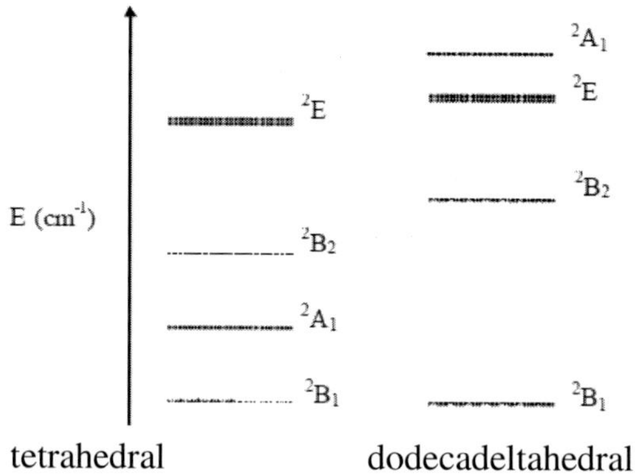

Figure 6.8. Distribution of energy levels for the tetrahedral and dodecadeltahedral coordination environments of the V^{+4} [14].

In agreement with Vicent et al. [17] the level of green color, and consequently the amount of V^{5+} in the sample, cannot be associated with microstructural parameters such as the shape or particle size of the pigment powder, the vanadium amount or its distribution in the particles which are similar in both the turquoise and green samples. However, by IR analysis the authors observe that IR spectra were similar for all fired samples: five bands at 900, 620, 420, 380 and 310 cm^{-1}, associated with $ZrSiO_4$ were detected. In addition, bands at 1400 associated with ZrO^- (O^- signifying non bridging oxygen ions), 1100 (Si-O), 980 and 800 cm^{-1} were observed. The bands at 1100 and 800 cm^{-1} appeared more intense and sharper for green samples (polymeric gel and impregnation method respectively), which could indicate the development of an Si-O-Si network in the green material associated with highly reactive zirconium precursors.

6.2. The Yellow of Praseodymium

Pr-zircon is a pigment associated with praseodymium oxide entrance in zircon discovered in 1961 [2]. The sample shown on Figure 6.1 has been prepared from quartz (SiO_2 content=95.5%, D_{50}=3.9 μm particle size), zirconia (ZrO_2 content=95.0%, D_{50}=5.6 μm particle size), and Pr_6O_{11}, with the composition $Pr_{0,05}Zr_{0,95}SiO_4$ and addition of a 7.5wt.%NaF+6w.%NaCl as flux, the precursors were homogeneously mixed in a ball milling in acetone medium and fired at 1000ºC/3h, the fired sample was washed and milled to optimized particle size by 5 μm.

As in the case of vanadium zircon, the presence of NaF is necessary for the formation of the pigment with sufficient saturation in the ceramic route. The reason is the inhibition of the volatilization of praseodymium at synthesis temperatures by formation of the intermediate compound $Na_2Pr_3(Si0_4)_5F_2$. This compound remains stable until the intense yellow coloration of the pigment associated with the crystallization of zircon (Pr-$ZrSiO_4$ pigment) [18].

Praseodymium yellow in zircon can be obtained at slightly lower calcination temperatures using the colloidal gel method (1100°C versus 1150°C for industrial ceramic pathways). In the colloidal method, 5 g. of final product were prepared starting with colloidal silica (Quimidroga) dispersed in 100 ml of water, then was added the precursor of zirconium ($ZrOCl_2.8H_2O$) maintaining temperature at 70ºC and with continuous stirring, and then the precursor of praseodymium (Pr_6O_{11} dissolved in the minimal quantity of concentrated HNO_3 (37wt.%), finally the correspondent mineralizer was added and maintained in stirring at 70ºC one hour, then the homogenous dispersion was gelled by slowly dropping ammonia (the gelling appears at pH around 6). The presence of praseodymium favors the stabilization of tetragonal zirconia detected in the gel pathways, which occurs up to 1200°C in the polymer gel sample without fluorides. The presence of fluorides destabilizes the tetragonal zirconia, allowing the reaction between monoclinic zirconia and silica to give zircon: in this sense, the presence of NaF makes it possible to obtain zircon at a lower temperature. In the samples obtained from gels, the stabilized tetragonal zirconia develops a yellow coloration that resists enameling in ceramic coating flux matrices (pale yellow Pr-$Zr0_2$ (t) pigment at 1000ºC in Figure 6.9).

Badenes et al. [18] study the pigment comparing the ceramic route with unconventional methods by XRD, UV-Vis-NIR diffuse reflectance and SEM-EDX in order to analyze the nature of the pigment. The main results are in accordance with the possibility of an inclusion mechanism in the pigment, based on the following arguments:

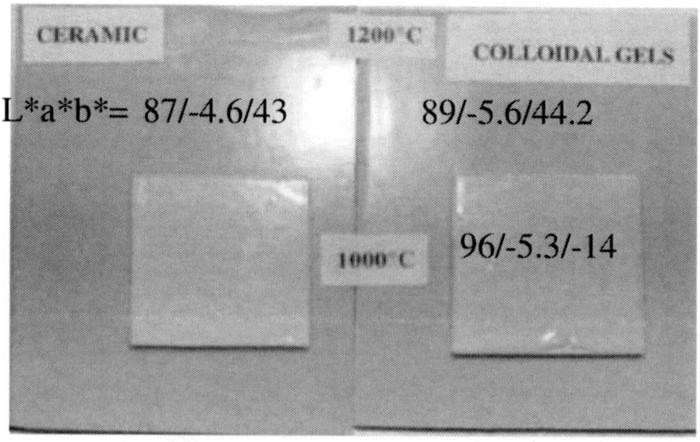

Figure 6.9. Ceramic and colloidal samples (($ZrSiO_4$)(Pr_2O_3)$_{0,05}$(NaF)$_{0.2}$) 5wt.% glazed in double firing frit (1000ºC) fired at 1000ºC (enclosed samples) and 1200ºC.

a) A visual loss of color is observed when firing the intense yellow powders at 1300°C (see Figure 6.10: visual view in optical microscopy x500, showing the discoloration of the microcrystals) without changing the cell parameters of zircon of ceramic pigment powder $((ZrSiO_4)(Pr_2O_3)_{0.05}(NaF)_{0.2})$. Figure 6.10 shows that the crystals do not have a uniform color; there are crystals with vivid yellow colored (which trap chromophore oxides) and other with light yellow color (those that do not have encapsulated coloring particles in agreement with the inclusion model).

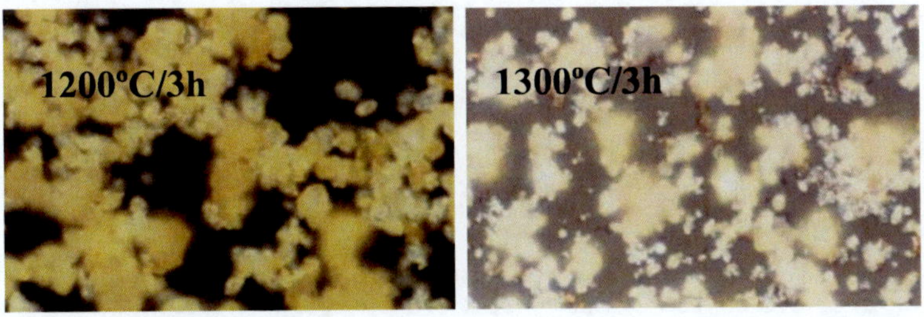

Figure 6.10. Ceramic pigment powder (visual view in optical microscopy x500) $((ZrSiO_4)(Pr_2O_3)_{0,05}(NaF)_{0.2})$ fired at 1200°C/3h and at 1300°C/3h showing the discoloration of the microcrystals.

b) The UV-Vis-NIR spectrum of $Pr\text{-}ZrSiO_4$ with NaF addition which is shown in Figure 6.11.a shows praseodymium ions in oxygenated environments similar to those of the unmineralized sample, as well as Pr^{3+} in some salts [18]. They show the bands associated with Pr^{3+} at the wavelengths indicated on Figure 6.11 described by Seeber et al. [19,20] from Judd-Ofelt model calculations [21] as it is shown in Figure 6.11.b. The appearance of the same bands in the UV-V spectrum is consistent with the presence of Pr^{3+} in this pigment.
It should be pointed out that the presence of Pr(IV) cannot be detected by UV-Vis-NIR optical spectroscopy as it does not present absorption in the visible and near infrared region of the spectrum [22,23]. Furthermore since 4f orbitals of praseodymium are shielded from the surroundings it is not possible to obtain information related to its geometry co-ordination sphere [20]. Obviously, an ion that

does not absorb in the visible range cannot act as a yellow chromophore absorber.

c) The EDX-mapping analysis (Figure 6.12) indicates the accumulation of the EDX-Pr signal ($L_{1\alpha}$5.034 keV) in some particles: this praseodymium heterogeneity distribution is in agreement with the inclusion mechanism and it cannot be explained by a solid solution mechanism.

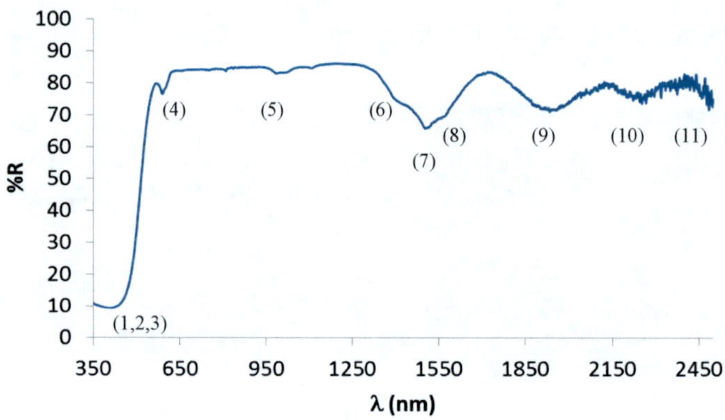

a

(1) 495 nm, $^3H_4 \rightarrow 3P_o$
(2) 481 nm, $3H4 \rightarrow {}^3P^1$
(3) 443 nm, $3H4 \rightarrow {}^3P_2$
(4) 590–600 nm, $^3H_4 \rightarrow {}^1D_2$
(5) 1020 nm, $^3H_4 \rightarrow {}^1G_4$
(6) 1423 nm, $^3H_4 \rightarrow {}^3F_4$
(7) 1525 nm, $^3H_4 \rightarrow {}^3F_3$
(8) 1600 nm, $^3H_4 \rightarrow {}^3F_2$
(9) 1900 nm, $^3H_4 \rightarrow {}^3H_6$
(10) 2200 nm, $^3H_4 \rightarrow {}^3H_5$
(11) 2400 nm, $^3H_4 \rightarrow {}^3H_4$

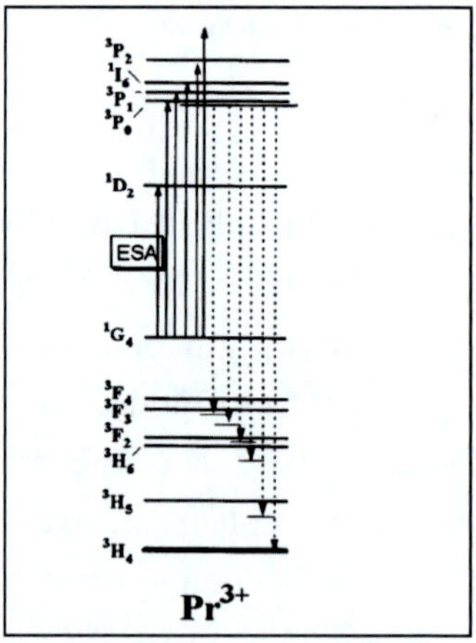

b

Figure 6.11. (a) UV-Vis-NIR diffuse reflectance spectra of powders $((ZrSiO_4)(Pr_2O_3)_{0,05}(NaF)_{0.2})$ fired at 1200°C/3h, (b)Absorbance bands of Pr^{3+} optical spectrum: ESA= excited state absorption [19].

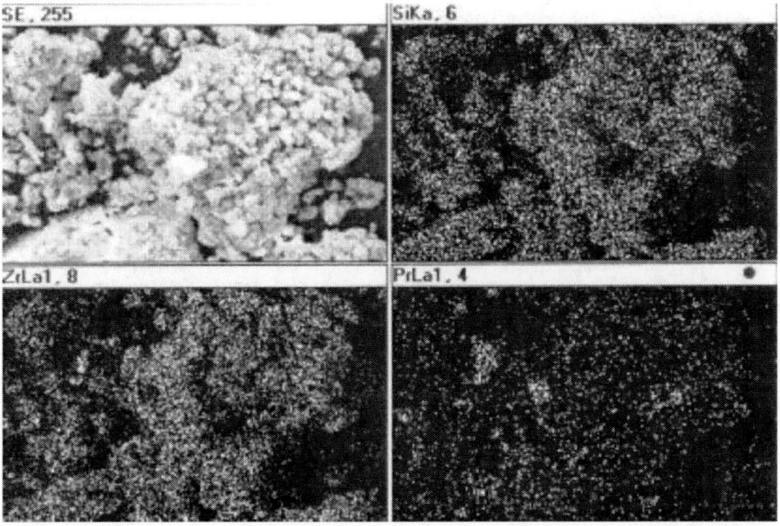

Figure 6.12. EDX composition mapping of of powders $((ZrSiO_4)(Pr_2O_3)_{0,05}(NaF)_{0.2})$ fired at 1200°C/3h.

d) The role of NaF is to stabilize the $Na_2Pr_8(SiO_4)_6F_2$ phase which retains praseodymium in the reaction media, hence avoiding its volatilization and generating the adequate praseodymium concentration necessary to be able to produce an intense yellow color.

e) In order to explain the above results, it is necessary to differentiate the zircon crystallization, the presence of praseodymium in solid solution and the color production: samples prepared following any of the methods ceramic, colloidal or polymeric sol-gel method but without mineralizers addition are not capable of producing intense yellow color although a small increase of cell parameters is observed (indicating that the solid solution of Pr-zircon has been reached). It is well known that using Sol-Gel techniques a uniform distribution of Pr^{3+} is achieved in the ceramic matrix, so that the colloidal or polymeric samples show a soft yellow. The intensity of color increases in samples with NaF when, considering the inclusion model, an adequate synchrony between the crystallization of zircon and the inclusion of praseodymium oxide microcrystals.

f) As above discussed, the yellow color cannot be associated to NaF addition exclusively. UV-Vis spectra of yellow samples, both with and without NaF addition (CG and PG), are similar but the former has higher absorbance.

Therefore, Pr-$ZrSiO_4$ yellow pigment shows a mixed nature: solid solution and encapsulated pigment. This pigment follows an inclusion mechanism when the trivalent praseodymium oxide is present but the cell parameters expansion do not exclude other secondary mechanism of coloring related to the praseodymium solid solution [18].

Ocaña et al. [24] prepared Pr-doped zircon by heating at 1020°C for 3h an equimolecular mixture of ZrO_2 and SiO_2 in the presence of Pr_6O_{11} (Pr/Zr atomic ratio=0.04) and alkaline halides (NaF and NaCl) used as fluxing agents. Chemical analysis of this powder carried out by ICP gave a Si/Zr atomic ratio of 0.96 and a Pr/$ZrSiO_4$ mole ratio of 0.04. According to X-ray diffraction data (not shown), this sample mainly consisted of zircon. The only additional phase detected was a small amount of monoclinic ZrO_2 due, at least in part, to the little excess of Zr revealed by the measured Si/Zr ratio, which was slightly lower than that required for stoichiometric zircon. The sample was analyzed by X-ray diffraction, XAS (XANES and EXAFS) and XPS spectroscopies. Based on this analysis, it is concluded that the Pr cations present a tetravalent state, but the color of the sample is not described or evaluated; the relatively low temperature (1020ºC) employed in the synthesis probably give to a pale yellow color and the sample does not have possibility of encapsulation. According to these authors, they have unambiguously demonstrated that the Pr cations in Pr-doped zircon pigments present a tetravalent state and form a solid solution in the zircon lattice, replacing Zr in its triangular dodecahedral positions. The XPS data show that cations are not homogeneously distributed in the pigment particles and are preferentially concentrated at their outer layers. But as above discussed, Pr (IV) does not absorbs in visible range and obviously for to give a yellow color is necessary to absorb the violet-blue range (see Figure 6.11).

Del Nero et al. [25] synthesized Pr doped zircon pigments by a Sol–Gel procedure for variable Pr concentrations and with the addition of several mineralizers. The presence of Pr was observed to promote the crystallization of the zircon phase, in agreement with the literature results and with the effects caused by other doping metals (V, Fe) [18]. However, the promoting effect did not increase regularly with the Pr amount but showed a maximum for intermediate concentrations (3%). The Pr amount relative to the best zircon promotion corresponded also to the largest sizes of zircon and tetragonal zirconia crystallites. At the same time, Vis–NIR reflectance spectra showed that in the region 400–500 nm the absorption intensity regularly increased to 3% Pr content and then decreased for the samples containing larger Pr amounts. Further, in the case of these latter samples, the presence of shoulders in the spectra at 495, 481 and 483 nm showed that Pr ions, besides being distributed into the host matrix, were also present in separate crystalline phases. The presence of separate Pr containing phases (encapsulated or not in the host lattice) was also noticeable in X-ray spectra in the case of the highest Pr amounts, at the same time, EDX analyses showed a non-homogeneous distribution of the metal.

The authors [25] report that the addition of mineralizers to the reacting Sol-Gel mixture provokes in any case a definite increase in the amount of zircon formed. The addition of NaCl and NaF, besides promoting the growth of $ZrSiO_4$, produces an intense ochre-yellow color. At 5% Pr addition the analysis by Vis–NIR shows that Pr is entirely distributed in the zircon lattice. The role of these two mineralizers can be that of extending the solubility range of Pr species in the zircon lattice. The samples obtained in the presence of LiCl contain the largest amount of zircon phase with respect to the other samples and present a brilliant canary-yellow color. These samples show the largest Pr(III) content and the total absence of segregated phases. The addition of Mg salts supports the formation of the zircon phase but gives rise to a whitish color. If as suggested by Ocaña et al. [20, 26] the intense yellow color of the NaF samples is the result of the stable formation of a Pr(IV)-zircon solid solution, then in the case of Mg salts a possible competition between Mg and Pr might cause the segregation of Pr(III) containing phases decreasing the color intensity.

In these works [20, 25, 26] all UV-Vis-NIR diffuse reflectance of powders show the presence of Pr (III) with a similar feature of the presented in Figure 6.10, the authors indicate that Pr(IV) is not evaluated because does not absorbs at Vis-NIR. CIEL*a*b* color evaluation of the powders, although not references to behavior of the pigments in glazes or ceramic matrices, are reported, and the developed yellows always show high lightness (L*>90) an the best yellow powder shade is the corresponding to the addition of 5% of LiCl as flux agent (L*a*b*=92.3/4.7/46.7) [25] or a pale yellow using alkoxides free of mineralizers at 1600ºC (L*a*b*=84.90/-7.03/25.03 for the composition $Pr_{0.06}$–$ZrSiO_4$) [26].

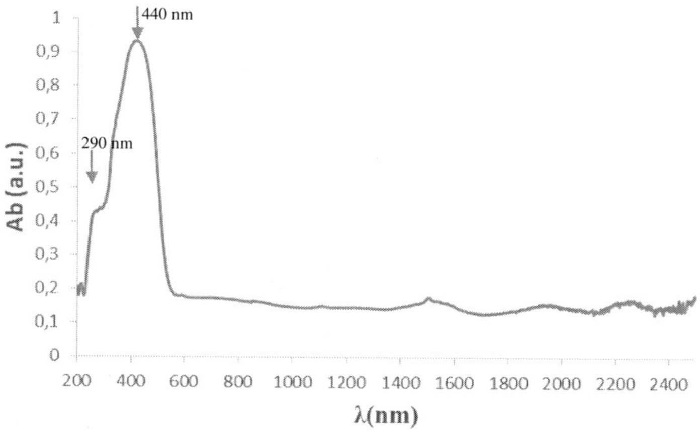

Figure 6.13. UV-Vis-NIR absorption spectrum of Pr-$ZrSiO_4$ yellow powders of Fig 6.1 showing the absorption in the violet-blue range.

The oxidation state IV for praseodymium is unknown in the solution chemistry of this element; it is not even possible to obtain this high valence by attacking the Pr (III) compounds (always yellow-green in color) with fluorine. However Pr_6O_{11} oxide (triclinic, *P-1*) is a black mixed valence oxide with a disordered structure and rich in lattice defects capable of absorbing the entire optical spectrum. However, the entrance of the Pr^{4+} ion in solid solution replacing Zr^{4+} into the dodecadeltahedral position of the zircon would allow stabilizing this valence.

In any case, either by the entrance of Pr^{4+} ion or by the entry of Pr^{3+} with the formation of cationic vacancies ($Zr^{4+}{}_{1-4x}Pr^{3+}{}_{4x}V^{+4}{}_{x}SiO_4$: V=cationic vacant), an additional charge transfer band is induced additionally centered at 440 nm (Figure 6.13, associated 2.6 eV band gap) added to that of zircon (which appears centered at 290 nm with 3.5 eV band gap), resulting in strong absorption of the violet-blue range of spectrum that reflects the yellow color (Figure 6.13).

6.3. The Pink Coral of Iron Zircon

The pink coral of iron in zircon, is considered an inclusion pigment of hematite (Fe_2O_3) in zircon crystals, it was discovered in 1965 [11, 9, 27]. The sample showed on Figure 6.1 has been prepared from quartz (SiO_2 content=95.5%, D_{50}=3.9 μm particle size), zirconia (ZrO_2 content=95.0%, D_{50}=5.6 μm particle size), and $FeSO_4.4H_2O$, with the optimized composition $[(SiO_2)(ZrO_2)_{0,97}(Fe_2O_3)_{0,15}][NaF]_{0,7}[NaCl]_{0,3}$ with NaF and NaCl added as mineralizers, the precursors were homogeneously mixed in a ball milling in acetone medium, pelletized (100 kg/cm^2) and fired at 1100°C/3h; the fired sample was washed and milled to optimized particle size around 5 μm.

The pink coral of iron in zircon ceramic pigment is accepted to be an inclusion of hematite microcrystals in $ZrSiO_4$. In this system, hematite acts as chromophore when it is encapsulated within the zircon crystals; therefore is crucial to have a synchrony between the zircon growth-coarsening and the encapsulation of hematite that also grows and if its coarsening is larger, the encapsulation is ineffective.

Therefore, the selected mineralizers for the control of synthesis of zircon allowing this phase to be formed and grown up for an effectively entrapping of hematite particles is crucial, also the selected precursor of hematite is very important in order to show the effective size to be entrapped in the synchronized growth of the two crystals hematite (guest) and zircon (host), following the synthesis steps drawn in Figure 6.14 [28] namely:

a) Homogenization of the mixture of iron oxide (hematite) and silica and zirconia crystals: in the model, for an efficient encapsulation the hematite should be surrounded by silica and zirconia crystals.
b) Silica and zirconia react to produce zircon crystals which trap the microcrystal of hematite which is also coarsening.

c) Finally if the coarsening of hematite is synchronized with the zircon growth, the inclusion of hematite in the zircon crystal is efficiently obtained.

As in their homologous zircon pigments, alkaline halides are widely used as mineralizers in the synthesis of the pigment, basically as flux agents in order to facilitate the ionic diffusion process to the progress of the hematite-zircon assembly.

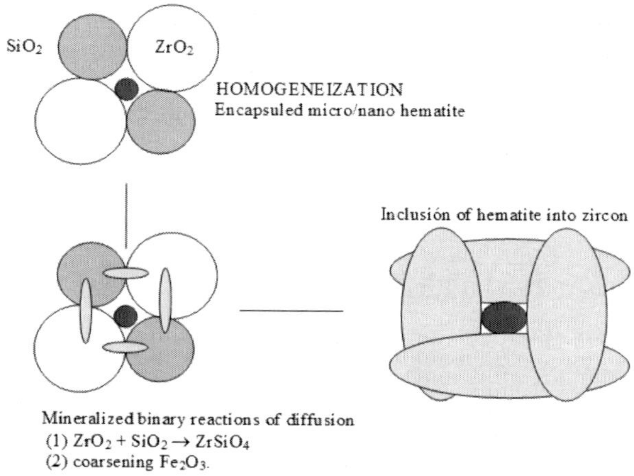

Figure 6.14. "Inclusion-growth" model for encapsulation of hematite in zircon crystals [28].

As above discussed, the optimization of the mineralizer system is very effective in the synthesis of ceramic pigments: the role of these mineralizers in this case is to maximize the formation of zircon and the effective encapsulation of hematite by properly crimping, both processes according to the encapsulation-coarsening model proposed for the pigment formation. In this sense, Llusar et al. [27, 28] study the effect of different mineralizers and related environmental aspects such as iron or sulfate leaching, they have found:

a) The species such as NaF, NaCl, BaF_2, favor the formation of the zircon phase, also allowing the obtaining of acceptable pink colorations, while other compounds, such as KNO_3, Na_3AlF_6,

Na_2SiF_6 and KI, can develop a collateral action promoting and regulating the states of growth and inclusion of hematite; therefore, on the effect of different mineralizer it is pointed out:

- Primary mineralizers: The addition of different mineralizing agents have a different behavior in the synthesis of the coral pink pigment of iron-zircon $(Fe,Zr)SiO_4$, depending on the iron precursor used (sulfate or goethite): with iron sulfate, compounds such as NaF, NaCl, and BaF_2 stand out as "zircon-forming mineralizers": with relatively low addition (3 wt.%), compositions with NaF and NaCl already develop zircon at 850°C (although always with residual monoclinic zirconia in the case of NaCl); for its part, with BaF_2 a yield of zircon of 25% is obtained at 1100°C. With goethite (FeOOH) the colorimetric yield obtained is poorer than with iron sulfate, highlighting in this case the mineralization with Na_2SiF_6 and Na_3AlF_6, which produce a good development of zircon at 950°C with additions of 5wt.%, while the BaF_2 and MgF_2 develop zircon with more difficulty (the first at high concentrations and the second at high temperature, 1050°C).
- Secondary mineralizers: another series of compounds, which have been called "secondary mineralizing or adjuvant agents" (KNO_3, Na_2SiF_6 and Na_3AlF_6 with sulfate, and KI and KNO_3 with goethite). They are not capable of developing zircon in their simple addition, but could have a collateral effect in the synthesis of the pigment, regulating the processes of growth and inclusion of hematite.
- The mineralized compositions on the binary systems (NaF-Na_2SiF_6), (NaF-Na_3AlF_6) or (NaF-KNO_3) were selected as optimal in pigment synthesis, allowing the obtaining of more intense reddish colorations.

b) From an environmental point of view, binary mineralization with NaF- Na_3AlF_6 allows to obtain a low level of emission and iron leaching, while the NaF-BaF_2 system minimizes the removal of sulfates by stabilization of the barite during calcination.

c) On the other hand due the crucial importance in the synthesis of this pigment to achieve an optimal synchrony in the processes of zircon growth and the encapsulation of the hematite, the use of Sol-Gel synthesis methods does not allow to obtain good results, with these unconventional methods it is possible to increase considerably the reactivity of the system in terms of the formation of the zircon phase, but this greater reactivity prevents an effective encapsulation.
d) The observations made by optical microscopy (Figure 6.15) confirm the hypothesis of encapsulation of hematite in zircon crystals as the origin of color, showing the existence of a correlation between the percentage of colored granules (with occluded hematite) and the intensity of the enameled colored samples.

However, García et al. [29] report that in order to achieve an effective encapsulation of the chromophore agent into the crystal host, the sintering and/or crystallization of the matrix must be synchronized with nucleation-growth of the included chromophore phase. The crystallization, sintering and inclusion processes are thus required to take place simultaneously. Consequently, control of particle size of the raw powder will go towards high levels in encapsulation. But traditional ceramic processing is limited by the diffusional barrier that prevents the control of the microstructural heterogeneity, and the nucleation and growth processes, and thereby the reactivity of the system. In order to avoid this problem, a microemulsion route was applied so as to improve the microstructural characteristics.

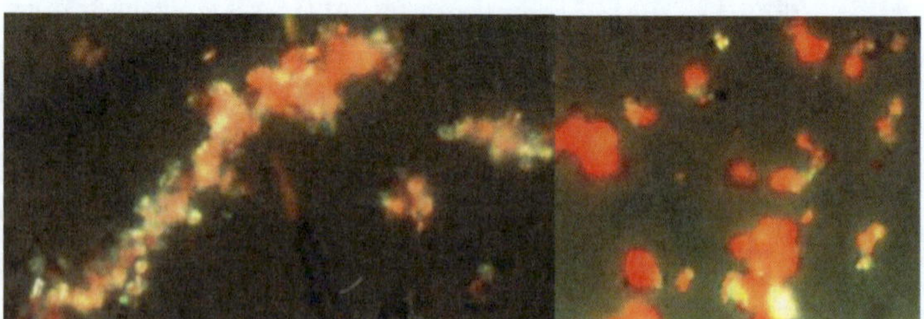

a b

Figure 6.15. Optical micrographs of iron-zircon powders of Figure 6.1: a) x500, b) x1000.

A microemulsion is a microheterogeneous system which provides a large interfacial area and low viscosity. Usually, microemulsion systems contain water, oil, a surfactant, and often a cosurfactant in order to stabilize the system. The microemulsion of the precursors is formed by forming droplets of water containing the soluble precursors of the product, suspended and stabilized in an organic non-polar solvent (usually iso-octane). Iron zircon samples of molar composition $0.4Fe_2O_3$:$ZrSiO_4$:0.2M (where M is the flux agent: NaF or the eutectic NaF.2NaCl), were prepared by microemulsion (ME) synthesis route and by the reference ceramic method (CE).

The microemulsion of the precursors was obtained by mixing 0.5 M aqueous solution with iso-octane (oil phase), the surfactant agent (3 wt.% of N-cetyl-N,N,N-trimethylammonium bromide, alkylbenzildimethyl-ammonium chloride or sodium lauryl sulfate have been used), and a co-surfactant (1 wt.% of 1-propanol). The mixture was stirred for 30 min and then was kept in ultrasonic agitation for 30 min. A similar procedure was followed to prepare the ammonia microemulsion. The microemulsion of precursors and the precipitant microemulsion were mixed by ultrasonic agitation for 1 hour. In the resulting microemulsion, three different phases could be distinguished: an upper organic (iso-octane) phase, an aqueous phase at the bottom, and the intermediate product layer. The iso-octane phase was removed, and the microemulsioned raw powder was obtained by drying the remainder at 110ºC in an oven. Samples were fired at 1000ºC in an electrical furnace, with a soaking time of 2.5 h. The fired samples were micronized and washed in a ball-mill in water media and dried at 110 ºC. The micronized powders were tested as ceramic pigments for glazes using a conventional double firing glaze (1080ºC) [29].

The results obtained by microemulsion procedure are in agreement with the encapsulation model of hematite into zircon, usually accepted for the pink coral ceramic pigment. The zircon reactivity and color of samples depends strongly on the conditions of the microemulsion such as surfactant, precipitant or flux agent:

- Alkyltrimethylammonium chloride produces the best zircon reactivity and reddish color using flux agent addition, but does not react without a mineralizer.
- When ammonia is used as precipitating agent produces amorphous powders that show high zircon reactivity and high hematite inclusion effectiveness giving red colors when glazed. NaOH 6M used as precipitating agent shows a very different behavior: it reacts and stabilizes $Na_4ZrSi_3O_{11}$ and Na_2ZrSiO_5 silicozirconates avoiding the zircon crystallization using N-cetylN,N,N-trimethylammonium bromide, but show amorphous powders that react and produce red colors when using alkyltrimethylammonium chloride; it gives sintered material when using lauryl sulfate as surfactant agent.
- NaF works better than NaF.2NaCl in order to encapsulate hematite, giving reddish colors. This fact is associated to the greater growth of zircon crystals when using NaF.2NaCl flux agent.

However, the red pigments obtained by microemulsion method are lower or similar to the shade obtained by the reference ceramic powder: García et al. [29] operates at high iron content (0.4 moles of hematite per mole of zircon) and a relatively low temperature, the ceramic route gives L*a*b* values of 63/11/12, 65/15/7 and 71/11/8 in unfluxed, NaF and eutectic NaF.2NaCl flux agent respectively; the best reddish shades using microemulsion method were 66/15/18 (sodium lauryl sulfate and NH_3), 65/13/6 (alkylbencildimethylammonium and NaOH) and 75/15/21 (sodium lauryl sulfate and NH_3). The use of pressure (pellets) and an adequate halide mixture in the industrial synthesis of this pigment is more efficient than non conventional methods.

Ozel et al. [30] study the synthesis of colored zircon ceramic pigments from intermediate products obtained by decomposing zircon sand with NaOH, avoiding the above described route by calcining a mixture of SiO_2, ZrO_2 and a dopant ion which produces the color (NH_4VO_3, Pr_6O_{11} or $FeSO_4.7H_2O$), and the mineralizer (a mixture of NaF and NaCl). After decomposing zircon with NaOH at around 850ºC, decomposed products (Na_2ZrSiO_5 and Na_2ZrO_3) (also detected by García et al. in the microemulsion mixtures precipitated with NaOH [29]), were used as coloring agents. As a result, Ozel et al. [30] shown that it is possible to obtain zircon pigments from zircon sand instead of using pure ZrO_2 and SiO_2. In effect, a V-zircon pigment comparable with a conventional synthesis (L*a*b*=79.9/-13.1/-11.6, powder previously washed with sulfuric acid, versus 70.1/-16.7/-19.6 of the conventional powder fired at 1000°C when enameled in a transparent glaze, both are not stable in porcelain glaze), also an interesting reddish pink coral (L*a*b*=60.5/26.1/24.2 and 57.3/11.1/8.4 in transparent glaze and porcelain glaze respectively, versus 61/1/20.9/23.5 and 59.8/10.3/9.7 of the conventional powder) and a good Pr-zircon yellow pigment (L*a*b*=84.5/-10.2/41.9 and 62.1/-1.6/23.1 in transparent glaze and porcelain glaze, versus 88.1/1/-7.3/51.2 and 75.5/-2.2/29.8 of the conventional powder).

6.4. Cadmium Sulphoselenide-Zircon

Cadmium sulphoselenide $CdS_{1-x}Se_x$ red pigment was synthesized in 1919 [31]. Further improvements were made to obtain pure cadmium sulphoselenide red pigment. The research results were major issued in patent form [32-35]. Cadmium sulphoselenide pigments could yield pure bright color with abundant hues and strong tinting capacity, and were widely used in ceramics, enamels, plastics, glasses and inks [36]. The main synthesis techniques to prepare cadmium sulphoselenide pigments are solid phase sintering method, chemical co-precipitation method, and hydro-thermal method. Cadmium sulphoselenide pigments are unstable, sensitive to calcination temperature and atmosphere, have poor thermal stability and could easily react with glassy phases and thus have been confined to on-glaze decoration for a comparatively long time [37].

There is evidence that the well-known Fe-zircon pink or pink coral are actually formed by the entrapment of hematite within zircon crystals, lending credence to the idea of the protection of unstable pigments by wrapping them in a protective envelope. The so called inclusion pigments in which the chromophore crystal is included into the host crystal that is completely alien to that of chromophore. Degussa (https://www.ceramic-glazes.com/degussa-ceramics) showed that this encapsulation mechanism is possible for the cadmium sulphoselenide [38]. By solid state reaction a reddish color appears when zirconia, quartz and the cadmium pigment are interdifussed but only 1-2% of cadmium was trapped and is unable to provide the true yellow-orange-red shades of cadmium sulphoselenide. Figure 6.16 shows the optical image (x500) of powders and the shade obtained 5wt.% glazed in a double firing frit (1080°C) of $ZrSiO_4$-$Cd(S_xSe_{1-x})$ synthesized by solid state method using CdS, S, Se, quartz, zirconia and a mixture of NaF and NaCl as mineralizer in a tubular kiln in nitrogen atmosphere.

In the later 1970s C.A. Seabright and the Harshaw Chemical Co. patent the encapsulation method [39] which made cadmium sulphoselenide pigments available for high temperature application.

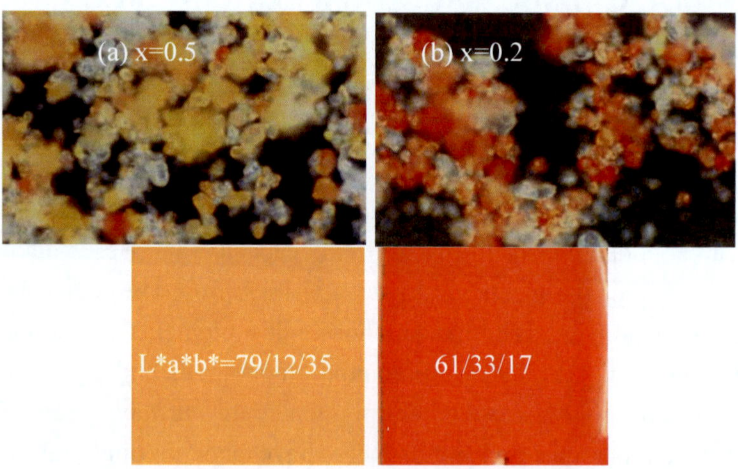

Figure 6.16. Optical image (x500) of $ZrSiO_4$-$Cd(S_xSe_{1-x})$ (from orange with x=0.5 to red with x=0.2) powders and the shade obtained 5wt.% glazed in a double firing frit (1080°).

However, the encapsulated cadmium sulphoselenide pigments produce a much lighter color than their original color. The principal factor which affects the coloring effect of encapsulated pigments is encapsulation efficiency and a minor factor is the opacifying effect of encapsulated layer that affects all zircon pigments. Following Castilone et al. [40], high levels of zircon are required to impart opacity in glazes, zircon additions of <3 wt.% dissolve into the glassy phase; additions of >12% cause all the zircon to crystallize from the glaze. Dynamic high-temperature X-ray diffraction has shown that, during heating, a portion of the zircon dissolves and subsequently recrystallizes, using undissolved zircon crystals as seeds. This recrystallization does not occur at lesser zircon levels, because of a lack of seed crystals. When zirconia is added to the glaze, zircon is formed as elongated crystals at the glaze surface and lesser zircon levels have been observed in the glazes that have zirconia additions.

Encapsulated pigments with bright color need to be implemented by improving encapsulation efficiency. The efficiency of cadmium inclusion improves up to 12% by a coprecipitation route from zirconium hydroxide and colloidal silica in presence of the cadmium pigments, then the obtained pigment has sufficient color saturation to match the bright colors of cadmium pigments and the inclusion remain stable to temperatures in excess of 1400ºC. The receipt for the red $ZrSiO_4$-$Cd(S_xSe_{1-x})$ reported by Airey is:

> "29 g of cadmium sulfate were dissolved in 200 cm^3 of water and then this solution was mixed with 140 g of a solution containing approximately 35% by weight of zirconium acetate. 3.6 g of selenium powder was dissolved in a sodiumsulphide solution (28 g dissolved in 1000 cm3 of water) and was then made alkaline by adding ammonium hydroxide solution. The two solutions were mixed together and the precipitate obtained was washed, filtered and dried. The dry powder was then mixed with 35 cm^3 of sodium silicate solution diluted to 200 cm^3 with water and neutralized with acid to produce a gel. The gel was washed with water, filtered and dried at 130° C. to give a fine powder. This powder was dry-mixed with 3% by wt. of lithium fluoride and fired in a closed crucible at 1050ºC for 30 minutes. The resulting red pigment composition was ground in a ball mill, and then washed with a mixture of concentrated sulphuric acid and concentrated nitric acid. This was followed by washing with water, then with 5M sodium hydroxide and finally with water. The resulting red protected pigment contained 5.9% by wt. of cadmium. The

red protected pigment was completely stable to acids and alkalis and to temperatures of 1200°C."

The formation mechanism of the red $ZrSiO_4$-$Cd(S_xSe_{1-x})$ was proposed by Lambies and Rincón in 1981 [41]. The formation process of $Cd(S_xSe_{1-x})$ solid solution and $ZrSiO_4$ crystal was analyzed. It was found that $Cd(S_xSe_{1-x})$ was encapsulated as a unit by $ZrSiO_4$, rather than trapped in the crystal lattice in the form of ions. The difference of the different crystal formed in the pigments was pointed out and a two stage formation mechanism was developed [28]. A small amount of microcrystalline $ZrSiO_4$ is formed in the initial stage after the formation of microcrystals of the solid solution $Cd(S_xSe_{1-x})$, then the microcrystals of $Cd(S_xSe_{1-x})$ were encapsulated in the $ZrSiO_4$ crystals previously formed in the main reaction step and the encapsulated pigment of $ZrSiO_4$- $Cd(S_xSe_{1-x})$ was developed. It is indicated that the nucleation rate of cadmium sulphoselenide (pigment core) should be lower than the crystal growth of zircon (see the formation process in Figure 6.17). There are two means to gain a better encapsulated pigment: decelerating the nucleation rate of cadmium sulphoselenide (pigment core) or accelerating the formation of zirconium silicate [36]. Chemicals used to prepare cadmium sulphoselenide should be carefully selected to control the reaction rate. To obtain a fully developed pigment core and completely crystallized zirconium silicate, the former method achieved better result.

But the inclusion mechanism of hematite and cadmium sulphoselenide in zircon it is possibly to expand and generalize to others systems such as Airey et al. report [43]:

> "pigments which are unstable at high temperatures are protected by inclusion in zirconium silicate crystals. Improved protected pigments are produced by the step of precipitating the constituents of the protected pigments from aqueous or non-aqueous solution. Especially applicable to pigments for ceramics such as those containing cadmium and pigments for use in plastics where contamination of food stuffs is to be avoided."

For example:

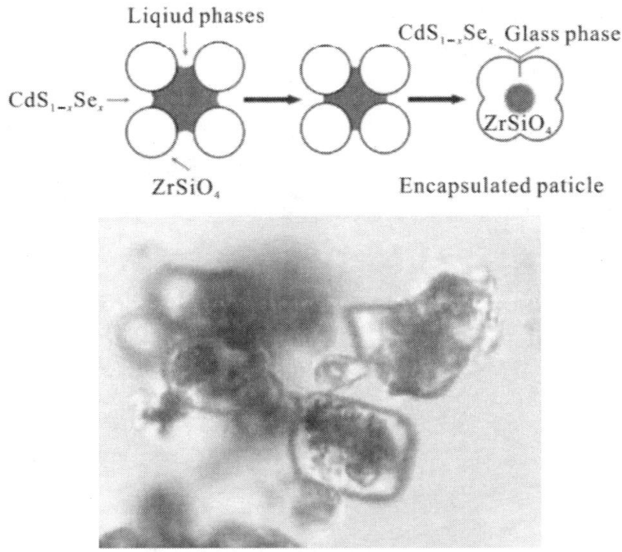

Figure 6.17. Formation process of $ZrSiO_4$-$Cd(S_xSe_{1-x})$ [41] and optical image of $ZrSiO_4$-$Cd(S_xSe_{1-x})$ inclusion pigment. (https://www.puttgarden.com/crystal/2008/1-25-08/inclusion_pigments_-_andreas_widhalm.html).

I. Gold pink associated to colloidal particles of gold entrapped into zircon crystals; in this case the main problem is the adjustment of the gold colloid particles, Airey et al. used hydrated zirconia, from a solution of zirconyl chloride and chloroauric acid. The precipitate was filtered and mixed damp with sodium silicate solution which was neutralized to precipitate silicic acid. The mixture was filtered, washed and dried to a powder which was then mixed with 3 wt.% LiF and fired in a closed crucible for ten minutes at 1000ºC [43].

II. Chrome green: is relatively easy to encapsulate eskolaite in zircon by coprecipitation route but the pigment has not been success.

III. Carbon black: adding an organic compound at the zirconia stage precipitation in the receipt for red cadmium in zircon pigment above described is possible to prepare a grey pigment comparable to $(Sn,Sb)O_2$ tin-antimony cassiterite pigments. The pigment is today investigated in order to obtain a carbon-zircon stable, avoiding pin-hole, for underglaze applications [44].

IV. Cobalt blue: all cobalt pigments show problems associated with the cobalt tendency to migrate or flow, thus it making it difficult to produce precisely defined designs. If cobalt could be prevented from moving by entrapping it in zircon then not problems of this type should be encountered [43], but this encapsulation as in the case of chromium have not been success.

REFERENCES

[1] Seabright C. A., Blue turquoise, *Ceramic Pigments*, U.S. patent, 2,441,407, May 1948.

(a) Seabright C. A., *Yellow of Praseodymium*, US Pat. 2,992,123, July 11, 1961, (b) Fujii, K., Sono, K., Jap. Patent, 5239 (1954).

[2] Seabright C. A., Pink Koral, *Iron Ceramic Pigment*, U.S. Pat., 3166430,1965.

[3] Minton, R. H., "Use of substitutes for tin oxide in glazes," *J. Amer. Ceram. Soc,* 3,1(1920): 6-12.

[4] Budnikov P. P., "Zircon for white glazes," *Tonind. Ztg.,* 47 (1923): 1734-1738.

[5] Budnikov P. P., Ginstling A. M., Shaw K., *Principles of Solid State Chemistry*, (Elsevier Science, 1968).

[6] Titanium Alloy Mfg. Co., "Zirconium glazes," *Tech. Reinew*, 1(1944): 45.

[7] Eppler, R. A., "Mechanism of formation of Zircon Stains," *J. Am. Ceram. Soc*, 53,8(1970):457-462.

[8] Li C. H., Eppler, D. R., Eppler, R. A., "Iron zircon pigments," *Ceram. Eng. Sci. Proc.,* 16, (1992): 109-118.

[9] Monrós, G., Carda, J., Escribano, P. Alarcón, J., "Synthesis of V-ZrSiO4 solid solutions," *J. Mater. Sci. Lett.,* 9, (1990):84-86.

[10] Matkovitch, V. I., Corbett, P. M., "Formation of zircon from zirconium dioxide and silicon dioxide in the presence of vanadium pentoxide," *J. Am. Ceram. Soc.,* 443(1961):128-131.

[11] Monrós, G., Carda, J., Tena, M. A., Escribano, P., Sales, M., Alarcón, J., "Diferent kinds of available solid solutions in V_2O_5-$ZrSiO_4$-NaF system by sol-gel processes and their characterization," *J. Eur. Ceram. Soc.,* 11, (1993): 77-86.

[12] Llusar, M., Vicent, J. B., Badenes, J., Tena, M. A., Monrós, G., "Environmental Optimisation of Blue Vanadium Zircon Ceramic Pigment," *Journal of the European Ceramic Society* 19 (1999): 2647-657.

[13] Xiaou, H., Gui-Ku, B., Ming-Guang, Z., "The study of the optical and the EPR spectra od V^{+4} in zircon-type crystals," *J. Phys. Chem. Solids,* 46(1985):719-726.

[14] Demiray, T., Nath, D. K., Hummel, F. A., "Zircon-vanadium blue pigment," *J. Amer. Ceram. Soc.,* 53(1970):1-9.

[15] Di Gregorio, S., Greenblatt, M., "An ESR and optical study of V^{+4} in zircon-type crystals," *J. Chem. Phys.,* 76(1981):1931-1940.

[16] Vicent, J. B., Badenes, J., Llusar, M., Tena, M. A., Monrós, G., "Microstructural diferentiation between the green and turkish blue solid solutions of V in a zircon lattice obtained by the sol-gel process," *J. Sol-Gel Sci. Tech.*, 13(1998):347-352.

[17] Badenes, J. A., Vicent, J. B., Llusar, M., Tena, M. A., Monrós, G., The nature of Pr-$ZrSiO_4$ yellow ceramic Pigment, *Journal of Materials Science* 37 (2002):1413–1420.

[18] Seeber, W., Downing, E. A., Hesselink, L., Fejer, M. M., Ehrt, D., Pr^{3+} doped fluoride glasses," *J. Non-Cryst. Solids* 189 (1995) 218-223.

[19] De, G., Licciulli, A., Nacucchi, M., "Uniformly dispersed Pr^{+3} doped silica glass by the sol-gel process," *J. Non-Crystalline Solids* 201 (1996):153-158.

[20] Ofelt, G. S., "Intensities of Crystal Spectra of Rare-Earth Ions," *J. Chem. Phys* 37 (1962):511-520.

[21] Asprey, L. B., Keenan, T. K., "Quadrivalent lanthanides. I. Sodium praseodymium (IV) fluorides" *J. Inorg. Nucl. Chem.* 16(1961):260-266.

[22] Anemburg, M., Burnham, A. D., Hamilton, J. L., "Quadrivalent praseodymium in planetary materials," *Am. Mineralogist*, in press, http://www.minsocam.org/MSA/Ammin/AM_Preprints/7325AnenburgPreprint.pdf, https://doi.org/10.2138/am-2020-207325.

[23] Ocaña, M., Caballero, A., González-Elípe, A. R., Tartaj, P., Serna, C. J., "Valence and Localization of Praseodymium in Pr-Doped Zircon," *Journal of Solid State Chemistry,* 4 (1998) 412-415.

[24] Del Nero, G., Cappelletti, G., Ardizzone, S., Fermo, P., Gilardoni, S., "Yellow Pr-zircon pigments The role of praseodymium and of the mineralizer," *Journal of the European Ceramic Society* 24 (2004) 3603–3611.

[25] Montoya, N., Herrera, G., Alarcón, J., "Synthesis and characterization of praseodymium-containing $ZrSiO_4$ solid solutions from gels," *Ceram. Int.*, 37(8) (2011)342-356.

[26] Llusar, M., Badenes, J., Calbo, J, Tena, M. A., Vicent, M., Monrós, G. "Environmental and colour optimisation of mineraliser addition in synthesis of iron zircon ceramic pigment," *British Ceramic Transaction*, 99, 1, (2000):14-22.

[27] Llusar, M., Badenes, J. A., Calbo, J., Tena, M. A., Monrós, G., "Analysis of the effect of different mineralizers in the ceramic synthesis of the "pink coral" iron-zircon Pigment," *Bol. Soc. Esp. Cerám. Vidrio*, 38,3(1999); 201-208.

[28] García, A., Llusar, M., Sorlí, S., Calbo, J., Tena, M. A., Monrós, G., "Effect of the surfactant and precipitant on the synthesis of pink coral by a microemulsion method," *Journal of the European Ceramic Society* 23 (2003) 1829–1838.

[29] Ozel, E., Turan, S., "Production of coloured zircon pigments from zircon," *Journal of the European Ceramic Society* 27 (2007):1751–1757.

[30] Huckle W. G., Swigert G. F., Wilerley S. E. "Cadmium Pigments. Structure and Composition." *Industrial & Engineering Chemistry Product Research and Development*, 5, 4, (1966): 362-366.

[31] Loeffler, J. *Process for the Production of Cadmium Red.* US Patent, 2 351 985. 1944-6-20.

[32] Wagner, H. B., Homburg B, Broll A. *Process for the Production of Firing Stable Yellow to Red Glaze mixtures*. US Patent, 4 008 901. 1977-2-15.

[33] Juredine, G. M., *Method of Making Cadmium Red Pigments.* US Patent, 2 248 408. 1941-7-8.

[34] Broll, A, Mann, H., *Inorganic Coloured Pigments and a Process for Their Production.* GB Patent, 1 440 718. 1976-6-23.

[35] Jianfeng, W., Kun, L., Xiaohong, X., Yaxiang, Z., Xiaoyang, X., Xinbin, L., "Research and Development of Cadmium Sulphoselenide Red Pigment," *Journal of Wuhan University of Technology-Mater. Sci.* 30(2015): 1247-1254. Ed. www.jwutms.net.

[36] Monrós, G., Garcia, A., Sorli, S., "Heteromorphic Pigment Synthesis Mechanisms." *Ceramica Acta*, 2002, 14(2): 12-27.

[37] Volker, W., Inclusion pigments and their use in various glaze systems "*Glazes and Glazing Techniques, Conference proceedings published by the Swedish Ceramic Society*." 1981.

[38] Seabright, C. A., *Ceramic Pigments*. US Patent, 2 441 447. 1948-5-11.

[39] Castilone, R. J., Sriram, D., Carty W. M., "Crystallization of Zircon in Stoneware Glazes," *Journal of the American Ceramic Society*, 82(10)(1999): 2819-2824.

[40] Lambies, V., Rincón J. M., "Study of the mechanism of formation of a zircon-cadmium sulphoselenide pigments," *Trans. & jour. British Ceramic Society*, 80 (1981):105-108.

[41] Vicent, J. B, Llusar, M., Badenes, J., "Oclusión De óxidos Cromoforos Mediante Métodos Sol-Gel: Aplicación a la Síntesis de Rojo Hematita Sílice [Occlusion of Chromophore Oxides by Sol-Gel Methods: Application to the Synthesis of Red Hematite Silica]." *Boletín de la Sociedad Española de Cerámica y Vidrio*, 39(1)(2000): 83-93.

[42] Airey, A. C., Spiller, A., *Method of protecting pigments with transparent crystalline zircon*, US Pat 4,482,390, 1984.

[43] Wang, Q., Chang, Q., Wang, Y., Wang, X, Zhou, J., "Preparation of Zircon-Encapsulated Carbon Black Ceramic Pigment Using the Collapsed Mesoporous-Structure," *Silicon*, 10(2018):2253–2262

Chapter 7

THE SCHEELITE PIGMENT FAMILY

As described in Chapter 1, the crystalline structure of scheelite or calcium tungstate $CaWO_4$ is integrated by the W^{6+} ions forming WO_4 units in tetrahedral coordination with the oxygens and separated in three dimensions by the Ca^{2+} ions in dodecadeltahedral coordination (IC=8) [1]. From the crystallographic point of view, scheelite is tetragonal and belongs to the Space Group *I41/a*, its lattice constants are a= 5.25 Å b = 11.40 Å, its color is white showing a biaxial refractive index (n=1.92-1.94) (see Table 1.1) The value of the refractive index n is very important when considering pigments [2]: a high refractive index allows to achieve high light scattering efficiency when the pigment is added to a matrix with a very different refractive index. The dispersion efficiency increases with the difference between the refractive index of the pigment and that of the substrate. The polymorphs of TiO_2 (anatase and rutile) show high refractive index (2.55 and 2.73 respectively, Table 1.1) making them the quintessential white pigments, hematite Fe_2O_3 (n = 2.7-3) or massicot PbO (n = 2.5) are excellent red and yellow pigments, zircon (n = 1.96) is the basis of the important family of ceramic pigments discussed in Chapter 6.

ABO_4 family of ternary oxides contain many compounds since cation A with valences +1, +2, +3, and + 4 can be accommodated with cation B with valences +7, +6, +5, and +4, respectively. They show interesting technological properties and have been used during the last years as solid-state scintillators ($CaWO_4$ and $PbWO_4$), laser-host or solid state lasers (YVO_4). A number of crystal structures in ABX_4 compounds, consist of BX_4 tetrahedra and AX_8 eight-coordinated polyhedra, which can be seen as two interpenetrating tetrahedra, known as bisdisphenoids or dodecadeltahedra [3, 4]. Among these structures some important mineral structures as scheelite or zircon are included where A and B cations form two interpenetrating diamond structures in a tetragonally distorted FCC array. Zircon is the mineral $ZrSiO_4$ and also the general name of the compounds crystallizing in the tetragonal *I41/amd* phase, like many ABO4 silicates, vanadates, phosphates and arsenates ($HfSiO_4$, YVO_4, YPO_4, and $YAsO_4$).

The Scheelite is the mineral $CaWO_4$ and the general name of the compounds crystallizing in the tetragonal *I41/a* space group, like other alkaline-earth tungstates ($SrWO_4$ and $BaWO_4$), $PbWO_4$, alkaline-earth molybdates, $PbMoO_4$, $CdMoO_4$, and a number of fluorites ($YLiF_4$, $LuLiF_4$, $CaZnF_4$). As described in Chapter 1 zircon and scheelite structures are superstructures of the rutile structure common in AX_2 compounds and the following pressure-driven phase transitions sequence can be predicted on the light of recent advances: *I41/amd* (zircon) → *I41/a* (scheelite) → *I2/a* (fergusonite) → *P21/n* ($BaWO_4$-II-type) → orthorhombic phases ($BaMnF_4$-type, *Cmca*) → amorphous [3]. Fergusonite $YNbO_4$ occurs as different polymorphs depending on the temperature of crystallization [4]: a high-T tetragonal polymorph (space group *I41/a*) isostructural with scheelite ($CaWO_4$) and powellite ($CaMoO_4$); and a low-T monoclinic polymorph with the fergusonite-type structure (known also in the mineralogical literature as fergusonite-beta, space group *I2/a*).

On the other hand, with the same thickness of paint film applied, not all paints are equally opaque (or transparent). We call this "hiding strength" or "covering power." This hiding ability (or opacity) of the paint is highly dependent on the difference in the refractive indices of the pigment and the medium, the size of the particles, and the dispersion of the pigment (i.e., the shape of the particles and the degree of aggregation of the particles), the proportion of pigment in the vehicle (that is, pigment volume concentration or PVC) and the thickness of the applied film. These factors determine the opacity of a particular painting [5].

Novel yellow ceramic pigments have been reported based on chromium doped scheelite or powellite solid solutions [6,7] such as ($CaCr_xW_{1-x}O_4$); the color is due to the replacement of tungsten by chromium in the scheelite structure, and only taxes of x=0.05-0.10 are required for obtaining a deep yellow color.

From the sequence of pressure-driven phase transitions above described for ABX_4 compounds [3] (*I41/amd* (zircon) → *I41/a* (scheelite) → *I2/a* (fergusonite)), it can be observed that the extremes (zircon and fergusonite) produce some known stable ceramic pigments but there are not have been detected in the literature any scheelite based ceramic pigments before the chromium doped scheelite or powellite solid solutions [6,7]. Only recently and related to powellite (isostructural with scheelite) ceramic pigments with general formula $Pr_{2-x}Ca_xMo_2O_{9-\delta}$ (x between 0 and 1) have been developed [8] which give colorations from yellow to green with easily application in plastics and in low reactive ceramic glazes. The performance of these pigments enameled in a lead free double firing ceramic glaze produces light yellow colors not better than b*=19. Using NH_4Cl, NaF and Na_2SiF_6 as flux agents in the $(Pr_{2-x}Ca_x)Mo_2O_9$ x=0.1 composition with the same molar addition of halides (0.84 mol per formula weight), a structural effect of fluoride ion is observed but the yellow color on enameled samples does not improve. Finally, using an ammonia coprecipitation method in the x=0.6 sample, more regular and higher crystal size crystallization are produced. The microstructure obtained from coprecipitation method gives more intense yellow colored powders and improve their resistance against glaze, producing significantly best yellow colors than samples prepared by ceramic method [9].

In the case of Cr-powellite the main characteristics of the pigment were analyzed by Galindo et al. [6]: a) the stoichiometry of the solid solution, b) the effect of the dopant concentration on the pigmenting properties, c) the effect of the chromium precursor and mineralizer's addition, d) the effect of the method of preparation. The composition $Cr_xCaMo_{1-x}O_4$ x=0.075 using $K_2Cr_2O_7$ or Cr_2O_3 as chromium precursors fired at 1000-1100°C produces a yellow color, slightly lower in intensity than the commercial praseodymium-zircon yellow pigment in double (1000°C) or single firing (monoporosa-1080°C or porcelain stoneware-1200°C) ceramic glazes. By ammonia coprecipitation method and firing at 950°C, the yellow ceramic pigment glazed in a conventional CaO-ZnO-SiO_2 glaze (double firing glaze, 1080°C) shows a yellow color similar to the commercial praseodymium-zircon yellow pigment. However (Cr-$CaMoO_4$) yellow ceramic pigment release Cr^{6+} when washed, and the leaching water should be treated as in other industrial ceramic pigments that present chromium in its composition. The observation of a homogeneous composition throughout the samples suggests an ideal solid solution behavior beneficial for the immobilization of Cr (VI) in powellite lattice as it is considered in literature for the immobilization of radionuclides in a waste repository scenario. However the release and possible mobilization of the Cr (VI) should be controlled in order to get its industrial implementation [6].

Regarding to the Cr-scheelite solid solution $Ca(Cr_xW_{1-x})O_4$, Cerro et al. [10] studied it up to x=0.2, showing only scheelite as crystalline phase at XRD, the lattice parameters for the solid solution x=0.2 shows a contraction of the lattice in agreement with the replacement of W^{6+} (0.56 Å) for Cr^{6+} (0.40 Å). Finally, the coloring behavior of Cr-scheelite and Pr-zircon in lead-free glazes (double firing or single firing for porcelain stoneware) was similar. The UV-Vis-NIR diffuse reflectance spectra evolution with x in $Ca(Cr_xW_{1-x})O_4$ solid solution shows that its coloring capacity saturates at x=0.1 in agreement with the L*a*b* data evolution.

Looking at a possible industrial application of the Cr-scheelite pigment, and considering that Cr^{6+} is toxic and moreover easily mobilized ion, a washing study was carried out for both Cr-scheelite and Pr-zircon yellow ceramic pigments: with this aim, 1 g of fired pigment placed in a funnel was carefully washed with water at 25ºC until obtaining 100 ml of leaching water. Then, the electrical conductivity, the Cr amount (in Cr-scheelite sample) and the fluoride amount (in Pr-zircon sample) were measured in the leaching water. Electrical conductivity was measured at 20ºC using a CRISON 540 conductivity-meter, the chromium amount was measured by colorimetric determination as dichromate ion in the acidified water (adding 10 ml of HNO_3 0.3M) at 350 nm, and finally the fluoride determination was carried out using a fluoride selective electrode (a lanthanum fluoride electrode, doped with europium supplied by CRISON). The leaching water of Cr-scheelite was yellow colored and the one of Pr-zircon colorless, the conductivity of leaching water was higher for Cr-scheelite (1368 μS versus 825 μS of Pr-zircon) and the measurement of Cr(VI) was 6 mg/g (lower than the amount measured in other pigments based on chromium). In any case, the fluorides released by Pr-zircon (16.8 mg/g) as well as the Cr(VI) of Cr-scheelite should be treated and controlled for industrial implementation.

The comparative study performed by Cerro et al. [10] with the four historical lead-based pigments (XVII-XVIII century), the modern Ni,Sb-rutile (1939) and Pr-zircon (1961) yellow pigments, and the novel yellow ceramic pigments based on Cr-scheelite and Mo-yttrium cerate [11] pigments shows that all powders have a wide band gap semiconductor behavior (E_g=2.1-2.5 except lead-tin yellow I; 3.1 eV) and high NIR reflectance ranging from 80% for Cr-Scheelite to 86% of Naples yellow, except Ni,Sb-rutile pigment that show a moderate 55%. However, the low color performance of lead based pigments, along with its toxicity; discard these pigments as eco-friendly cool pigments. Ni,Sb-rutile, with moderate NIR reflectance and toxicity associated to the Ni presence is also discarded. The Pr-zircon, Cr-scheelite and Mo-yttrium cerate pigments exhibit similar NIR reflectance (Rnir=80-83%) and yellow coloration (b*=57.6-45.3). However, only Pr-zircon and Cr-scheelite show color stability in the pigment applications; maintaining relative high b* values in alkyd paint and in the lead-free double firing glaze, and also the L* values enter into the Munsell range (around 80). Finally regarding to results of Orange II photodegradation test, lead based and Ni,Sb-rutile yellow pigments inhibits the photolysis of the UV irradiation, acting as protecting agents from irradiation. The Mo-cerate shows a slight photocatalytic activity ($t_{1/2}$ =301 min) and both Cr-scheelite and Pr-zircon pigments show a moderate activity ($t_{1/2}$ =198 and 184 min. respectively vs. 60 min. for anatase).

The effect of Cr in scheelite (space group *I41/a*), which has been above described integrated by a central Ca^{2+} ion coordinated by eight singly-bound tungstate WO_4^{2-} tetrahedral groups, is explained by the entrance of Cr^{6+} replacing W^{6+} which modifies the charge transfer band of scheelite giving a wide band gap semiconductor (2.3 eV calculated by Tauc plot). Doping Scheelite with the main transition metals different colors can be developed depending on the used chromophore: white by doping with Zr or Ti, yellow with Cr, ocher with Co, green with Co-Cr or black-brown with Fe-Cr, which are discussed later [7].

7.1. WHITE PIGMENT BASED ON ZR AND TI DOPED SCHEELITE

Figure 7.1 shows the white B10 ($Ca(Zr_{0.2}W_{0.8})O_{4-\delta}$) scheelite doped with zirconium and the B20 ($Ca(Ti_{0.2}W_{0.8})O_4$ M-$CaWO_4$ scheelite doped with titanium pigments, fired at 100ºC/3h and prepared starting with $CaCO_3$, WO_3 and ZrO_2:

a) The white powders (Figure 7.1.a) show L*a*b* values with lightness L* around 90 and very low values of a* and b*. Both samples show similar XRD pattern, with scheelite as the only crystalline phase detected, indicating that the solid solution of zirconium or titanium replacing tungsten has been achieved.

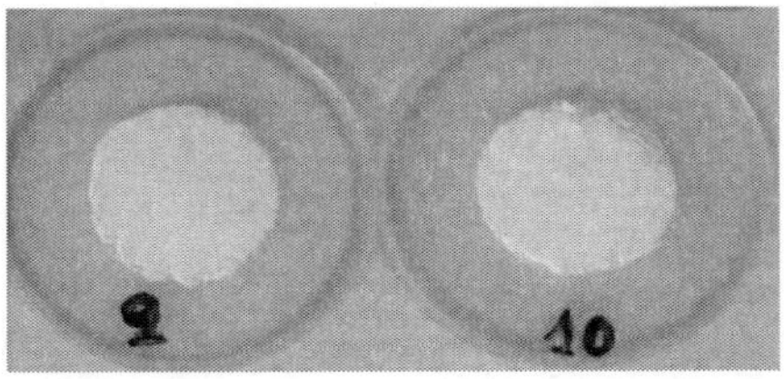

B10 ($Ca(Zr_{0.2}W_{0.8})O_{4-\delta}$) **B20** ($Ca(Ti_{0.2}W_{0.8})O_{4-\delta}$) **1000ºC/3h**

L*a*b*= 88.85/-0.58/2.67 89.94/-0.88/2.27

Rvis/Rnir/R=88/86/87 88/85/86

a

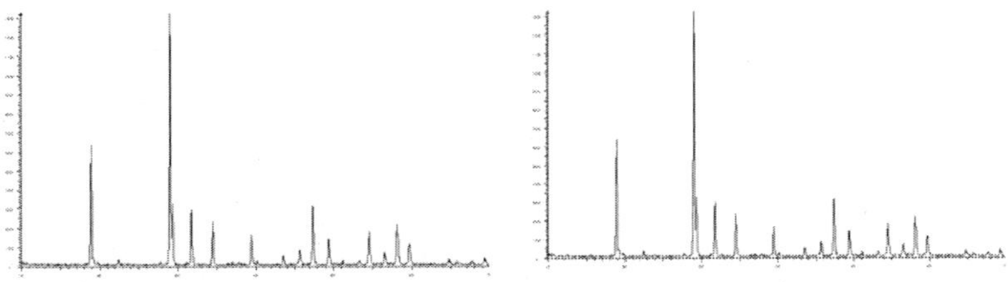

b

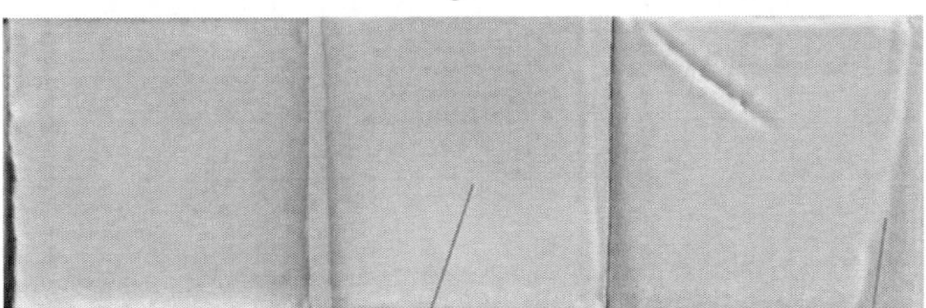

B10 3wt% in: Double firing (1025) Single firing (1080) Single firing (1190ºC)

L*a*b*= 87.1/0.8/-2.00 85.54/1.06/3.4 83.91/0.20/2.5

c

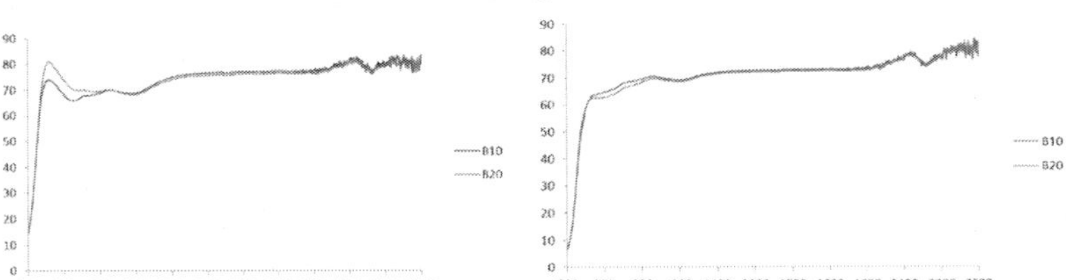

Rvis/Rnir/R= 71/74/72 (B10) 69/70/69 (B10)

72/74/73 (B20) 68/70/69 (B20)

d

Figure 7.1. White M-$CaWO_4$ (M=Zr (B10) and M=Ti(B20)) pigments: (a) powders, (b) XRD of powders, (c) B10 images of 3wt.% glazed samples, (d) UV-Vis-NIR diffuse reflectance spectra of the indicated glazed samples.

b) The 3 wt.% glazed samples of the pigments in several transparent glazes (double firing (1025ºC), single firing (1080ºC) and a porcelain single firing (1190ºC) (Figure 7.1.c) indicate the development of white shades, as can be observed at the images, with lightness around L*=84-87 and very low values of a* and b*.
c) The UV-Vis-NIR diffuse reflectance spectra of glazed samples (Figure 7.1.d) shows a similar feature of both B10 and B20 pigments in the checked pigments with Rnir around 69-73 depending of the transparent glaze (slightly higher in the double firing glaze fired at low temperature).

Figure 7.2 shows the 10 wt.% applications of B10 and B20 white samples in alkyd paint over wood and their UV-Vis-NIR spectra. The white coating developed shows an L*a*b* with lightness L* around 87-90 and very low values of a* and b*. These wood samples were irradiated with UVA lamp (PHILIPS TL-K 40W/05 lamp, λ=365 nm), and the evolution of L*a*b* color parameters with the time of irradiation (in hours) are shown in Table 7.1 indicating a good stability of the color to the UVA rays.

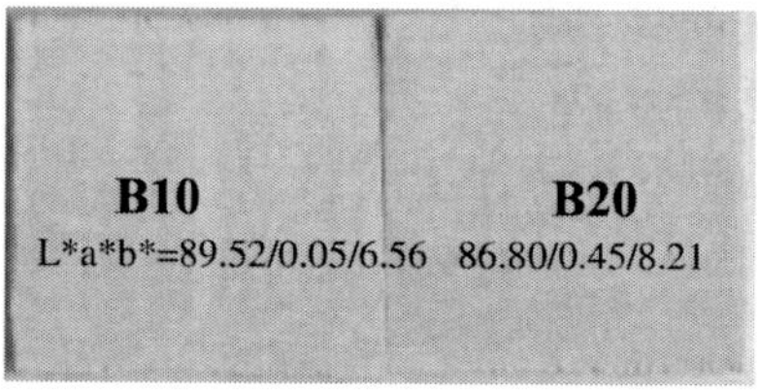

a

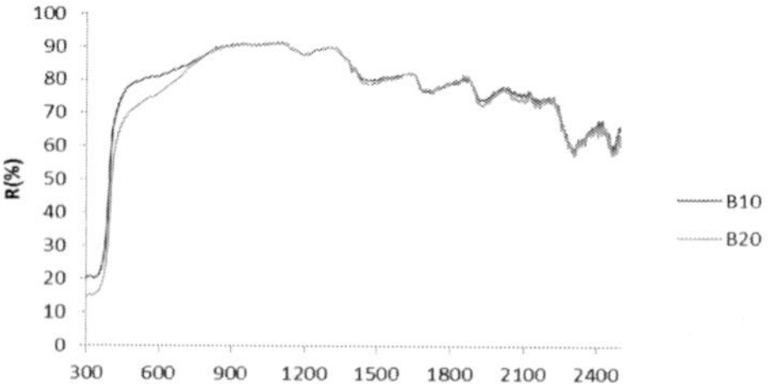

b

Figure 7.2. B10 and B20 white samples 10wt.% added in alkyd paint (a) and their UV-Vis-NIR reflectance spectra (b).

Table 7.1. Evolution of L*a*b* color parameters with the time of irradiation to UVA rays of white samples 10wt.% in alkyd paint

SAMPLE	0 h	24 h	48 h	120 h
B10 (1000°C)	89.52/0.05/6.56	89.39/-0.13/6.39	89.64/-0.14/6.56	89.69/-0.24/6.20
B20 (1000°C)	86.80/0.45/8.21	86.32/0.57/7.92	86.86/0.56/7.99	86.99/0.43/7.47

7.2. YELLOW PIGMENT BASED ON CR DOPED SCHEELITE

The yellow of Cr-scheelite discussed above based on the comparative study performed by Cerro et al. [10] with the four historical lead-based pigments (17-18th c.), the modern Ni,Sb-rutile (1939) and Pr-zircon (1961) yellow pigments, and the novel yellow ceramic pigments based on Cr-scheelite and Mo-yttrium cerate show interesting characteristics described below.

a) Effect of chromium amount and temperature.

Figure 7.3 shows the effect of temperature and doping amount x over the $Ca(Cr_xW_{1-x})O_4$ pigment obtained from $CaCO_3$, WO_3 and $K_2Cr_2O_7$ as starting precursors 5wt.% glazed in a double firing frit (1025°C): the L*a*b* values of glazed samples (Figure 7.3.a) are very similar in all cases, therefore the more simple and cheaper sample (x=0.075 fired at 1000°C) is considered the optimal sample, even it shows the better yellow b* parameter. XRD patterns of powders fired at 1000°C/3h at different x, showed in Figure 7.3.b, only scheelite peaks are detected, showing that the solid solution of chromium replacing tungsten is achieved in this range of composition.

In effect, the SEM-EDX (mapping and spectrum) study of the optimized $Ca(Cr_{0.075}W_{0.925})O_4$ pigment fired at 1000°C/3h (Figure 7.4) shows fine particles of 1-2 μm forming agglomerates of 10-20 μm which show homogenous dispersion of chromium content.

b) Effect of chromium precursor and mineralizer addition.

Figure 7.5 shows the effect of chromium precursor amount in the optimized $Ca(Cr_{0.075}W_{0.925})O_4$ pigment using Cr_2O_3, $Cr(NO_3)_3.9H_2O$ or $K_2Cr_2O_7$ as chromium precursor. It is observed that Cr (VI) precursor (potassium dichromate), employed in the samples above described, shows the better yellow L*a*b* color parameters with higher b* parameter (b*=62.1), the use of chromium oxide Cr_2O_3 gives green shades (a*=-1.5).

a

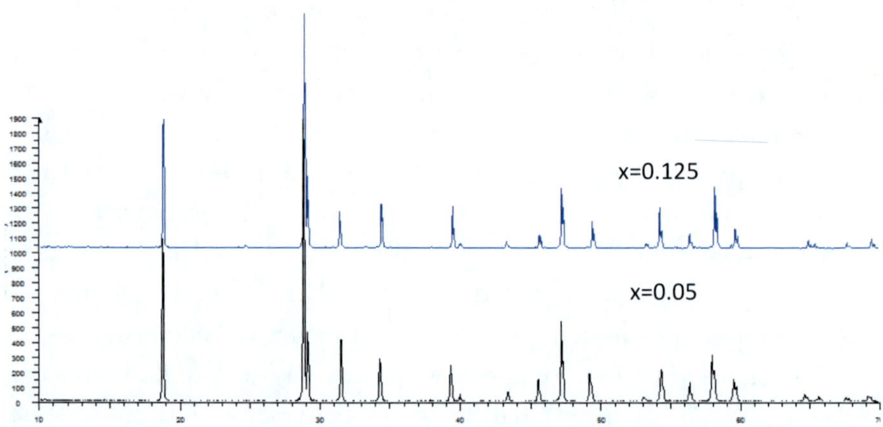

Figure 7.3. Effect of temperature and doping amount x over $Ca(Cr_xW_{1-x})O_4$ pigment 5wt.% glazed in a double firing frit (1025°C) (a), and XRD of powders fired at 1000°C/3h (only scheelite peaks are detected) (b).

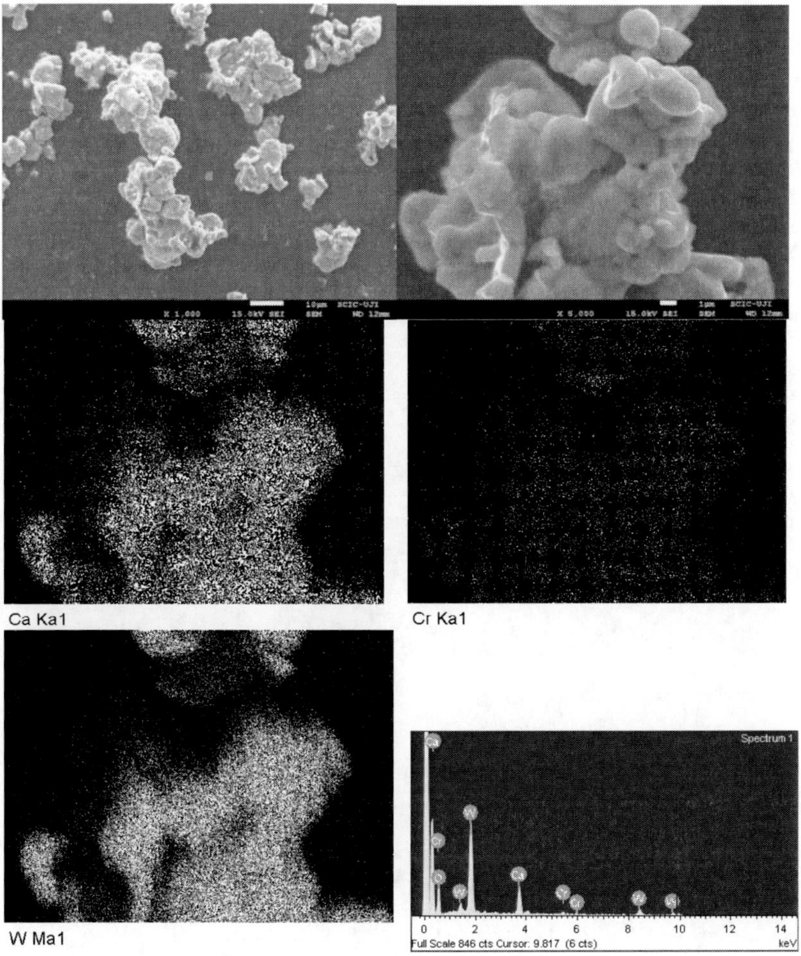

Figure 7.4. SEM-EDX (mapping and spectrum) of $Ca(Cr_{0.075}W_{0.925})O_4$ pigment ($K_2Cr_2O_7$,1000ºC/3h).

Likewise, Figure 7.6 shows the effect of mineralizers addition (3wt.% of NH_4Cl, 3wt.% of ($3NaF.2MgF_2.Li_2CO_3$) mixture or 2wt.%($CaCO_3$.KCl) + 1wt.%($NaF.Na_4B_2O_7$) mixture) on the $Ca(Cr_{0.075}W_{1.925})O_4$ pigment. The results, both in powder, double firing or single firing frit show that the best yellow color is developed without mineralizer addition, therefore the use of mineralizers is discarded in this pigment.

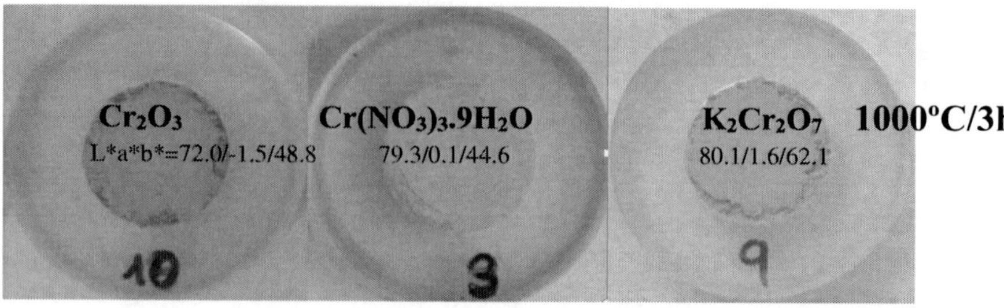

Figure 7.5. Effect of chromium precursor in $Ca(Cr_{0.075}W_{0.925})O_4$ pigment.

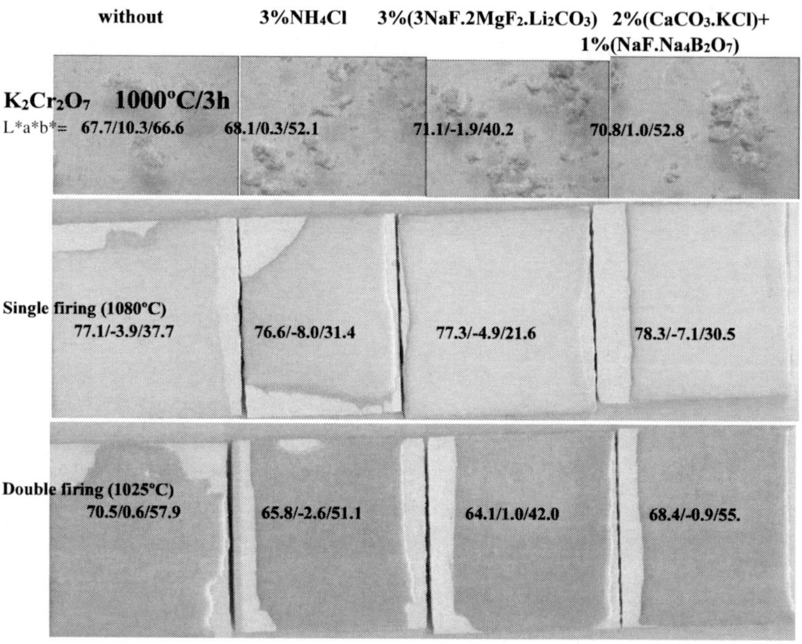

Figure 7.6. Effect of mineralizers in $Ca(Cr_{0.075}W_{0.925})O_4$ pigment.

c) Effect of method of preparation: use of MOD (Metal-Organic Decomposition) method.

Figure 7.7 shows the obtained results using an alternative colloidal method based on the MOD (Metal Organic Decomposition) methodology with citric acid addition to the $Ca(Cr_{0.075}W_{0.925})O_4$ pigment composition. The following steps are carried out in this methodology:

I. The WO_3 is dispersed in 40 ml of 1:1 HNO_3 by slowly addition of the powder over the acid, in the case of not complete dissolution, as is the case, keep stirring for 1 h to impregnate the solid.
II. The nitrate precursors (Ca, Cr) are dissolved in 160 ml of water.
III. The polycarboxylic acid (oxalic or citric acid) with a molar ratio (acid:sum of cations) = 0, 0.25, 1, 1.5 and 2.4 is added to the solution, then 1:1 (v/v) ammonia is added drop by drop until gelation or pH 8, and is maintained 8 h. in stirring. Direct application by screen printing (48 threads) on unfired enameled porcelain single firing stoneware is carried out with the prepared colloidal solution.
IV. Drying at 110ºC in an oven.
V. Firing at 500ºC/1h ("Charring" treatment) and screen printing (48 threads) of the ink obtained by the mixture (charred powder:DEG) = 2:5 in weight).
VI. Firing at 1000°C/3 h.
VII. Grinding and washing with water.

Figure 7.7 shows the results obtained by the above described MOD method using citric addition of the optimized $Ca(Cr_{0.075}W_{0.925})O_4$ pigment. The sample without citric addition (x=0) shows the worst yellow result in 5wt.% glazed sample (Figure 7.7.b) in a double firing glaze (1080ºC) with b*=26.9; the best results were obtained with x=0.25 (b*=45.5) and x=1 (b*=51.4 but greenish b*=-3.4) which are similar to the obtained by the simple ceramic route.

XRD of samples of fired samples at 500ºC/1h (Figure 7.7.c) shows the crystallization of WO_3 in samples x=0, 0.25, 1 (decreasing the intensity of peaks in that same order), the sample x=2 only shows scheelite; likewise at 1000ºC/3h MOD powders and the ceramic CE powders only show scheelite (Figure 7.7). The peaks of scheelite at 500ºC broadening (associated to nanometric particle size of powders) and shows low intensity in comparison with fired scheelite in agreement with the fine particle size observed at SEM images showed in Figure 7.8.

In effect, Figure 7.8 shows SEM (Scanning Electronic Microscopy) images (x1000 and x5000) of MOD samples without citric addition after the "charring" treatment at 500ºC/1h and after the final firing at 1000ºC/3h compared with the ceramic optimized powder of $Ca(Cr_{0.075}W_{0.925})O_4$ pigment. It is observed the low particle size of the colloids in the MOD treated at 500ºC versus the 200-300 nm of particle size in MOD fired at 1000ºC and the 1-2 μm showed by the ceramic sample fired also at 1000ºC/3h.

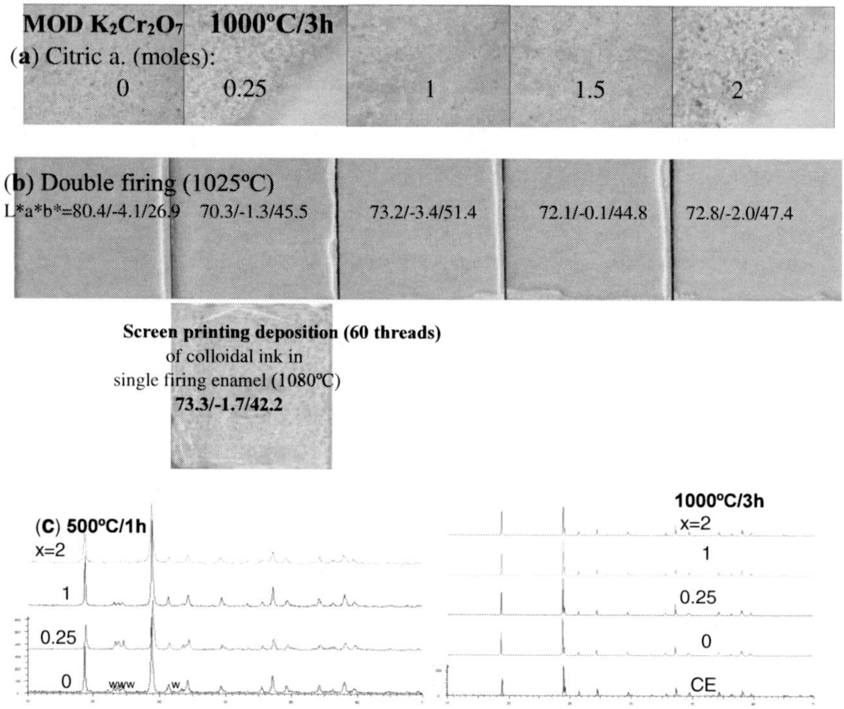

Figure 7.7. Obtained results by a MOD method using citric acid addition $Ca(Cr_{0.075}W_{0.925})O_4$ pigment: (a) powders, (b) 5 wt.% double firing frit (1025ºC), (c) XRD of samples fired after "charring" treatment and after firing at 1000ºC.

In order to analyze the particle size of MOD powders, Figure. 7.9 shows TEM (Transmission Electronic Microscopy) images and electron diffraction pattern of MOD samples with citric addition x=0 and 0.25 of $Ca(Cr_{0.075}W_{0.925})O_4$ pigment. The particle size of the x=0.25 with minor presence of WO_3 at XRD is lower than the particle size of x=0 (20-40 nm versus 25-60 nm) and the electron diffraction pattern shows a lattice distribution compatible with scheelite.

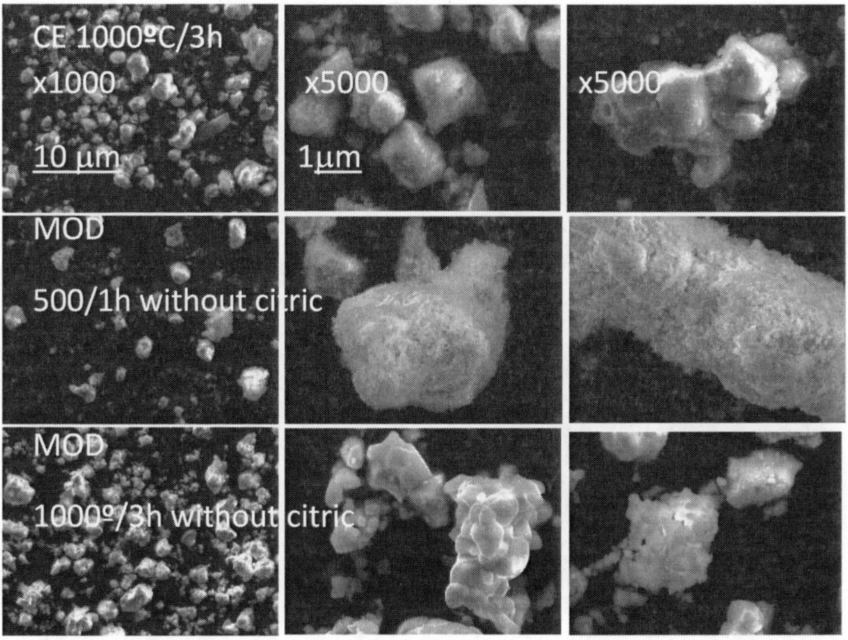

Figure 7.8. SEM images (x1000, x5000) of ceramic and MOD without citric addition of $Ca(Cr_{0.075}W_{0.925})O_4$ pigment.

More interesting is the application of the direct colloidal ink on the porcelain single firing stoneware which gives an intense yellow color (L*a*b*=73.3/-1.7/42.2) in order to use the ink on inkjet applications by control of pH and viscosity (Figure 7.7).

Figure 7.10 presents MOD sample with citric addition x= 0.25 of $Ca(Cr_{0.075}W_{0.925})O_4$ pigment 10 wt.% in alkyd paint over wood; its UV-Vis-NIR spectrum and the L*a*b* evolution with time of UVA (365 nm) irradiation. The yellow coating obtained shows L*a*b*=81.81/-4.1/48.03. The wood samples was irradiated with UVA lamp (PHILIPS TL-K 40W/05 lamp, λ=365 nm), and the evolution of L*a*b* color parameters with the time of irradiation are shown in Table 7.2 indicating a good stability of the color to the UVA rays (L* does not change and b* decreases 2 points at 24 h of irradiation but shows no change at 120 hours).

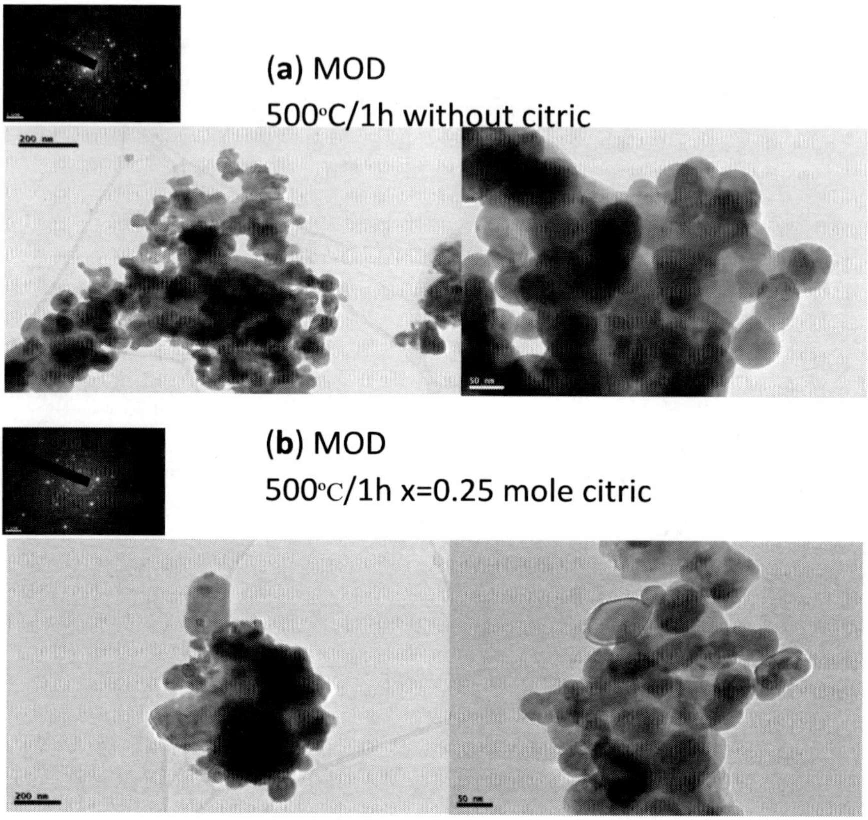

Figure 7.9. TEM images and electron diffraction pattern of $Ca(Cr_{0.075}W_{0.925})O_4$ MOD samples 500°C/1h with citric addition: (a) x=0 , (b) 0.25 moles.

Table 7.2. Evolution of L*a*b* color parameters with the time of irradiation to UVA rays of MOD sample (citric addition x= 0.25 of $Ca(Cr_{0.075}W_{0.925})O_4$ pigment) 10wt.% in alkyd paint

Sample	0 h	24 h	120 h
MOD x=0.25 (1000°C/3h)	81.81/-4.1/48.03	82.23/-4.59/45.96	82.18/-4.49/46.06

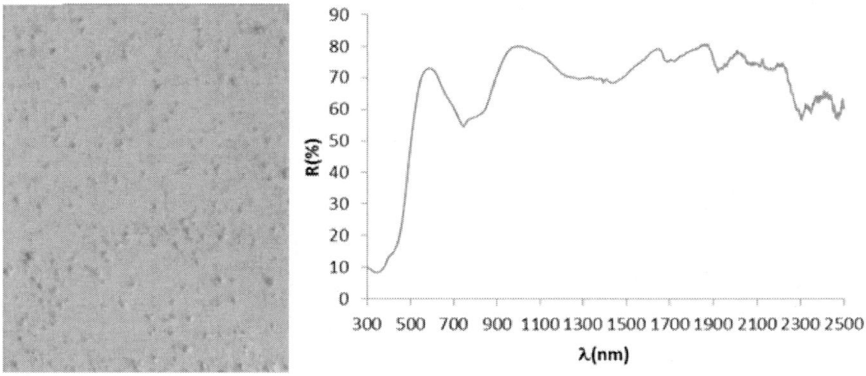

Figure 7.10. MOD sample with citric addition x= 0.25 of $Ca(Cr_{0.075}W_{0.925})O_4$ pigment 10wt.% in alkyd paint over wood and its UV-Vis-NIR spectrum.

7.3. Ochre Pigment Based on Co Doped Scheelite

Figure 7.11 shows the characterization of the ochre pigment based on Co doped scheelite $Ca(Co_{0.2}W_{0.8})O_{4-\delta}$ prepared from $CaCO_3$, WO_3, Cr_2O_3 and Co_3O_4 as starting precursors by solid state method and fired at 1000ºC/3h. The powder shows a dark ochre color (L*a*b*=54.90/3.83/ 19.25) and by XRD only shows scheelite peaks indicating that all cobalt has entered into scheelite lattice in solid solution.

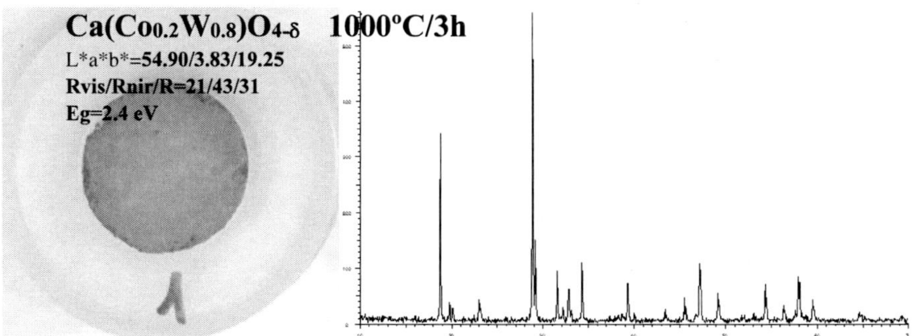

a b

Figure 7.11. (Continued).

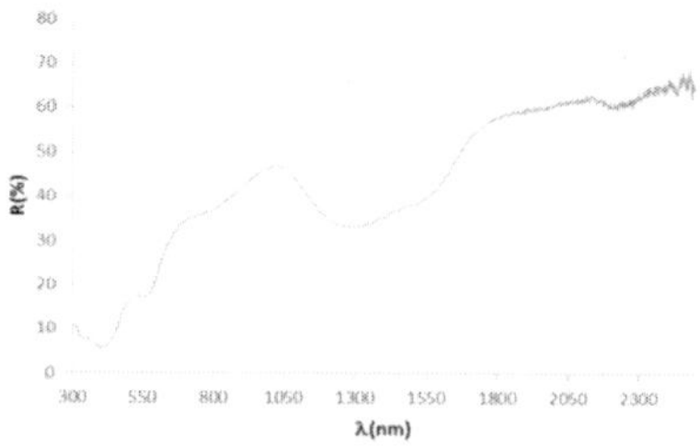

c

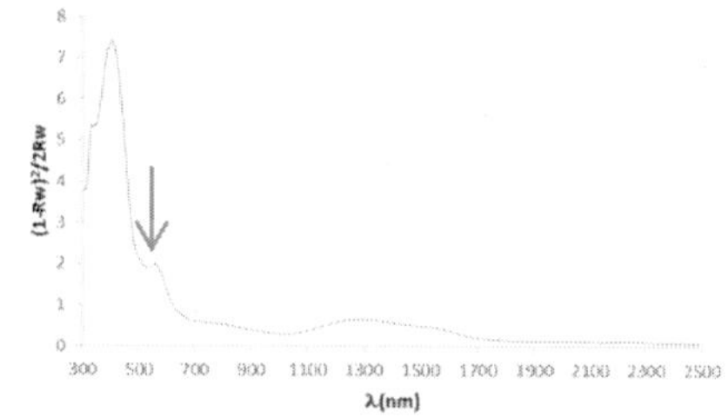

d

Figure 7.11. Characterization of the ochre pigment based on Co doped scheelite: image of the powder (a), XRD of powder (b) (only scheelite peaks detected), UV-Vis-NIR spectrum reflectance (c) and Kubelka-Munk function (d)).

a

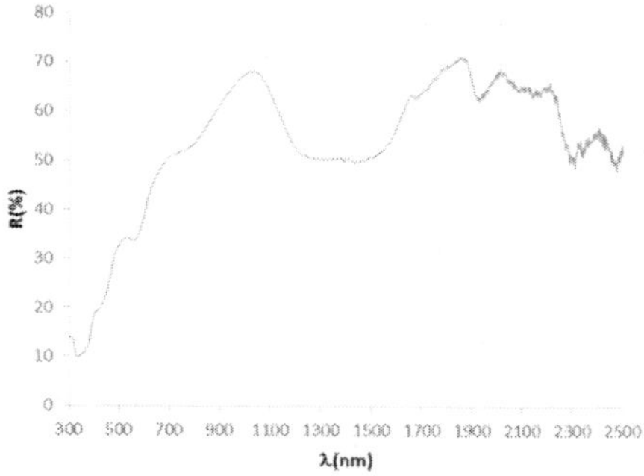

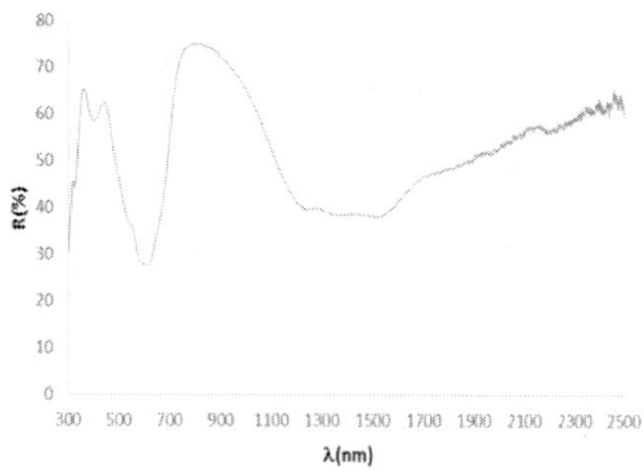

b

Figure 7.12. (Continued)

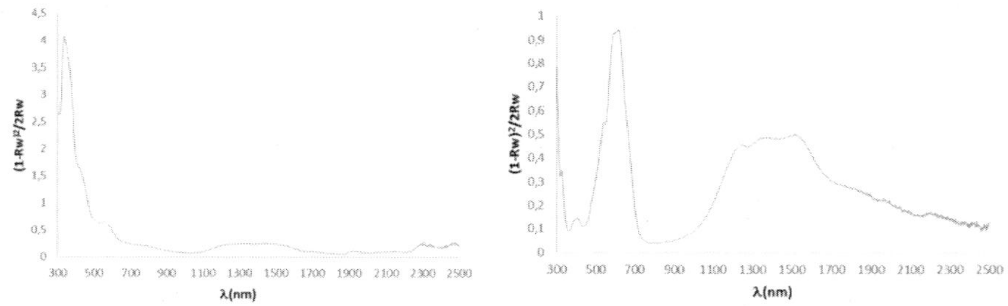

c

Figure 7.12. Characterization of the applications of ochre pigment based on Co doped scheelite in alkyl paint over wood and in single firing glaze (1080ºC): image of the application (a), and UV-Vis-NIR spectrum reflectance (b) and Kubelka-Munk function (c) of the corresponding application.

UV-Vis-NIR diffuse reflectance spectrum shows light absorptions at 420, 570, 800, and two broad and intense bands at NIR region (1300 and 1600 nm) that are not compatible with Co (II) neither in tetrahedral nor octahedral coordination discussed in chapter 4: in effect, the crystalline structure of scheelite or calcium tungstate $CaWO_4$ is integrated by the W^{6+} ions forming WO_4 units in tetrahedral coordination with the oxygens and separated in three dimensions by the Ca^{2+} ions (green) in dodecadeltahedral coordination (IC=8); then Co^{2+} enters replacing Ca^{2+} ions in dodecadeltahedral coordination giving an UV-Vis-NIR absorption response different. The dark pigment shows a solar reflectance Rvis/Rnir/R=21/43/31 with high Rnir component. From the Kubelka-Munk function a band gap of 2.4 eV can be measured.

Figure 7.12 shows the characterization of the applications of the ochre pigment based on Co doped scheelite in alkyl paint over wood and in a single firing glaze (1080ºC).

The 10 wt.% in alkyl paint over wood sample shows a dark ochre color (L*a*b*=67.00/2.41/18.00) and its UV-Vis-NIR diffuse reflectance spectrum is similar to the powder pigment, but with lower intensity, and the band at 1600 nm is not observed. The coating shows a solar reflectance Rvis/Rnir/R=37/59/46 with high Rnir component as in the powder pigment. From the Kubelka-Munk function a band gap of 2.4 eV, the same of the powder, is measured.

The 5 wt.% glazed sample in a single firing glaze (1080ºC) shows a light blue color (L*a*b*=67.13/-1.03/-20.39) indicating that Co (II) dissolves in the glaze (the called cobalt leaching effect). Its UV-Vis-NIR diffuse reflectance spectrum is different to the powder an can be associated with Co (II) (d^7) in tetrahedral environment, at least partially, since the maintenance of the dodecadeltahedral Co (II) bands of the powder previously discussed, cannot be discarded due to overlapping of bands. It is detected an intense absorption band at 600 nm that gives the blue color (which can be associated with the overlap of the triple band centered at 600 nm of the $^4A_2(F) \rightarrow {}^4T_1(P)$ transition split by a Jahn-Teller effect of tetrahedral vacancies of tetrahedral Co(II)) and a broad NIR band between 1200-1600 nm (associated with υ_1: $^4A_2(F) \rightarrow {}^4T_2(F)$ and υ_2: $^4A_2(F) \rightarrow {}^4T_1(F)$ in Figure 4.1.a).

The glaze coating shows a solar reflectance Rvis/Rnir/R= 48/58/53 with high Rnir component as in the powder pigment. From the Kubelka-Munk function a band gap of 1.72 eV is measured showing a shift to lower values associated with the intense absorption band at 600 nm (with low intensity in the case of powder in Figure 7.11.d signalized by an arrow).

7.4. Dark Green Pigment Based on Co-Cr Doped Scheelite

Figure 7.13 shows the characterization of the dark green pigment based on Co-Cr doped scheelite obtained from the composition $Ca(Co_{0.2}Cr_{0.2}W_{0.6})O_{4-\delta}$ fired at 1000ºC/3h. The powder shows a dark green color (L*a*b*=47.87/0.84/18.81) and in XRD patterns intense scheelite and weak $Ca_3Co_2O_6$ hexagonal *R-3c* (G in Figure 7.13) peaks are detected, indicating that the solid solution is not completed.

UV-Vis-NIR diffuse reflectance spectrum shows light absorptions at 350, 470, 550, 800, and a broad and intense band in NIR region at 1300 nm associated with Co (II) and Cr (III) overlapping absorption bands (see Chapter 4). The dark pigment shows a solar reflectance

Rvis/Rnir/R=13/33/22 with relative high Rnir component. From the Kubelka-Munk function a band gap of 2.3 eV can be measured.

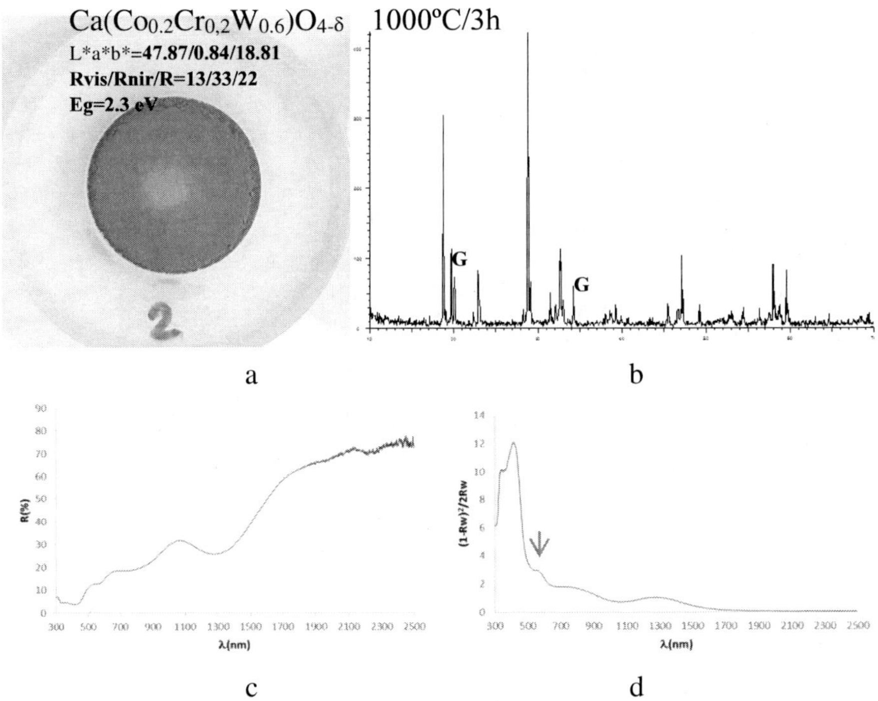

Figure 7.13. Characterization of the dark green pigment based on Co-Fe doped scheelite: (a) image of the powder, (b) XRD of powder (intense scheelite and weak $Ca_3Co_2O_6$ hexagonal *R-3c* (G) peaks are detected), (c) UV-Vis-NIR reflectance spectrum, (d) Kubelka-Munk function.

Figure 7.14 shows the characterization of the applications of the dark green pigment based on Co-Cr doped scheelite in alkyl paint over wood and in a single firing glaze (1080ºC).

The 10 wt.% in alkyl paint over wood (Figure 7.14.a) shows a dark green color (L*a*b*=64.12/0.35/28.50) and its UV-Vis-NIR diffuse reflectance spectrum is similar to the powder pigment on visible range but with lower intensity (weak absorption band associated to the binder of alkyd paint are detected also). In agreement with the corresponding UV-Vis-NIR diffuse reflectance spectrum, the coating shows a solar reflectance

Rvis/Rnir/R=31/60/44 with higher Rnir component than in the powder pigment (Rnir =33). From the Kubelka-Munk function, a 2.3 eV band gap is measured, the same of the powder.

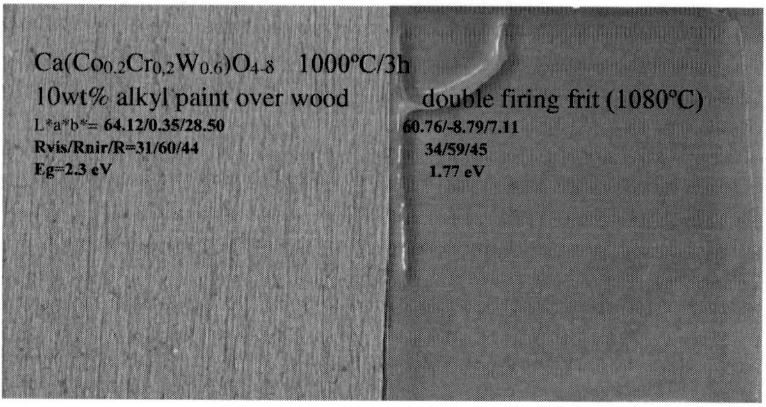

a

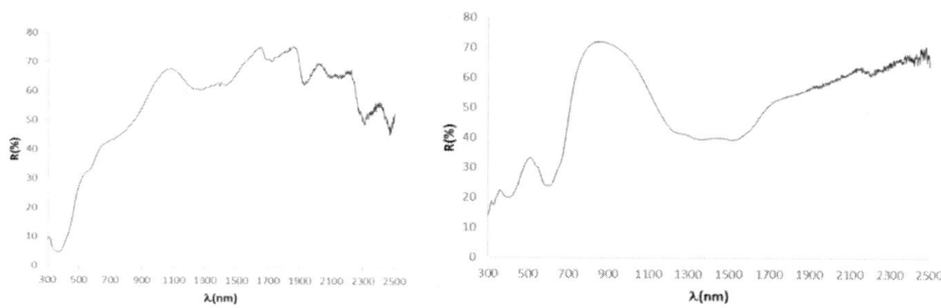

b

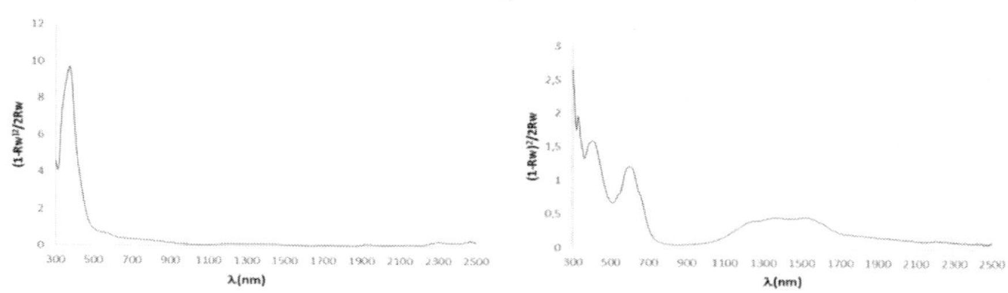

c

Figure 7.14. Characterization of the applications of dark green pigment based on Co-Cr doped scheelite in alkyl paint over wood and double firing glaze (1080ºC): (a) image of the applications, (b) UV-Vis-NIR reflectance spectra, and (c) Kubelka-Munk function of the corresponding application.

The 5wt.% glazed sample in a single firing glaze (1080ºC) shows a dark green color (L*a*b*=60.76/-8.79/7.11) with relative high b*. The corresponding UV-Vis-NIR diffuse reflectance spectrum is different to the powder; it shows a relative intense absorption band at 400 and 600 nm (giving green color) and a broad NIR band between 1100-1600 nm that can be associated to a partial cobalt leaching effect. The coating shows a solar reflectance Rvis/Rnir/R= 34/59/45 with high Rnir component as in the wood application. From the Kubelka-Munk function a 1.77 eV band gap is measured, showing a shift to lower values associated with the absorption band at 600 nm (with low intensity in the case of powder in Figure 7.13.d signalized by an arrow as in the case of ochre composition based on cobalt, and associated with the mentioned cobalt leaching effect).

7.5. Black Pigment Based on Co-Fe Doped Scheelite

Figure 7.15.a shows the characterization of the black pigment based on Co-Fe doped scheelite. The powder shows a soft black color (L*a*b*=47.87/0.84/18.81) and by XRD intense scheelite and weak peaks of $Ca_{4.26}Co_{10}O_{30}$ (tetragonal *P4/mbn*, B in Figure 7.15) are detected, indicating that the solid solution has not been achieved.

The UV-Vis-NIR diffuse reflectance spectrum of the black powders in Figure 7.15.c shows no absorptions bands; the reflectance increases progressively with wavelength. The dark pigment shows a solar reflectance Rvis/Rnir/R=13/33/22 with high Rnir component for a black pigment. From the Kubelka-Munk function a band gap of 2.1 eV is measured.

Figure 7.16 shows the characterization of the applications of the black pigment based on Co-Fe doped scheelite in alkyl paint over wood and in a single firing glaze (1080ºC), showing the image of the application and below the UV-Vis-NIR spectrum (reflectance and Kubelka-Munk function) of the corresponding application.

The 10 wt.% in alkyl paint over wood shows a black color (L*a*b*=42.82/0.41/7.95) with a better black feature than the powder. The UV-Vis-NIR diffuse reflectance spectrum is different to the powder pigment: weak absorption bands associated with the binder of alkyd paint in the UV and NIR range are detected and a weak band at 550 nm is now observed. The coating shows a solar reflectance Rvis/Rnir/R=13/30/20 with similar Rnir component to the powder pigment. From the Kubelka-Munk function, a band gap of 1.39 eV is measured, the same as the powder, associated with the band at 550 nm marked with an arrow.

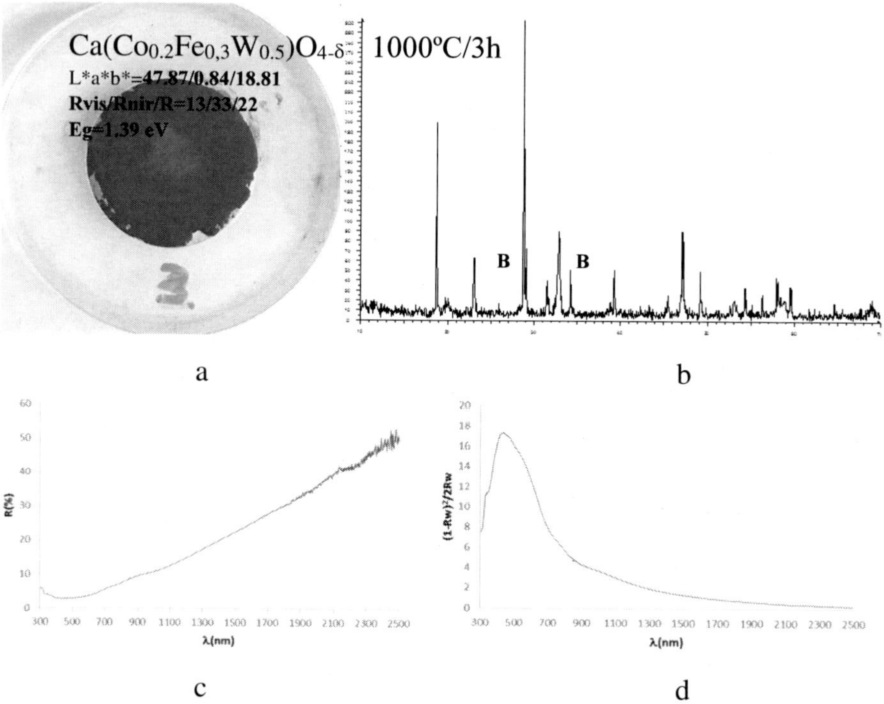

Figure 7.15. Characterization of the black pigment based on Co-Fe doped scheelite: (a) image of the powder, (b) XRD of powder (intense scheelite and weak $Ca_{4.26}Co_{10}O_{30}$ tetragonal *P4/mbn* (B) peaks are detected), (c) UV-Vis-NIR reflectance spectrum, (d) Kubelka-Munk function.

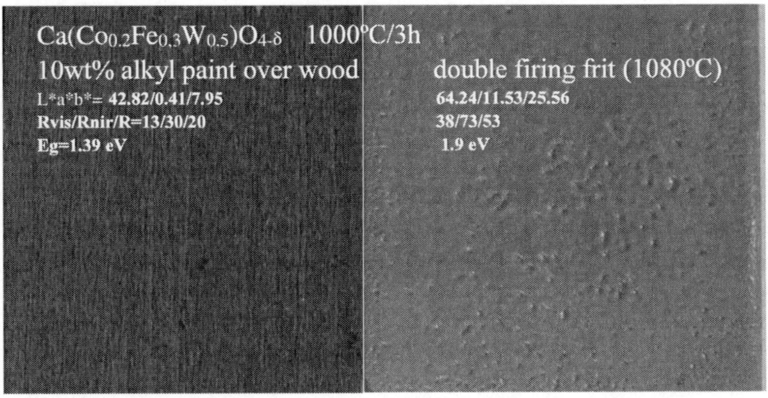

a

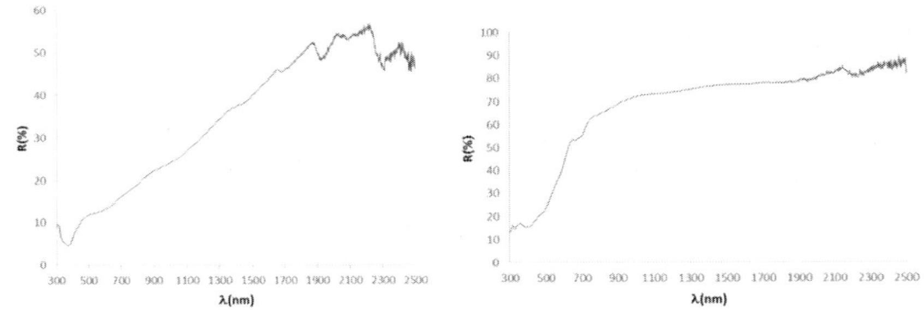

b

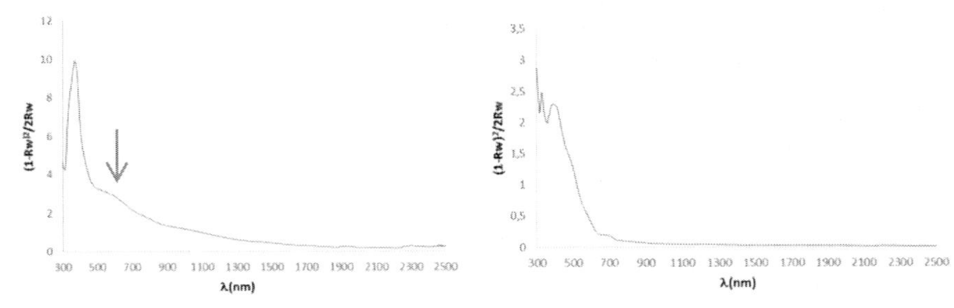

c

Figure 7.16. Characterization of the applications of black pigment based on Co-Fe doped scheelite in alkyl paint over wood and double firing glaze (1080°C): (a) image of the applications, (b) UV-Vis-NIR reflectance spectra, (c) Kubelka-Munk function of the corresponding application.

The 5 wt.% glazed in a single firing glaze (1080°C) shows a brown color (L*a*b*=64.24/11.53/25.56). Its UV-Vis-NIR diffuse reflectance spectrum shows an inflection of absorbance at 600 nm and a continuous absorption in the NIR region. The coating shows a solar reflectance Rvis/Rnir/R=38/73/53 with excellent Rnir component higher than in the powder pigment. From the Kubelka-Munk function, a band gap of 1.9 eV is measured, showing a shift to higher values in reference to alkyd application (1.39 eV) but slightly lower in reference to the powder (2.1 eV), associated with the sigmoidal absorbance appearance of the reflectance spectrum.

REFERENCES

[1] Zalkin, A., Templeton, D. H. "X-ray diffraction refinement of the calcium tungstate structure." *Journal of Chemical Physics*. 40 (1964): 501–504. doi: 10.1063/1.1725143.

[2] Singh, S., "Refractive Index Measurement and its Applications," *Physica Scripta*, 65 (2002): 167-180).

[3] Manjon, F. J., D. Errandonea, "Pressure-induced structural phase transitions in materials and Earth sciences," *Physica Status Solidi* 246 (2009) 9–3.

[4] Hazen, R. M., Finger, L. W., Mariathasan, J. E. E., "High-pressure crystal chemistry of scheelite-type tungstates and molybdates," *Journal of Physics and Chemistry of Solids*, 46, 2 (1985): 253–263.

[5] Feller, R. L., editor, *Artists Pigments: A Handbook of Their History and Characteristics*, Vol. I, (Oxford University Press, pp. 289–291, New York, 1993). https://www.naturalpigments.com/artist-materials/transparent-opaque-paints/.

[6] Galindo, R., Gargori, C., Fas, N., Llusar, M., Monrós, G., "New chromium doped powellite (Cr–$CaMoO_4$) yellow ceramic pigment," *Ceram Int* 41 (2015): 6364–6372.

[7] Monrós, G., Gargori, C., Esteve, V., Llusar, M., Jaume I University, *Spanish Pat.* ES 2616347 A1, 2018.

[8] Giable, G., Sandhya Kumari, L., Vishnu, V. S., "Synthesis and characterization of environmentally benign calcium-doped $Pr_2Mo_2O_9$ pigments: Applications in coloring of plastics," *J. of Solid State Chem.*, 181 (2008) 487-489.

[9] Galindo, R., Gargori, C., Llusar, M., García, A., Badenes, J., Monrós, G., "Ecopigmentos cerámicos verdes y amarillos de $Pr_2Mo_2O_9$ dopados con calcio obtenidos en presencia de mineralizadores y por coprecipitación química [Green and yellow ceramic eco-pigments of Pr2Mo2O9 doped with calcium obtained in the presence of mineralizers and by chemical coprecipitation]." *Bol. Soc.Esp. Ceram Vidrio*, 29 (2011) 219-228.

[10] Cerro, S., Llusar, M., Gargori, C., Monrós, G., "Cool and photocatalytic yellow ceramic pigments; from lead-tin to Cr doped scheelite pigments," *Ceram Int*, 45,4 (2019): 4613-4625.

[11] Vishnu, V. S., Reddy, M. L., "Near-infrared reflecting inorganic pigments based on molybdenum and praseodymium doped yttrium cerate: Synthesis, characterization and optical properties," *Sol Energy Mater Sol Cells*, 95(2011)2685–2692.

Chapter 8

Comparative Behavior as Cool and Photocacatalytic Pigments

Inorganic pigments are generally more stable against heat and aging than their organic counterparts and are consequently the materials of choice for outdoor applications and in particular for the construction industry.

The solar spectrum, which reaches the Earth's surface, contains about 50% NIR radiation, 45% visible light, and 5% UV radiation, and thus NIR radiation accounts for most of the produced heat from solar light. Hence, when solar radiation strikes a surface that is not coated with a NIR reflective material, some of the energy is absorbed and transformed into heat. By this, the exposed building walls become hot and cause heat inside, as well as higher peripheral temperature (urban heat island effect).

The most used material for NIR reflective purposes is the white pigment TiO_2. Unfortunately, for colored outdoor applications, only few materials are available. Moreover, they often contain toxic metals such as Co and Sb, and their reflective performance is still in need of improvements. But it is necessary to clarify the concept of cool Pigment and its evaluation.

On the other hand the passive protection against air pollution (NO_x, COVs) by building envelopes with photocatalytic activity are of interest in order to contribute to achieving optimal urban air quality.

In order to perform an adequate characterization of the cool pigments characteristics, the reflectance characterization of the powders only gives a partial vision, making it necessary to analyze its pigmenting yield in different matrices.

Moreover, high values of R and R_{nir} must be considered only relatively and not as a definitive parameter for the evaluation of colored powders as cool pigments. In effect "poor" pigments with light chroma and high lightness L* usually show high reflectance, but in this case they are poor pigments (with low pigmenting performance) rather than cool pigments. In this sense, authors have associated the high reflectance or SRI (Solar Reflectance Index, see section 3.2.3) with the lightness L* or Y factor. For instance, L. M. Schabbacha et al. analyzed the thermal and optical properties related to solar reflectance index in pigmented glazed ceramic roof tiles in Brazil [1]: the results show that most of the colored glazed tiles did not meet the 2013 LEED requirement, that is, SRI≥39 for roof tiles with a steep slope. A polynomial relationship between the Y factor (or L*) and the SRI (or R) was found (a correlation factor R^2 value > 0.95 is reported), and thus the tile colors that would probably not qualify for SRI certification could be estimated. Cool pigments with higher reflectance should be used to improve the performance of the glazes with a low SRI.

Therefore, in order to assure a high performance as cool pigments the following conditions or rules should be considered, either for powders or in its coating applications:

- Rule 1. Tinctorial strength index.

Molinari et. al. [2] introduce the *tinctorial strength* index because must be emphasized that the concept of color intensity for ceramic applications involves both color saturation and brightness [3], so it should not be merely expressed by the chromatic coordinates. In order to overcome this limit, the *tinctorial strength* index is proposed:

$$T^* = (100 - L^*) + C^* = (100 - L^*) + (a^{*2} + b^{*2})^{1/2} \tag{8.1}$$

Where L* is lightness (0 = white, 100 = black) and C* is chroma. It should not be confused with the *tinting strength* (used in paints), the *color strength* (applied in the textile sector) or the tinctorial *power* (utilized in the food industry) all based on the Kubelka-Munk equation. From this stand point, the industrial pigments and dyes exhibit tinctorial strength that, in most cases, is in between T*~70 and T*~100. These values can be devised as another basic requirement for ceramic pigments and dyes to enter in the industrial practice [2].

- Rule 2. Range of chroma C*.

The chroma C* should be high, except for black hues, the chroma of which tends to zero. This value tends to the absolute value of a* parameter for red and green hues and to the absolute value b* in blue and yellow hues. The value of the lightness L* should be enclosed within the range of high chromaticity for the different hues: i.e., high for yellow (80-90), cyan (75-85) and green (70-80), middle for magenta (50-60) and red (45-55) and low for blue (25-35) and black (30-0) hues (see Figure 3.1).

- Rule 3. NIR reflectance.

Obviously for a cool pigment the R_{NIR} should be high.

- Rule 4. Comparing colors of pigments.

In order to compare the performance between two cool pigments of the same hue (as powders or in its coating applications), their L* values should be the same. In this paper we propose as a useful recommendation the evaluation of R85 (R at L*=85) for yellow and cyan hues, R75 for green, R55 for brown-magenta and R30 for blue and black hues. These evaluations can be obtained by gradual addition of the pigment to the matrices and using a binder (e.g., $CaCO_3$ in the case of powder as will be discussed later).

If a pigment does not accomplish conditions 1 and 2, it is obvious that its pigmenting capacity is poor and should not be considered a pigment, while in the case that condition 3 is not met, the powder should not be considered a cool pigment; finally, the comparison with traditional cool pigments, using condition 4, allows to evaluate the quality level as cool pigment of any new proposed pigment.

As above discussed (Chapter 3) novel pigments with high NIR reflectance based in rare earths chromophores or components have been proposed, but its evaluation as a pigment is in many respects questionable in light of the above conditions.

8.1. Comparative Evaluation as Cool Pigments of Classic and Novel Pigments

A comparison between a palette based on rare earths and a palette obtained from classical cool pigments optimized by a procedure based on "Dry Powder Coating" (DPC) methods is presented in this section by application of the four conditions above described.

In order to improve the NIR reflectance, decrease the amount of chromophore and the price of the pigment, DPC methods produce "decorated" pigments over a high reflectance ceramic base (e.g. TiO_2, ZnO). DPC coating method has been utilized for producing special chemicals and ceramics under wet or dry conditions for improving its functionalities. In the dry method, host particles (1–500 μm) can be coated mechanically with guest particles (0.1–50 μm) in order to improve wettability, solubility, and other characteristics [5]. As the guest particles are too fine, the Van der Waals interactions are sufficiently strong to keep them firmly attached to the host particles. Thus, either a discrete or continuous coating of guest particles can be achieved depending on the choice of equipment or operating conditions such as processing period, mechanical action, weight fraction of guest to host particles and physical properties of the particles used [6].

8.1.1. Samples Preparation

The cool pigments palette based on rare earths (except black) is integrated by the following pigments:

a) Blue of La,Li doped $SrCuSi_4O_{10}$ wesselsite [7] (structure of gillespite $BaFeSi_4O_{10}$, tetragonal, space group *P4/ncc* [8]. 50 g. of the composition $Sr_{0.3}La_{0.7}Cu_{0.3}Li_{0.7}Si_4O_{10}$ were prepared from carbonates ($SrCO_3$ and Li_2CO_3) and oxides (La_2O_3, CuO, quartz SiO_2). This pigment is based in the ancient Egyptian Blue Family optimized for NIR reflectance improvement by La introduction.

b) Red-brown of Pr-CeO_2 cerianite (fluorite structure, cubic, space group *Fm3m*) [9]. 50 g. of the composition $Ce_{0.95}Pr_{0.05}O_2$ were prepared from oxides (Pr_6O_{11}, CeO_2). Precursors were mechanically homogenized in an electric grinder (20000 rpm) during 5 min. and the mixture fired at 1400°C/6h.

c) Yellow of Mo-$Y_2Ce_2O_7$ ytrium cerate (fluorite structure, cubic, space group *Fm3m*) [10]. 50 g. of the composition $Y_2Ce_{1.5}Mo_{0.5}O_7$ were prepared from $(NH_4)_2MoO_4$ and oxides (Y_2O_3, CeO_2).

d) Black-brown of $Sr_4Mn_2CuO_9$ (hexagonal perovskite, Space Group *P321*) [11]. The structure of this free rare earth pigment consists of octahedra and trigonal prisms that form a chain structure by face-sharing. The Mn^{4+} ions occupy octahedral sites and Cu^{2+} ions randomly fill the center of the square faces of the trigonal prisms. 50 g. of the composition $Sr_4Mn_2CuO_9$ were prepared from carbonates ($SrCO_3$) and oxides (MnO_2, CuO).

The coolest pigment palette based on traditional pigments with transition metals chromophores is integrated by the following pigments (see section 3.2.2):

a') Blue of Co-Zn_2SiO_4 willemite-decorated ZnO (trigonal, space group *R-3*) [12]. 50 g. of the composition $[Co_{0.6}Zn_{1.4}SiO_4].0.2ZnO$ were prepared from oxides (Co_3O_4, ZnO, quartz SiO_2).

b') Red-brown of Cr-$ZnFe_2O_4$ franklinite (CPMA number 13-37-7, spinel, cubic, space group *Fd3m*) [13]. 50 g. of the composition $ZnFe_{1.8}Cr_{0.2}O_4$ were prepared from oxides (Cr_2O_3, Fe_2O_3 hematite, ZnO).

c') Yellow of Ni,Sb-TiO_2-decorated TiO_2 (CPMA number 11-15-4, tetragonal, space group *P42/mnm*) [14]. 50 g. of the composition $[Ni_{0.25}Sb_{0.25}Ti_{0.5}O_2].TiO_2$ were prepared from oxides (Sb_2O_3, TiO_2 anatase, NiO).

d') Black-green of $Co(Cr_{2-x}Co_x)O_4$-decorated TiO_2 (spinel CPM number 13-29-2, cubic, space group *Fd3m*) [15,16]. 50 g. of the composition $[0.1Co_3O_4.0.1Cr_2O_3].TiO_2$ were prepared from oxides (Co_3O_4, Cr_2O_3, TiO_2 anatase).

All precursors were supplied by ALDRICH SA. Precursors were mechanically homogenized in an electric grinder (20000 rpm) during 5 min. and the mixture fired at the corresponding temperature and soaking time: 950ºC/16h (pigment a), 1400ºC/6h (pigment b), 1500ºC/6h (pigment c), 1000ºC/6h (pigment d) and 1100ºC/3h (pigments a', b', c' and d').

Samples were characterized by: (a) X-Ray Diffraction (XRD) using Cu K_α radiation (10-70 º2θ range, scan rate 0.02 º2θ/s, 4 s per step and 40 kV and 20 mA conditions), (b) L*a*b* and C*h* color parameters of glazed samples measured following the CIE-L*a*b* ("Commission Internationale de l'Éclairage") colorimetric with standard lighting D65 and 10º observer. On this method, L* is a measure of lightness (100=white, 0=black) and a* and b* of chroma (-a*=green, +a*=red, -b*=blue, +b*=yellow), (c) UV-Vis-NIR spectra of fired powder samples and also of the applications of the pigments samples were collected through diffuse reflectance technique, which gives data in absorbance using arbitrary units (A) or in reflectance units (R(%)) using the Kubelka-Munk transformation model. As above discussed, the Kubelka-Munk (K-M) function model has a particularly simple solution in the case of reflection of radiant energy at boundary surfaces of mat (dull scattering) surfaces. The optical reflectance spectra were scanned in the range of 300 to 2500 nm for optimized pigments, and the total solar reflectance R and the NIR solar reflectance R_{NIR} were obtained as above discussed (eq. 3.6), (d) Microstructure characterization of powders was carried out by Scanning Electron Microscopy (SEM) following conventional preparation and imaging techniques. The chemical composition and homogeneity of the samples was determined by semi-quantitative elemental analysis with an EDX analyzer attached to the microscope. (e) The pigmenting capacity of the pigments was studied in three matrices: (i) 10 wt.% in colorless alkyd paint, (ii) 3 wt.% in a conventional double-firing frit of the SiO_2-CaO-ZnO system (1025ºC) and (iii) 2% wt.% in porcelain powder (1190ºC).

8.1.2. Results and Discussion

8.1.2.1. X-Ray Diffraction (XRD) and SEM-EDX Characterization

XRD patterns of the powders are shown in Figure 8.1, and the SEM images of powders (at low and high magnification) in Figure 8.2; while additional SEM images with its corresponding EDX mappings of pigment powders are shown in Figure 8.3 (for both families of pigments in all cases).

a) In the case of blue pigments, XRD shows the crystallization of wesselsite $SrCuSi_4O_{10}$ (E in Figure 8.1) as the main crystalline phase although several weak peaks associated to residual quartz (Q) and La_2O_3 (L) are observed. For the classical Co-willemite-decorated ZnO, willemite and ZnO are the main phases, but weak XRD peaks of unreacted quartz and Co_3O_4 are also detected. The SEM images (Figure 8.2) show that the La,Li-wesselsite blue powder presents aggregates of 10 μm of mean size highly sintered, and sometimes decorated by little orthohedra with particle size between 0.2-1 μm. The microstructure of Co-willemite blue pigment is homogeneous, showing aggregates between 5-25 μm integrated with fine spherical particles of around 1μm of particle size. The SEM-EDX results (Figure 8.3) are in agreement with XRD observations: the lanthanum $L_{\alpha 1}$ signal shows the presence of unreacted lanthanum oxide in the La,Li-wesselsite blue powder, while unreacted quartz along with a little Co inhomogeneity are also detected in the mapping of the Co-willemite powder.

b) Regarding to red-brown pigments, cerianite (P in Figure 8.1) is the only crystalline phase detected in Pr-cerianite red-brown pigment, and franklinite (F) in the case of Cr-franklinite pigment (Figure 8.1). The microstructure of Pr-cerianite red-brown pigment is homogeneous, showing aggregates between 5-15 μm, integrated by homogeneous and very fine particles of 0.5 μm in particle size (Figure 8.2); similarly, Cr-franklinite powders show high homogeneity with aggregates between 0.5-3 μm, constituted by fine spherical particles of 0.3 μm of particle size. SEM-EDX results in Figure 8.3 are in agreement with the above observations, and all cation EDX signals are homogeneous in the corresponding composition mappings.

c) For the yellow pigments, the yttrium cerate $Y_2Ce_2O_7$ is the single crystalline phase (Y) detected by XRD (Figure 8.1). In the case of Ni,Sb-TiO_2 decorated rutile, only very weak peaks of residual $NiTiO_3$ (N) are detected accompanying the main rutile phase (R). Highly sintered particles 10-50 μm-sized are observed by SEM in the case of the yellow of Mo-yttrium cerate pigment (Figure 8.2), while in the case of Ni,Sb-rutile, the microstructure consists of aggregates between 1-5 μm-sized, integrated by finer spherical particles of 0.2-2 μm of particle size. The SEM-EDX results in Figure 8.3 correlate well with the above observations, since all cation EDX signals are homogeneous in the case of Mo-yttrium cerate yellow pigment, while the Ni $K_{\alpha 1}$ signal show very scarce inhomogeneities associated with the presence of residual $NiTiO_3$ detected by XRD.

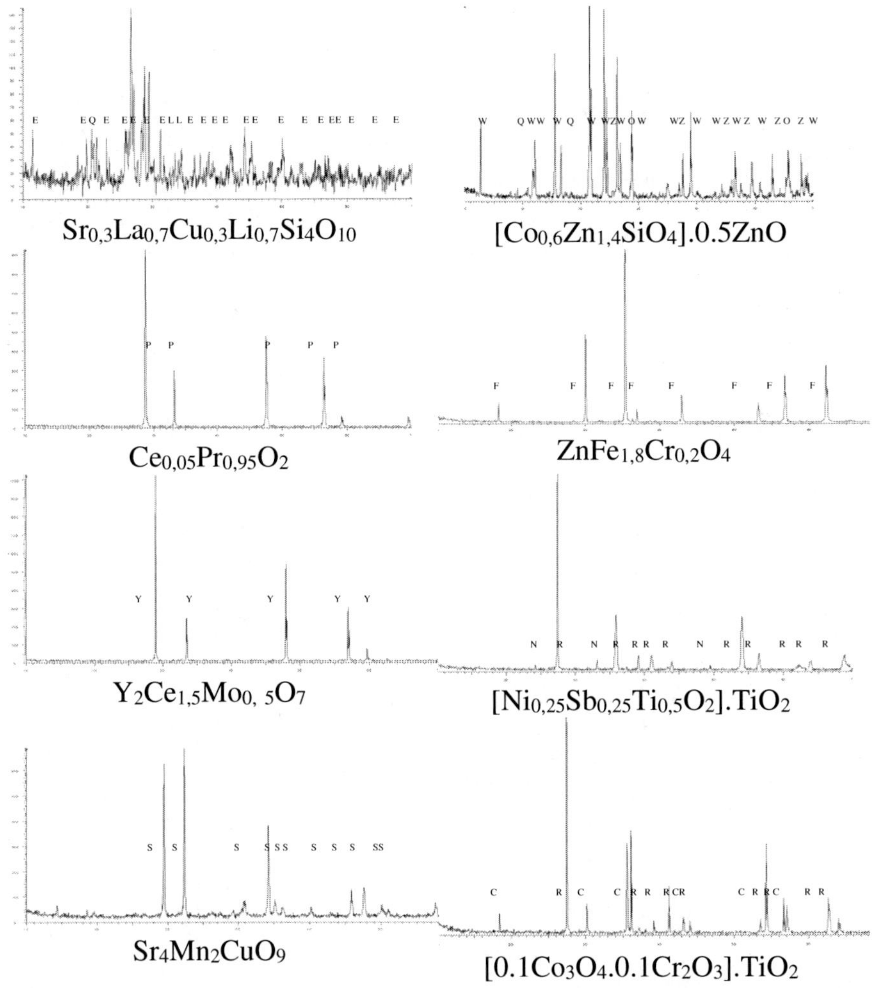

Figure 8.1. XRD patterns of powders. CRYSTALLYNE PHASES: E(wesselsite $SrCuSi_4O_{10}$), L (La_2O_3)W(willemite Zn_2SiO_4, Q(quartz),Z(ZnO),O(Co_3O_4), P(CeO_2) F(franklinite $ZnFe_2O_4$), Y($Y_2Ce_2O_7$), R(rutile), N($NiTiO_3$), S($Sr_4Mn_2CuO_9$), C($CoCr_2O_4$).

$Sr_{0,3}La_{0,7}Cu_{0,3}Li_{0,7}Si_4O_{10}$
950ºC/16h $R_{Vis}/R_{NIR}/R$=27/44/35

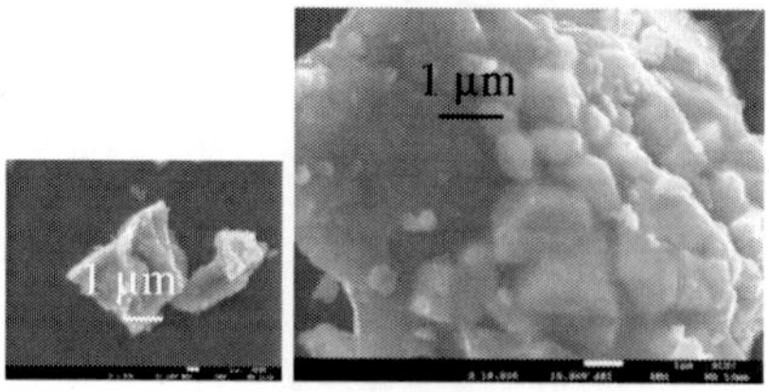

$Co_{0,6}Zn_{1,4}SiO_4.0,5ZnO$
1100ºC/3h $R_{Vis}/R_{NIR}/R$=13/30/21

$Ce_{0,05}Pr_{0,95}O_2$
1400ºC/6h $R_{Vis}/R_{NIR}/R$=17/65/40

$ZnFe_{1,8}Cr_{0,2}O_4$
1100ºC/3h $R_{Vis}/R_{NIR}/R$=14/48/30

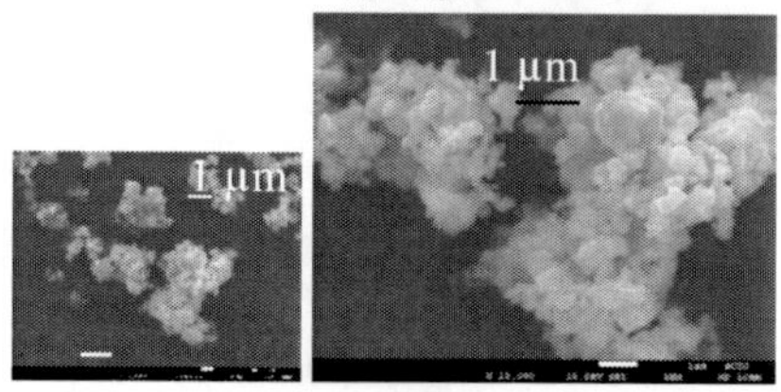

$Y_2Ce_{1,5}Mo_{0,5}O_7$
1500ºC/6h $R_{Vis}/R_{NIR}/R$=67/81/74

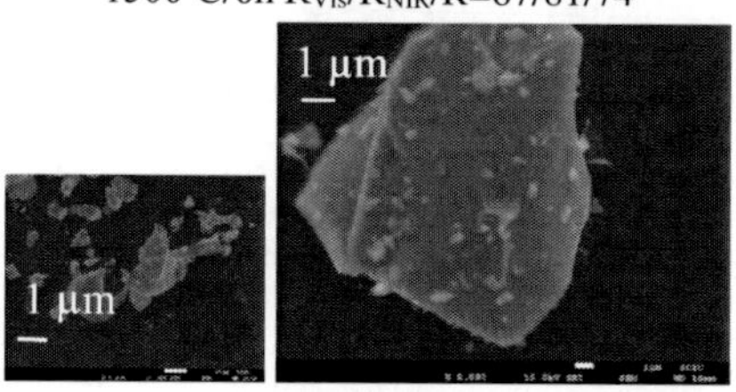

$Ni_{0,25}Sb_{0,25}Ti_{0,5}O_2.TiO_2$
1100ºC/3h $R_{Vis}/R_{NIR}/R$=40/55/47

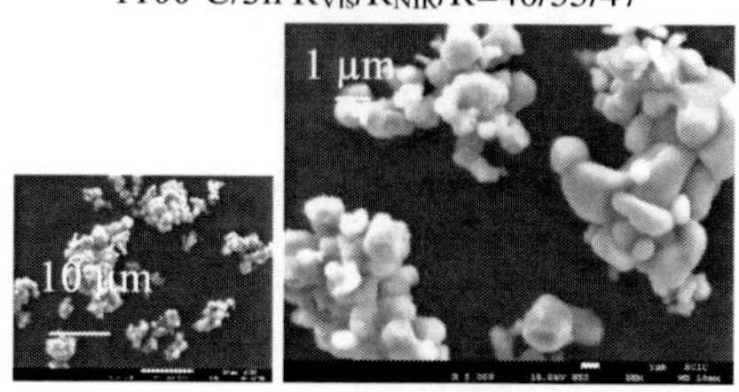

$Sr_4Mn_2CuO_9$
1000ºC/6h $R_{Vis}/R_{NIR}/R$=7/51/29

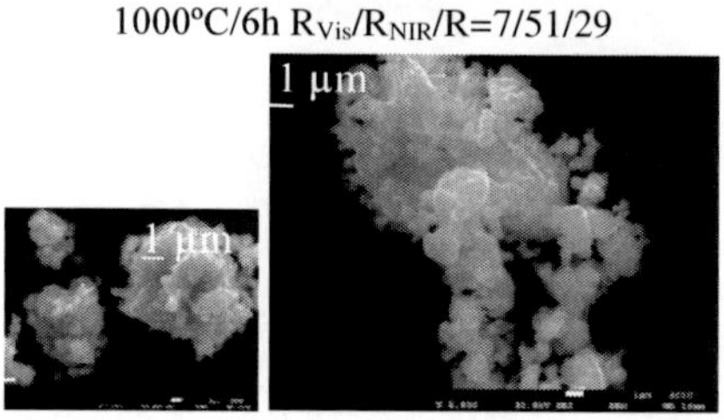

$0.1Co_3O_4.0.1Cr_2O_3.TiO_2$
1100ºC/3h $R_{Vis}/R_{NIR}/R$=7/21/14

Figure 8.2. SEM images of powders of the both families of pigments.

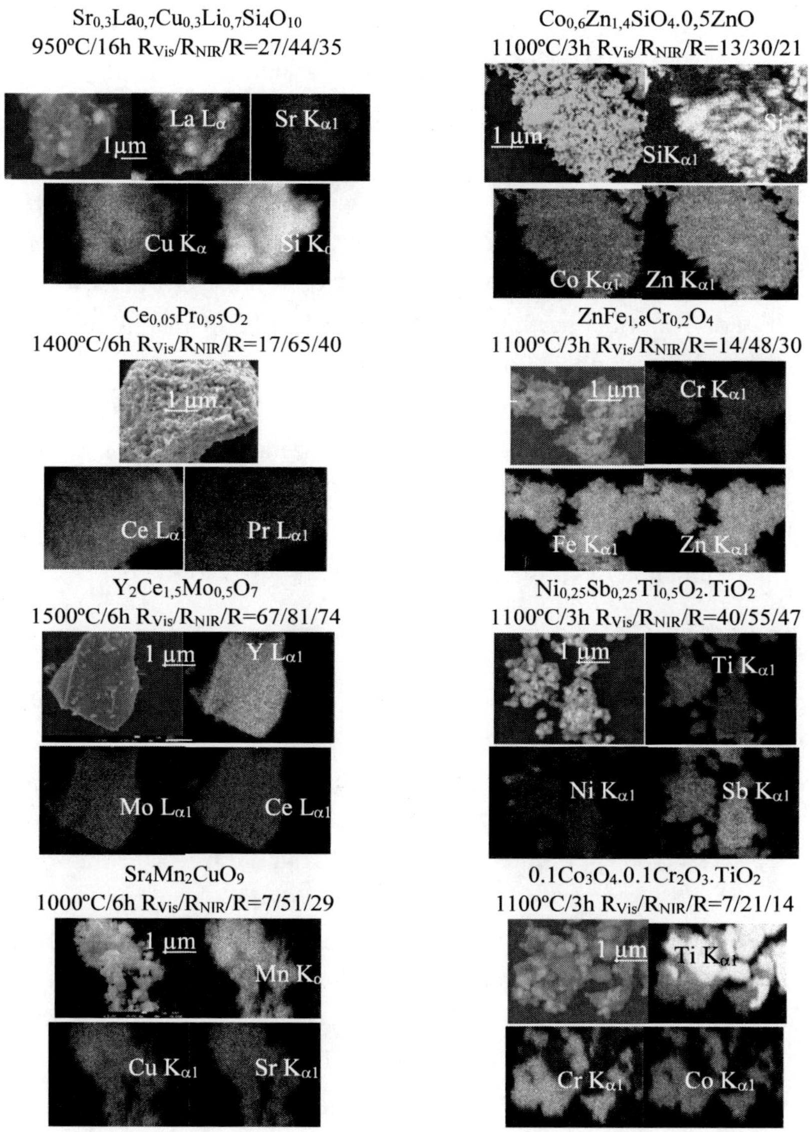

Figure 8.3. SEM and EDX mapping of powders of the families of pigments.

d) Finally, concerning the studied black pigments, the hexagonal perovskite $Sr_4Mn_2CuO_9$ (S) is the only crystalline phase detected for the $Sr_4Mn_2CuO_9$ black-brown pigment (Figure 8.1). Co-chromite $CoCr_2O_4$ (C) and intense peaks of rutile (R) are detected in the case of $Co(Cr,Co)_2O_4$-decorated rutile black green pigment. Regarding to morphology, homogeneous microstructures are observed by SEM images (Figure 8.2) in both cases; the $Sr_4Mn_2CuO_9$ black-brown pigment consists of aggregates between 2-10 μm integrated by fine spherical particles of 1 μm. In the case of $Co(Cr,Co)_2O_4$-decorated rutile black green pigment, fine aggregates between 1-5 μm in size are observed with two kinds of particles: i.e., cubic particles of 1 μm and spherical particles of 0.2 μm. SEM-EDX results in Figure 8.3 agree with the above observations, all cation EDX signals are homogeneous in the case of $Sr_4Mn_2CuO_9$ black-brown pigment, while the Ti $K_{\alpha 1}$ signal corroborates the presence of rutile phase associated with the largest cubic particles (around 1 μm) observed in SEM images in the case of $Co(Cr,Co)_2O_4$- rutile black green pigment.

8.1.2.2. UV-Vis-NIR Reflectance Spectra

The UV-Vis-NIR reflectance spectra of powders for both families of pigments are shown in Figure 8.4.

a) In the case of blue pigments, the structure of La,Li-$SrCuSi_4O_{10}$ wesselsite is composed of rings of four SiO_4 tetrahedra which are linked together by square-planar coordinated copper, thus forming double copper silicate layers which are joined by alkali-earth cations in a distorted cubic geometry. In the corresponding optical spectrum (Figure 8.4), three broad transitions are observed in the visible region originated from weakly-allowed d-d transitions of copper (despite a symmetry close to D_{4h} at the copper centre) due to vibronic coupling. The three transitions, observed as minimums of reflectance, occur at 800 nm ($^2B_{1g} \rightarrow ^2B_{2g}$), 630 nm ($^2B_{1g} \rightarrow ^2E_g$) and 540 nm ($^2B_{1g} \rightarrow ^2A_{1g}$) [17]. As for the classical Co-Zn_2SiO_4 willemite, its structure is an orthosilicate with all atoms in general position and composed by a framework of tetrahedra accommodating zinc and silicon in three different fourfold crystallographic sites; a deep blue color is obtained by doping the willemite structure with cobalt, which replaces zinc ions in tetrahedral positions. The advantage of willemite pigments comes from the lower amount of Co needed to obtain suitably saturated blue hues with respect to spinel $CoAl_2O_4$ or olivine; moreover, the use of an excess of ZnO protects the pigment from the attack of ZnO-rich glazes [18]. The blue color is associated with Co^{2+} in tetrahedral coordination environment, which shows a strong absorbance with three bands at 540, 590 and 640 nm (responsible for the color) associated to the split of ${}^4A_2(F) \rightarrow {}^4T_1(P)$ transition due to Jahn-Teller distortion. An additional strong and broad absorption appears at 1200-1600 nm (due to the coupling of ${}^4A_2(F) \rightarrow {}^4T_1(F)$ and ${}^4A_2(F) \rightarrow {}^4T_2(F)$ d-d transitions bands at 1400 nm and 1600 nm, respectively) (Chapter 4), which hinders NIR reflectance in this range of wavelength.

b) Regarding to the red-brown pigments, it has been discussed above that Pr^{4+} doping in Pr-CeO_2 introduces an additional conduction band, and its associated charge transference band absorbs from 610 nm (Figure 8.4) [9]. In the case of Cr-franklinite, Cr^{3+} enters replacing Fe^{3+} in the normal spinel $ZnFe_2O_5$, and the following absorption bands are detected in the UV-Vis-NIR reflectance spectra of fired powders: (a) intense bands at 300-520 nm, (b) weak band centered at 680 nm, (c) medium band at 780 nm, (d) intense band at 1200 nm. It is difficult to distinguish between the presences of Cr^{3+} or Fe^{3+} ions in octahedral sites in the UV-Vis-NIR spectra of samples containing simultaneously both ions in these positions, because its bands are overlapped to a high extent. However, the main characteristic of Fe^{3+} ions in octahedral sites is the strong absorption associated to the transitions ${}^6A_{1g}(S) \rightarrow {}^4A_{1g},{}^4E(G)$ which overlaps with the charge transference band at 400-500 nm that dominates the absorption spectrum of the powders. The bands at around 770 and 1200 nm, are characteristics of the zinc ferrite, which are attributed to ${}^6A_{1g}(S) \rightarrow {}^4T_{2g}(G)$ and $6A_{1g}(S) \rightarrow {}^4T_{1g}(S)$ for d-d electron transition of Fe^{3+}. These Fe^{3+} bands overlap with those of Cr^{3+} in octahedral coordination, namely: (a) three main parity-forbidden transitions ($^4A_2(^4F) \rightarrow {}^4T_2(^4F)$ at 570 nm and $^4A_2(^4F) \rightarrow {}^4T_1(^4F)$ at 445 nm, which are partially overlapped, and $^4A_2(^4F) \rightarrow {}^4T_2(^4P)$ at 235 nm, which overlaps with $Cr^{3+} \leftrightarrow O^{2-}$ charge transfer), (b) two weak Cr^{3+} spin-forbidden transitions ($^4A_2(^4F) \rightarrow {}^2T_1(^2G)$ and $^4A_2(^4F) \rightarrow {}^2E(^2G)$ that overlap [19, 20].

c) As for the yellow pigments, in the case Mo-$Y_2Ce_2O_7$ solid solutions the doping of Mo^{6+} replacing Ce^{4+} in $Y_2Ce_2O_7$ needs the formation of cation vacancies ($Y_2Mo_xCe_{2-x}O_{7+x}$) in order to satisfy the lattice charge balance [10]; Mo^{6+} doping introduces an additional conduction band and its associated charge transference band O_{2p}–Mo_{4d} absorbs from 505 nm (band gap of 2.5 eV), inducing a yellow color. All wavelengths above 505 nm are reflected, except the weak absorption due to 4f transitions at 1900 nm. In the case of Ni,Sb-TiO_2 decorated rutile yellow pigment, the following spin-allowed bands of Ni^{2+} in octahedral coordination can be observed in Figure 8.4: $^3A_{2g}(3F) \rightarrow {}^3T_{2g}(3F)$ at 1200 nm, $^3A_{2g}(3F) \rightarrow {}^3T_{1g}(3F)$ at 800 nm (which overlaps with other spin-forbidden bands) and $^3A_{2g}(3F) \rightarrow {}^3T_{1g}(3P)$ at 450 nm (which overlaps with $Ni^{2+} \leftrightarrow O^{2-}$ band transfer) [21].

d) Concerning to black pigments; in the case of $Sr_4Mn_2CuO_9$ black-brown pigment, the broad optical absorption band at the shorter wavelength (400 nm) is attributed to the partial overlap of charge transfer transition and $^4A_{2g} \rightarrow {}^4T_{1g}$ d-d transition of Mn^{4+} in octahedral coordination. The above described weak transitions from weakly-allowed d-d transitions of copper, that fill the center of the square faces of the trigonal prisms, can be observed as minimums of reflectance at a at 800 nm ($^2B_{1g} \rightarrow {}^2B_{2g}$, very weak band), 715 nm ($^2B_{1g} \rightarrow {}^2E_g$) and 540 nm ($^2B_{1g} \rightarrow {}^2A_{1g}$) [8,17], but the band at 715 nm is associated by Bae at al. [22] to $^2T_2 \rightarrow {}^2E$ transition of Cu^{2+} in a trigonally disordered tetrahedral crystal field [23]. In the case of Co,Cr decorated rutile black green pigment, the absorption bands of Co^{2+} in tetrahedral coordination and Co^{2+} in octahedral coordination in the spinel are overlapped, and the following absorption bands are detected: a charge transfer band at 350 nm, a band at 450 nm (associated to ${}^4A_2({}^4F) \rightarrow {}^4T_1({}^4F)$ of Cr^{3+} in octahedral coordination), a strong absorbance with three bands a 540, 590 and 640 nm (associated to the splitting of ${}^4A_2(F) \rightarrow {}^4T_1(P)$ transition of Co^{2+} in tetrahedral coordination, due to Jahn-Teller distortion), and a strong absorption at 1200-1600 nm range (due to the coupling of ${}^4A_2(F) \rightarrow {}^4T_1(F)$ and ${}^4A_2(F) \rightarrow {}^4T_2(F)$ d-d transitions bands in tetrahedral Co^{2+}) [18, 19].

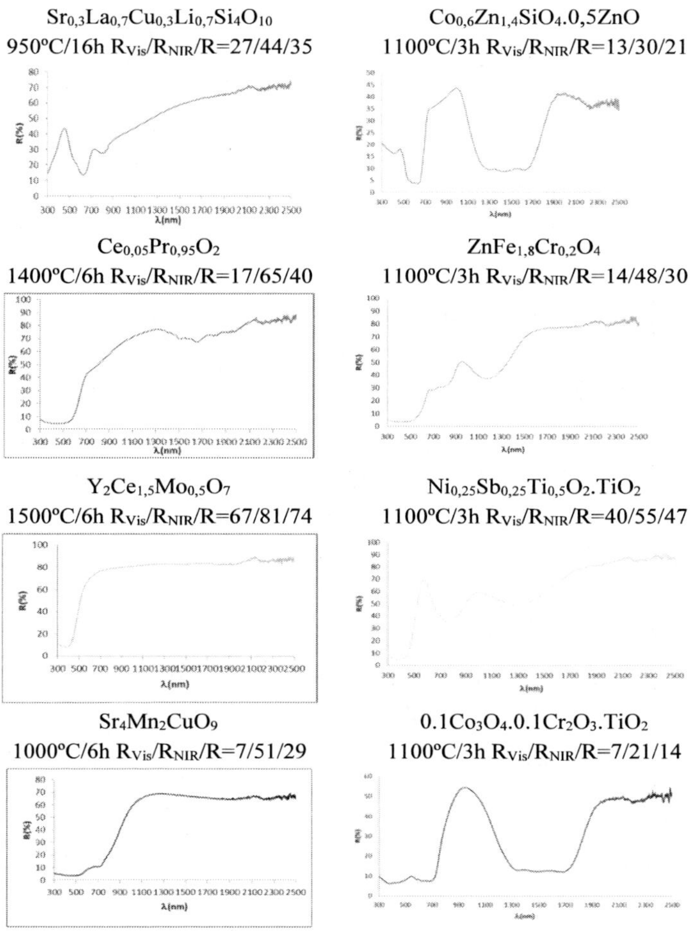

Figure 8.4. UV-Vis-NIR Reflectance spectra of powders of the both families of pigments.

*8.1.2.3. Color, Colorimetry (L*a*b* and C*h*) and Reflectance Parameters ($R_{Vis}/R_{NIR}/R$) of the Powders*

The image of powders, showing their color and its colorimetry (L*a*b* and C*h*) and reflectance parameters ($R_{Vis}/R_{NIR}/R$) are collected in Figure 8.5. Considering the rules described above it can be pointed out:

a) As for blue pigments the chroma C* is higher for the classical Co-willemite, and in both cases it is very similar to its b* value. The lightness L* of the rare earth pigment is high and falls out of the blue range (L*=25-35), but its R_{NIR} is higher. For the comparison of the performance as cool pigments, $CaCO_3$ was added as binder to the pigments and the evolution of colorimetry an reflectance data are shown in Table 8.1; likewise, the representation of R_{NIR} versus lightness L* of the blue pigments with gradual addition of the binder $CaCO_3$ is shown in Figure 8.6. According to this Figure 8.6 it is observed that rare earth pigment shows always higher L* and R_{NIR} values, and the values of both pigments (classical and rare earth-based) follow a linear relationship with R^2 value > 0.92. Therefore the $R_{NIR}30$ value (reflectance associated to a L*=30, which is proposed for blue hue evaluation) may be estimated and it is 30% for willemite and 7% for wesselsite blue. Indeed, for the blue range of chromaticity, the Co-willemite shows better performance as cool pigment.

b) In the case of red-brown pigments, the chroma C* is higher for Cr-franklinite (C*=30.2 versus 21.7 for Pr-cerianite), being this value closer to its absolute value of a* in the case of Cr-franklinite spinel pigment. The L* value is higher for Pr-cerianite (50.6 vs 36.2 for Cr-franklinite) and then shows higher R_{NIR} (65% vs. 48% for Cr-franklinite). The evolution of colorimetry and reflectance data with the gradual addition of $CaCO_3$ binder to the pigments is shown in Table 8.2 and the representation of R_{NIR} versus lightness L* follows a linear relationship in Figure 8.7. The $R_{NIR}55$ (reflectance associated to a L*=55, considered for red hues) is 69% for rare-earth Pr-cerianite pigment and 59% for classical Cr-franklinite spinel pigment. Therefore Pr-cerianite powder shows better performance as cool pigment.

c) Regarding to yellow pigments, the chroma C* is higher for the Ni,Sb-rutile pigment (C*=54.3 versus 45.3 for Mo-cerate pigment), and the lightness L* is similar for both yellow pigments and included in the proposed yellow range in rule 2 (L*=80-90). The Mo-cerate yellow powder shows higher R_{NIR} (81% versus 55% for Ni,Sb-rutile). The evolution of colorimetry an reflectance data with the gradual addition of $CaCO_3$ binder to the pigments are shown in Table 8.3 and the representation of R_{NIR} versus lightness L* follows a linear relationship in Figure 8.8. The $R_{NIR}85$ (reflectance associated to a L*=85, proposed for yellow hues in rule 4) is 80% for rare-earth Mo-cerate pigment and 63% for classical Ni,Sb-TiO_2 decorated rutile pigment. Therefore Mo-cerate shows better performance as cool pigment.

Table 8.1. Colorimetry and reflectance parameter for $CaCO_3$ addition to blue pigments

$Sr_{0.3}La_{0.7}Cu_{0.3}Li_{0.7}Si_4O_{10}$ 950°C/10h	R_{VIS} (%)	R_{NIR} (%)	R (%)	L a*b*
0% $CaCO_3$	26.6	44.3	35.2	62.3/-4.4/-19.9
20% $CaCO_3$	45.1	60.6	52.0	74.1/-5.1/-13.2
38.5% $CaCO_3$	53.9	66.0	59.3	79.6/-4.5/-8.4
50% $CaCO_3$	56.1	66.5	60.7	81.1/-4.1/-6.4
$[Co_{0.6}Zn_{1.4}SiO_4].0.2ZnO$ 1100°C/3h	R_{VIS} (%)	R_{NIR} (%)	R (%)	L a*b*
0% $CaCO_3$	13.1	29.9	21.3	30.1/-3.8/-24.6
20% $CaCO_3$	17.9	33.8	24.6	49.7/-3.9/-14.7
38.5% $CaCO_3$	21.3	35.2	27.2	53.2/-4.3/-13.9
50% $CaCO_3$	24.2	37.5	29.8	56.4/-4.1/-13.1

$Sr_{0.3}La_{0.7}Cu_{0.3}Li_{0.7}Si_4O_{10}$ 950ºC/16h	$Ce_{0,95}Pr_{0,05}O_2$ 1400ºC/6h	$Y_2Ce_{1,5}Mo_{0,5}O_7$ 1500ºC/6h	$Sr_4Mn_2CuO_9$ 1000ºC/6h
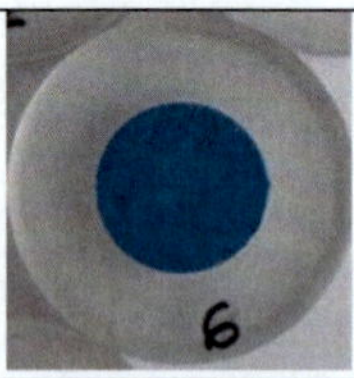	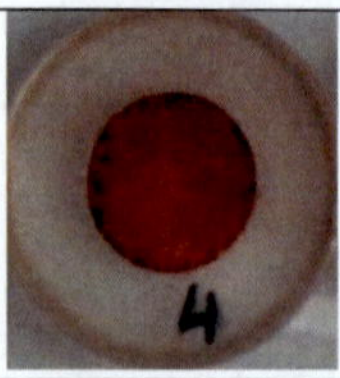		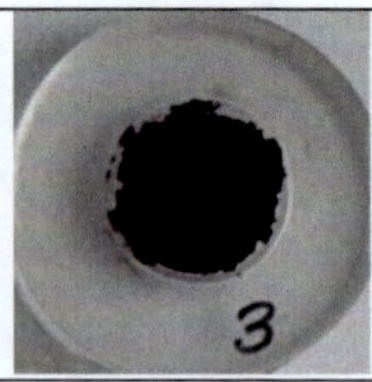
L*a*b*=62.3/-4.4/-19.9	50.6/17.5/12,8	85.2/0.2/45.3	43.9/7.2/6.3
C*h*=20.4/257.5	21.7/36.2	45.3/89.8	9.6/41.2
$R_{Vis}/R_{NIR}/R$=27/44/35	17/65/40	67/81/74	7/51/29
$R_{NIR}30$=7	$R_{NIR}55$=69	$R_{NIR}85$=80	$R_{NIR}30$=50
$[Co_{0.6}Zn_{1.4}SiO_4].0.2ZnO$ 1100ºC/3h	$ZnFe_{1,8}Cr_{0,2}O_4$ 1100ºC/3h	$Ni_{0,25}Sb_{0,25}Ti_{0,5}O_2.TiO_2$ 1100ºC/3h	$[0.1Co_3O_4.0.1Cr_2O_3].TiO_2$ 1100ºC/3h
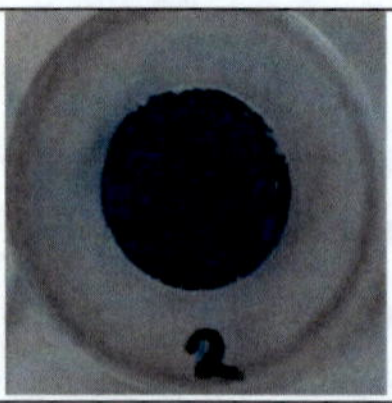		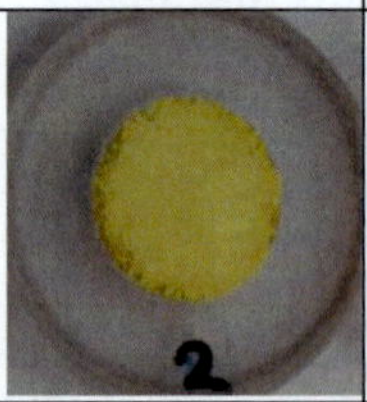	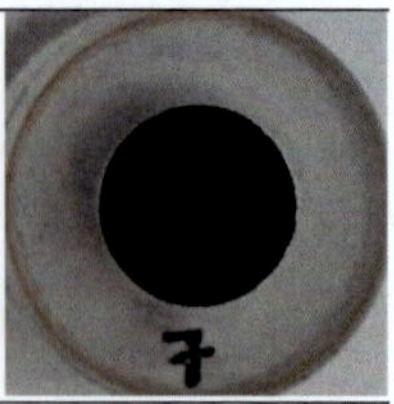
L*a*b*=30.1/-3.8/-24.6	36.2/22.1/20.6	84.3/-4.3/54.1	49.3/-2.9/5.8
C*h*=24.9/261.2	30.2/43	54.3/94.5	6.5/116.6
$R_{Vis}/R_{NIR}/R$=13/30/21	14/48/30	40/55/47	7/21/14
$R_{NIR}30$=30	$R_{NIR}55$=59	$R_{NIR}85$=63	$R_{NIR}30$=13

Figure 8.5. Image of the powders, color and its colorimetry (L*a*b* and C*h*) and reflectance parameters ($R_{Vis}/R_{NIR}/R$).

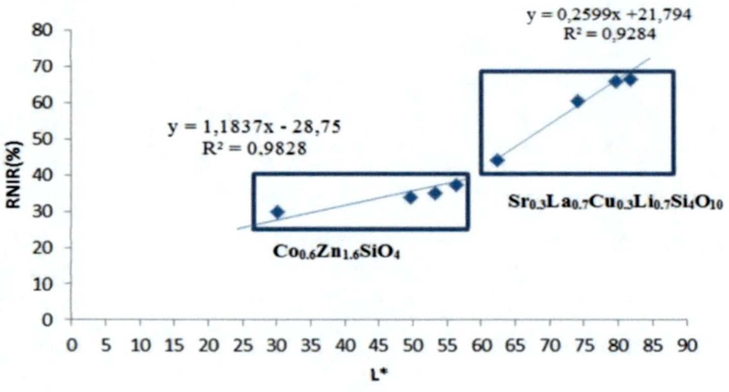

Figure 8.6. R_{NIR} versus lightness L* of blue pigments with gradual addition of the $CaCO_3$ binder.

Table 8.2. Colorimetry and reflectance parameters for $CaCO_3$ addition to red-brown pigments

$Ce_{0,95}Pr_{0,05}O_2$ 1400ºC/6h	R_{VIS} (%)	R_{NIR} (%)	R (%)	L *a*b*
0% $CaCO_3$	17	65	40	50.6/17.5/12.8
20% $CaCO_3$	25.1	71.1	45.1	54.6/14.1/9.3
38.5% $CaCO_3$	30.1	70.0	47.4	60/13.3/9.0
50% $CaCO_3$	32.6	70.3	49.0	62.0/13.0/9.0
60% $CaCO_3$	37.9	73.1	53.1	61.7/11.7/8.1
80% $CaCO_3$	48.3	75.3	60.0	70.8/8.1/6.4
95% $CaCO_3$	63.8	80.0	70.7	81.1/4.4/5.6
100% $CaCO_3$	79.3	83.0	80.8	88.2/0.8/7.4
$[ZnFe_{1.8}Cr_{0.2}O_4].ZnO$ 1100ºC/3h	R_{VIS} (%)	R_{NIR} (%)	R (%)	L*a*b*
0% $CaCO_3$	14.0	48.1	30,2	36.2/22.1/20.6
20% $CaCO_3$	18.2	46.0	29.7	41.5/19.0/12.6
38.5% $CaCO_3$	19.1	46.6	30.7	42.1/19.2/13.0
50% $CaCO_3$	19.8	47.7	31.9	45.2/18.7/10.1
60% $CaCO_3$	23.1	52.0	36.1	43.2/19.0/11.1
80% $CaCO_3$	29.2	56.1	43.0	48.1/17.2/12.5
95% $CaCO_3$	49.0	64.2	55,4	59.3/11.2.3/13.0
100% $CaCO_3$	79.3	83.0	80.8	88.2/0.8/7.4

Table 8.3. Colorimetry and reflectance parameters for $CaCO_3$ addition to yellow pigments

$Y_2Ce_{1.5}Mo_{0.5}O_7$ 1500ºC/6h	R_{VIS} (%)	R_{NIR} (%)	R (%)	L* a*b*
0% $CaCO_3$	54.0	81.1	67.3	85.2/0.2/45.3
20% $CaCO_3$	66.7	80.4	72.20	87.13/-2.60/30.71
38,5% $CaCO_3$	68.11	82.7	74.40	87.56/-3.12/24.73
50% $CaCO_3$	67.26	87.0	75.76	88.8/-3.02/19.66
100% $CaCO_3$	79.26	82.92	80.83	88.23/0.83/7.37
$[Ni_{0.25}Sb_{0.25}Ti_{0.5}O_2].TiO_2$ 1100ºC/3h	R_{VIS} (%)	R_{NIR} (%)	R (%)	L a*b*
0% $CaCO_3$	39.6	54.7	47	84.28/-4.29/54.05
20% $CaCO_3$	44.11	57.9	50.23	84.1/-4.05/44.50
38,5% $CaCO_3$	45.59	58.6	51.36	84.2/-3.97/43.83
50% $CaCO_3$	46.90	59.5	52.48	84.4/-3.98/41.78
100% $CaCO_3$	7.26	82.92	80.83	88.23/0.83/7.37

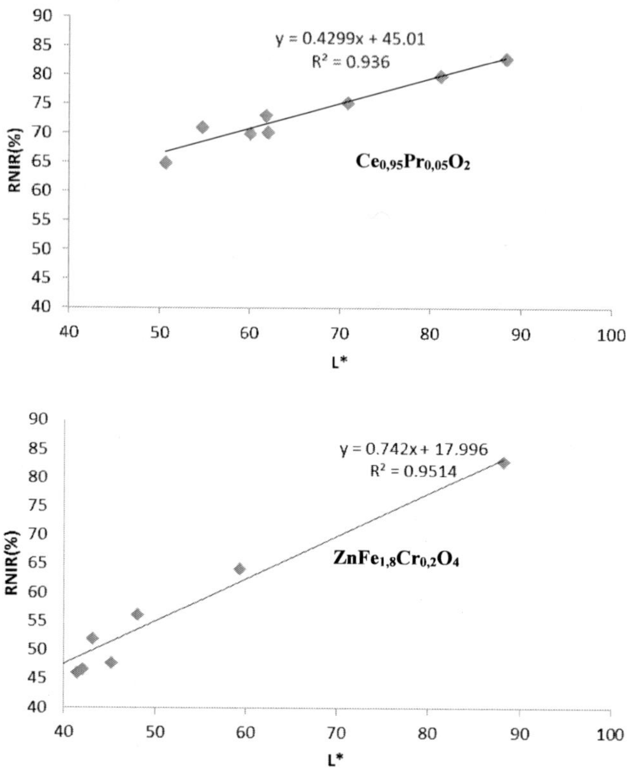

Figure 8.7. R_{NIR} versus lightness L* of red-brown pigments with gradual addition of the $CaCO_3$ binder.

d) Concerning to black pigments, from the data on Figure 8.5 the chroma C* is lower for chromite-rutile black-green (C*=6.5 versus 9.6 for $Sr_4Mn_2CuO_9$) indicating a better black color for the wesselsite pigment, and the hue is on the green range for chromite-rutile black-green (h*=116.6), while the hue of black-brown $Sr_4Mn_2CuO_9$ falls into the brown range (h*=41.2). Lightness is low and similar for both dark pigments (44-50), although it is high for pure black pigments, since they do not enter into the range proposed on rule 3 (L*=30-0). Moreover, the $Sr_4Mn_2CuO_9$ black-brown shows higher R_{NIR} (51% versus 21% for chromite-rutile black-green). Finally the evolution of colorimetry and reflectance data with the gradual addition of $CaCO_3$ binder to the pigments are shown in Table 8.4, and the representation of R_{NIR} versus lightness L* follows a linear relationship in Figure 8.9. The $R_{NIR}30$ (reflectance associated to L*=30, considered for black hues in rule 4) is 50% for $Sr_4Mn_2CuO_9$ black-brown pigment and 13% for chromite-rutile black-green. Therefore, $Sr_4Mn_2CuO_9$ shows better performance as cool pigment.

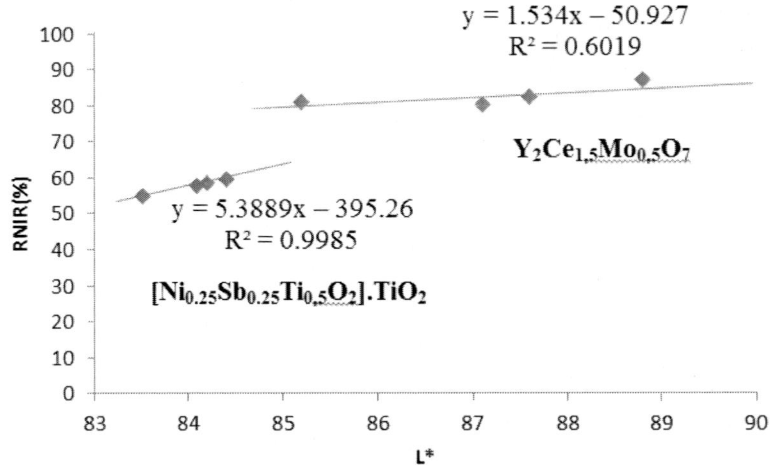

Figure 8.8. R_{NIR} versus lightness L* of yellow pigments with gradual addition of the $CaCO_3$ binder.

8.1.2.4. Application of the Pigments: Stability and Color Yield in Different Media

As above discussed, a pigment should be stable and exhibit an acceptable colorimetric performance when added (and thermally treated) to different media such as paints, glazes or ceramics. The coloring behavior of the pigments was studied in three different matrices: (a) 10 wt.% in colorless alkyd paint, (b) 3 wt.% in a conventional double-firing frit of the SiO_2-CaO-ZnO system (1050 ºC), and (c) 2% wt.% in porcelain powder (1190ºC). From this study we can use a modified CPMA classification (see section 3.1.1) based on:

- Category I: pigment of high temperature, stable in a, b and c applications.
- Category II: pigment of low temperature, stable only in a and b applications.
- Category III: pigment thermally unstable, stable only in application a.

The L*a*b* values of the different applications of the pigments are showed in Table 8.5, while its reflectance spectra and reflectance evaluation (R_{Vis}/R_{NIR}/R) are given in Figure 8.10 for rare earth pigments, and in Figure 8.11 for the classical cool pigments.

Table 8.4. Colorimetry and reflectance parameters for $CaCO_3$ binder addition to black pigments

$Sr_4Mn_2CuO_9$ 1000ºC/6h	R_{VIS} (%)	R_{NIR} (%)	R (%)	L *a*b*
0% $CaCO_3$	7	50.9	28.5	43.91/7.15/6.29
20% $CaCO_3$	10.16	51.0	27.63	47.79/6.32/5.45
38.5% $CaCO_3$	12.38	51.1	28.83	50.57/5.98/5.08
50% $CaCO_3$	14.63	51.3	30.76	52.77/5.85/5.21
[$0.1Co_3O_4.0.1Cr_2O_3$].TiO_2 1100ºC/3h	R_{VIS} (%)	R_{NIR} (%)	R (%)	L* a*b*
0% $CaCO_3$	7	21	14	49.3/-2.9/5.8
20% $CaCO_3$	9	20	14.5	50.1/-2.5/5.0
38.5% $CaCO_3$	10	21	15.5	51.6/-2.3/4.9
50% $CaCO_3$	11	22	16.5	52.7/-2.2/4.1

a) As for the blue pigments, the application of La,Li-$SrCuSi_4O_{10}$ wesselsite develops blue hues in alkyd paint (b*= -3.0) and in double firing frit (b*= -5.6), but it is colorless in porcelain powder application. In agreement to the color, the UV-Vis-NIR spectra of the applications (Figure 8.10) show similar features to the powder spectrum in the case of alkyd paint, indicating its stability in the paint. In the double firing frit application, only a strong absorption band centered at 750 nm is observed, which decreases the R_{NIR} dramatically (46%), although its L* is only 80.5). This band is associated to Cu^{2+} ions solved in the alkaline glaze and located in high distorted octahedral sites (due to Jahn Teller distortion) [24], showing that the pigment is not stable in the glaze. Finally, no bands are detected in the application in porcelain powder, also indicating that the pigment is not stable in this medium. In the case of classical Co-willemite decorated ZnO, the blue color is obtained in the three applications (Table 8.5) and the features of the spectra are similar in all cases (Figure 8.11), maintaining the spectrum of the powder associated to tetrahedral Co^{2+}. Nevertheless, in the case of double firing frit the absorption at 1200-1600 nm range [25] due to the coupling of ${}^4A_2(F) \rightarrow {}^4T_1(F)$ and ${}^4A_2(F) \rightarrow {}^4T_2(F)$) shows an unusual low intensity and the NIR reflectance increases in this application; this behavior has been observed in nanopowders of rare earth doped cobalt aluminate [25]. Following the proposed classification, La,Li-$SrCuSi_4O_{10}$ wesselsite pigment belongs to Category II and Co-willemite decorated ZnO to Category I.

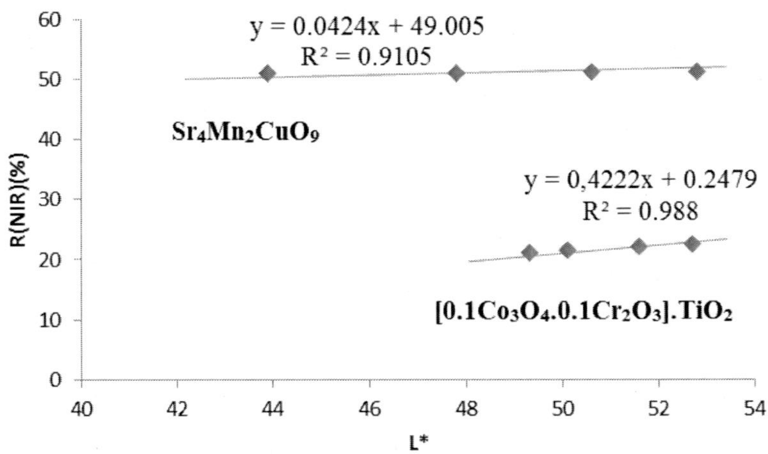

Figure 8.9. R_{NIR} versus lightness L* of black pigments with gradual addition of the $CaCO_3$ binder.

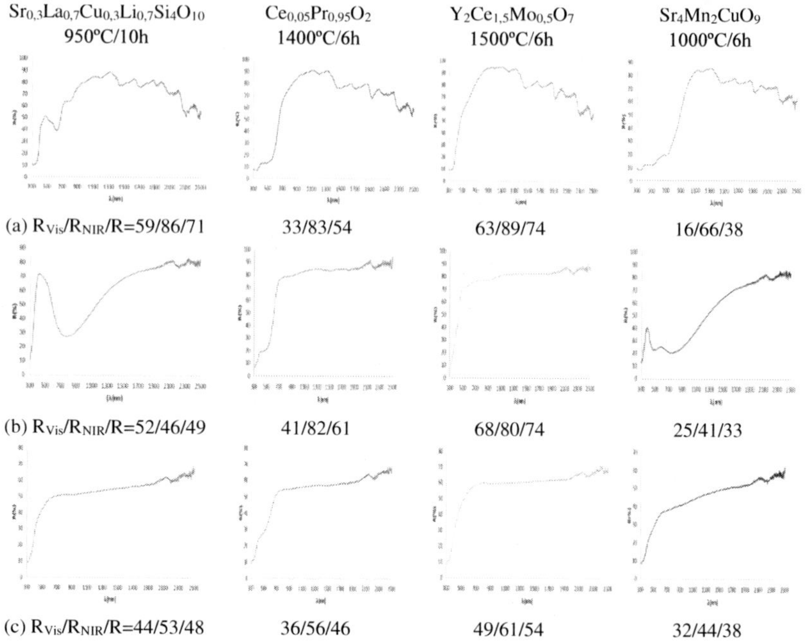

Figure 8.10. UV-Vis-NIR reflectance spectra of rare earth pigment applications: (a) 10 wt.% in colorless alkyd paint, (b) 3 wt.% in double firing frit (1000ºC), (c) 2 wt.% in porcelain powder (1190ºC).

b) In the case of red-brown pigments, the color of Pr-cerianite is maintained in the three applications, with a* values around 19 in the alkyd paint and double firing frit, although in porcelain powder a* value decreases to 6.8 (Table 8.5). The optical spectrum of the powder is maintained in all applications (Figure 8.10), and only in alkyd paint the spectrum shows weak absorptions in the NIR range, which can be observed always associated in all the applications with the alkyd resin. Concerning the classical Cr-franklinite spinel pigment, the red brown color is maintained in alkyd paint and in double firing frit (a* between 18 and 20), but it darkens in the porcelain powder application. The reflectance spectra of alkyd and double firing frit applications indicate the stability of the pigment (Figure 8.11), showing similar features to the powder spectrum; instead, the absorption bands observed at 300 and 500 nm produce the darkening of the sample of the porcelain powder application, moreover, a band at 900 nm and a strong absorption between 1200-1700 diminish the NIR reflectance of this application, probably associated to the decomposition of franklinite and partial reduction of iron (III) to iron (II). Therefore Pr-cerianite belongs to Category I and Cr-franklinite spinel to Category II, respectively.

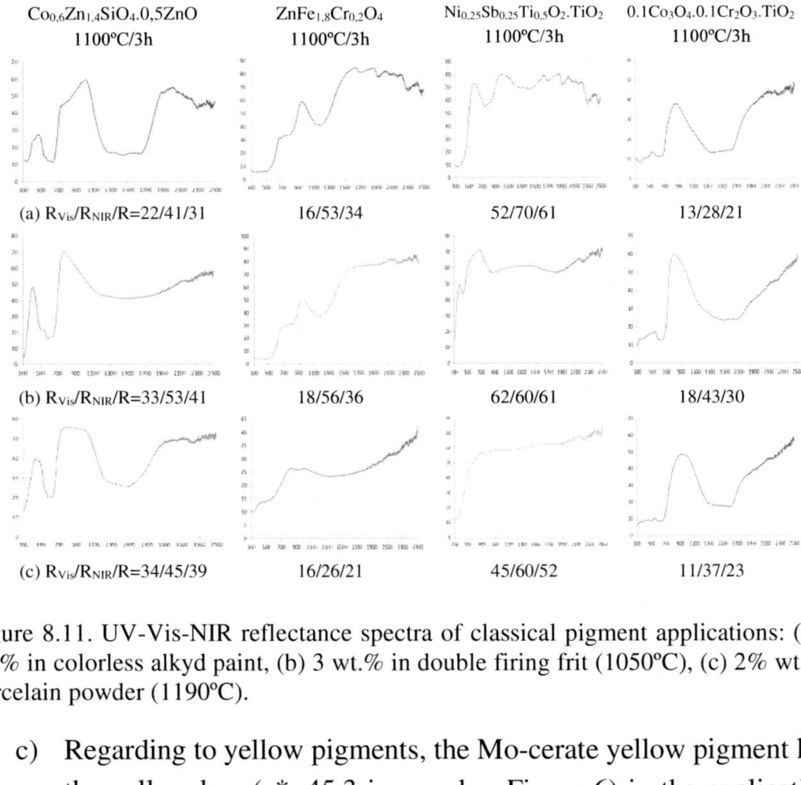

Figure 8.11. UV-Vis-NIR reflectance spectra of classical pigment applications: (a) 10 wt.% in colorless alkyd paint, (b) 3 wt.% in double firing frit (1050ºC), (c) 2% wt.% in porcelain powder (1190ºC).

c) Regarding to yellow pigments, the Mo-cerate yellow pigment loses the yellow hue (a*=45.3 in powder, Figure 6) in the applications, and only in alkyd paint shows an indifferent b*=18.9 (Table 8.5). The UV-Vis-NIR reflectance spectra of the applications (Figure 8.10) only indicate moderate absorption at 505 nm in the alkyd application, while in the case of double firing frit and porcelain powder there is no evidence of absorption at 500 nm, indicating that the pigment is not stable in these media. The Ni,Sb-rutile pigment, which shows b*=54.1 in powder (Figure 8.5), maintains b*=48 and

L*=81.8 in alkyd paint (into the range of high chromaticity for yellow), and even higher b* value in the double firing frit (52.6). However, b* decreases to only 17.4 and L* to 79.0 (out of the range) for porcelain powder indicating that the pigment is not stable and this application is colorless. The UV-Vis-NIR reflectance spectra of the applications (Figure 8.11) indicate similar features in the case of alkyd paint, while the spectrum is different for the double firing application, showing low absorption in UV range and a broadening of the bands associated with Ni^{2+} in octahedral coordination. In the case of porcelain powder, no absorption bands are detected. From the proposed classification, Mo-cerate belongs to Category III and Ni,Sb-rutile pigment to Category II of pigments, respectively.

d) Concerning the black pigments, the $Sr_4Mn_2CuO_9$ black-brown pigment maintains the color in alkyd paint and double firing frit, while only a light darkening is observed in the case of porcelain powder. The corresponding UV-Vis-NIR reflectance spectra of the applications (Figure 8.10) indicate a similar spectrum to the powder in alkyd paint, while in the double firing application shows a broadening of the absorption bands of Mn^{4+} in octahedral coordination and lower intensity of reflectance (particularly the band at 715 nm associated with $^4A_2(^4F) \rightarrow {}^4T_2(^4F)$ transition), that implies a decrease of reflectance in the 700-1000 nm, and thus decreasing its R_{NIR} performance. As for the application in porcelain powder, only a light absorption in the 700-900 nm range is observed, indicating the low stability of the black-brown pigment in this medium. In the case of Co,Cr decorated rutile black green pigment, the color is maintained in all applications, with UV-Vis-NIR spectra similar to the powder spectrum, associated with Co^{2+} in tetrahedral coordination and Cr^{3+} in octahedral coordination; This confirms the stabilization of the pigment in the three applications, which also show a R_{NIR} higher than the powder. Therefore, following the proposed classification $Sr_4Mn_2CuO_9$ black-brown belongs to Category II and Co,Cr decorated rutile black green to Category I of pigments, respectively.

In all applications the NIR reflectance increases with respect to powder, especially in alkyd paint, but this increase, as it is discussed above, does not necessarily prove a good performance as cool pigment, because in some cases the lightness and chromaticity of the color is out of range.

In summary the high NIR reflectance of colored powders is a necessary, but not sufficient condition to be considered as cool pigments with good performance. Usually, a satisfactory color evaluation of colored powders by a* or b* parameters, along with a high reflectance in the NIR range, measured by the UV-Vis-NIR (300-2500 nm) diffuse reflectance spectroscopy of powder, are considered as sufficient arguments to infer its good performance as cool pigments. However, this assumption may be erroneous sometimes. As a guidance to avoid this confusion, a simple method is proposed for a valid evaluation of the performance as cool pigments of powders, based on the rules established above.

The application of the rules to the proposed pigments (a palette of classical pigments and a palette based on rare earths) shows that in the case of the blue pigments, although the R_{NIR} of the powders are 44% and 30% for La,Li-wesselsite and Co-willemite respectively, the application of rule 1 by gradual addition of $CaCO_3$ as binder indicates that normalized $R_{NIR}30$ (NIR reflectance for mixture L*=30) are 7% and 30% respectively, thus having an inverse result. Regarding to red-brown pigments, Pr-cerianite shows the higher $R_{NIR}30$ (69% versus 59% for the classical Cr-franklinite spinel) but the difference is lower than for R_{NIR} direct measurements (65% for Pr-cerianite and 48% for Cr-franklinite). In the case of yellows, the rare earth pigment based on Mo-$Y_2Ce_2O_7$ ytrium cerate shows higher $R_{NIR}85$ (80%, and 63% for the Yellow of Ni,Sb-TiO_2) but the difference is also lower than for direct R_{NIR} measurements (81% and 55% respectively). Finally in the case of the black pigments, the black-brown of $Sr_4Mn_2CuO_9$ shows higher normalized $R_{NIR}30$ (50%, and 13% for classical $Co(Cr_{2-x}Co_x)O_4$-decorated TiO_2), showing in this case that the difference is greater than for direct R_{NIR} values (51% and 21% respectively). Finally, the pigmenting behavior and stability in different media is checked in order to evaluate the performance as cool pigments. Three applications of the pigments were carried out: in alkyd paint (10 wt.%), in double firing frit (3 wt.%, 1050ºC) and in porcelain powder (2 wt.%, 1190°C). If the pigment shows stability in the three media it is classified I, if it is only stable in alkyd paint an double firing glaze is classified II, and if it is only stable in alkyd paint its classification is III. The results show that the stability of the non-classical cool pigments (all II except Pr-cerianite, classified I, and Mo-cerate yellow classified III) is lower than that of classical pigments (Category I for blue and black, and Category II for yellow and brown pigments), except in the case of Pr-cerianite red brown pigment.

Table 8.5. L*a*b* values for the applications of pigments: (a) 10 wt.% in colorless alkyd paint, (b) 3 wt.% in double firing frit (1000°C), (c) 2% wt.% in porcelain powder (1190°C)

	$Sr_{0.3}La_{0.7}Cu_{0.3}Li_{0.7}Si_4O_{10}$ 950°C/10h	$Ce_{0.05}Pr_{0.95}O_2$ 1400°C/6h	$Y_2Ce_{1.5}Mo_{0.5}O_7$ 1500°C/6h	$Sr_4Mn_2CuO_9$ 1000°C/6h
(a)	72.4/-3.8/-3.0	51.6/19.9/15.5	81.3/0.4/18.9	46.6/7.8/7.5
(b)	80.5/-7.0/-5.6	65.6/18.2/20,0	87.6/-1.8/11.7	57.9/2.8/-0.2
(c)	76.8/1.2/9.8	70.3/6.8/11.4	79.2/1.5/11.5	73.4/1.7/11.6
	$[Co_{0.6}Zn_{1.4}SiO_4].0.2ZnO$ 1100°C/3h	$[ZnFe_{1.8}Cr_{0.2}O_4].ZnO$ 1100°C/3h	$[Ni_{0.25}Sb_{0.25}Ti_{0.5}O_2]$ TiO_2 1100°C/3h	$[0.1Co_3O_4.0.1Cr_2O_3]$ TiO_2 1100°C/3h
(a)	47.5/-6.2/-15.0	39.39/20.22/18.92	81.81/-4.41/48.03	39.6/-3.4/4.3
(b)	56.2/5.5/-21.1	48.5/18.1/17.9	78.7/-0.7/52.6	51.4/-5.2/1.0
(c)	64.8/-3.4/-13.1	46.3/3.4/4.7	79.0/-0.6/17.4	64.6/0.8/15.2

8.2. Evaluation as Cool and Photocatalytic Pigments

Although some commercial photocatalytic products show interesting efficiencies under controlled laboratory conditions (e.g., under ISO 22197-1 or the Italian UNI 11247: 2007 standard conditions), used to study their NO_x removal capacity, another very different issue is that their effects translate into measurable changes in air quality in the areas where these products can be applied. For a photocatalytic product applied on a horizontal surface in an outdoor environment to present a macroscopically detectable

NOx removal efficiency (which can positively affect the air quality in that environment), it is necessary to establish a vertical gradient of the concentration of the gas studied as a consequence of the sink effect induced by the photocatalytic surface and that this gradient is observable, that is, it can be measured [26].

Validation systems in real conditions are costly and ultimately ineffective systems, since they are still modeling [26]. For the validation of the materials, a comparative measurement installation has been designed using the two KUNAK measuring stations powered by a photovoltaic panel with temperature, humidity, CO, NO and NO_2 sensors installed on 60x120 platforms with the realizations of paint on wood or 5wt.% glazed tiles in double firing frit (1080°C) (Figure 8.12). KUNAK systems allow wireless monitoring and control, with environmental instrumentation, sensor networks and operational intelligence that guarantee the transmission and use of information and make it possible to integrate it into other systems [27].

8.2.1. Evaluation of Temperature Decrease with Cool Pigments

8.2.1.1. Alkyl Paints

Figure 8.12 shows the two KUNAK stations loaded with white paints (anatase white versus Zr-scheelite white described in Chapter 7) and yellow paints (organic diarylided yellow 12 versus Cr-scheelite described also in Chapter 7). Previously, studies of each one of the colors were carried out against the rutile base paint used (CONTROL or unpigmented alkyl base paint). Figure 8.13 shows the loaded stations with the white colors (anatase white vs. Zr-scheelite white) and yellow colors (diarylided yellow 12 vs. Cr-scheelite) paints.

The temperature evolution in full day (from 12 to 12 hours of the next day approximately) of pigments versus CONTROL is shown in Figure 8.14. It is found that there are only differences during daylight hours and always lower in white pigments than in yellow with respect to the control. Figure 8.15 shows the comparisons between 12 and 18 hours (daylight period).

Figure 8.16 shows the comparison of daytime temperatures from 12 to 18 hours in whites (anatase versus Zr-scheelite) and yellows (diarylided versus Cr-scheelite).

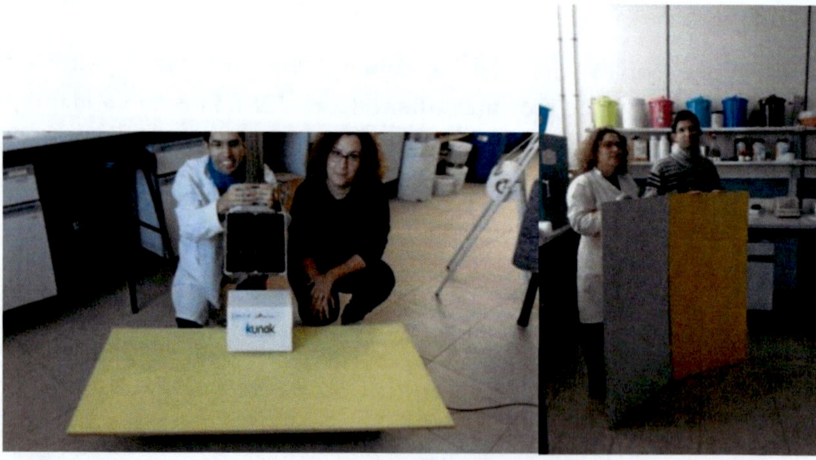

Figure 8.12. Measuring stations powered by a photovoltaic panel with installed on 60x120 platforms with the coatings with paint on wood or 5wt.% glazed tiles in double firing frit.

Figure 8.13. Stations loaded with the white (anatase white vs. Zr-scheelite white) and yellow (diarylided yellow 12 vs. Cr-scheelite) paints.

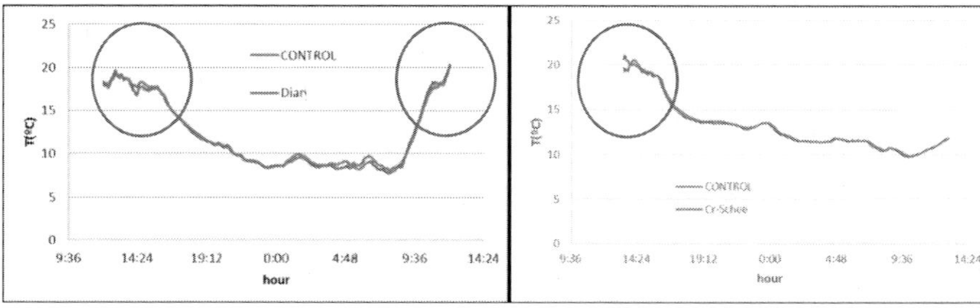

Figure 8.14. Daily evolution of temperatures.

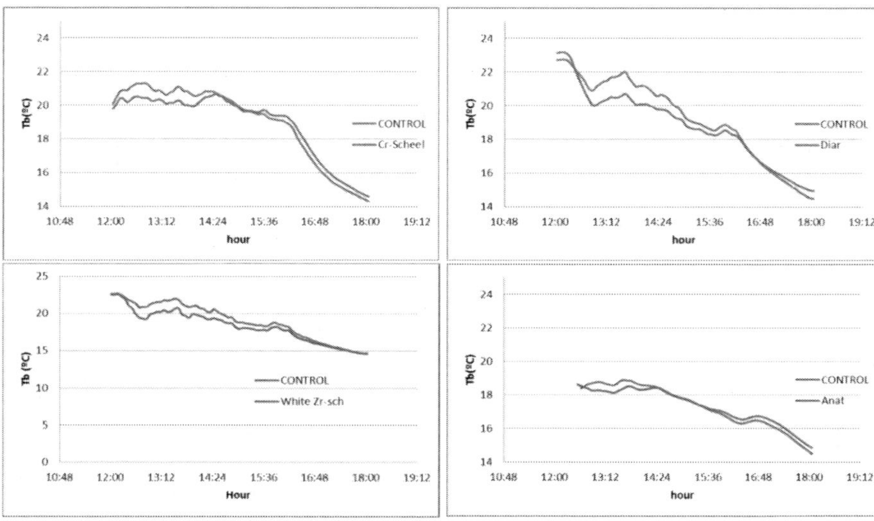

Figure 8.15. Evolution of daytime temperatures from 12 to 18 hours compared with CONTROL paint.

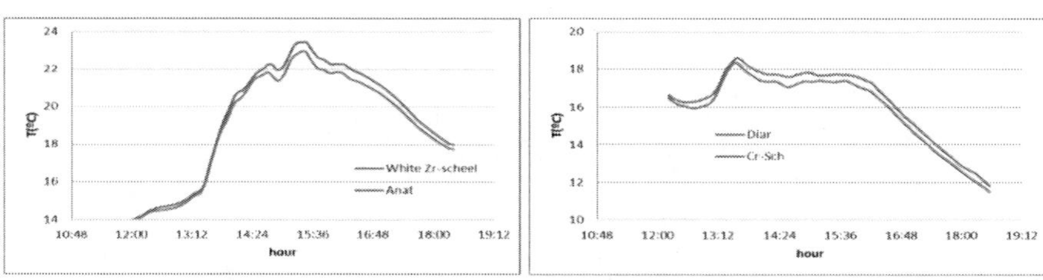

Figure 8.16. Comparison of daytime temperatures from 12 to 18 hours in white and yellow paints.

Table 8.6. Results of the mean of the differences observed in Figure 8.16 (ΔT_{kunak}) with the characteristics of the alkyl paint coatings

Pigment	L*/a*/b*	band gap (eV)	R_{NIR}(%)	R(%)	ΔT_{kunak} (mean ºC)
CONTROL	88.6/1.7/6.8	3.03	82.1	76.5	-
White Zr-sch	88.4/1.3/6.7	3.11	86.0	86.0	-1.5
Anatase	90.9/0.5/4.8	2.96	86.8	82.6	-0.7
Cr-sch	81.9/-2.8/46.1	2.59	81.2	66.5	-0.5
Diarylide 12	81.1/8.7/84.6	2.39	83.4	66.3	-0.2

The results of the mean of the differences observed in Figure 8.16 (ΔT_{kunak}) with the characteristics of the paint coatings are summarized in Table 8.6 and are consistent with the total reflectance values R of the surfaces: very high in whites (exceeding 82% and slightly higher for Zr-scheelite) and around 66% in yellows. White surfaces obviously always achieve greater decreases than the yellow ones with respect to the white rutile control.

8.2.1.2. Glazed Tiles in a Double Firing Frit (1025ºC)

Figure 8.17 shows the two stations loaded with the glazed tiles (zircon white vs. Zr-scheelite white) and yellow (Pr-zircon vs. Cr-scheelite); the zircon white is a commercial white frit with zircon used as reference, the corresponding pigments are 3 wt.% added to the zircon-white frit. Figure 8.18 shows the comparison of day time temperatures from 12 to 18 hours in whites (zircon white vs. Zr-scheelite white) and yellows (Pr-zircon vs. Cr-scheelite).

Figure 8.17. Stations loaded with the glazed tiles (zircon white vs. Zr-scheelite white) and yellow (Pr-zircon vs. Cr-scheelite).

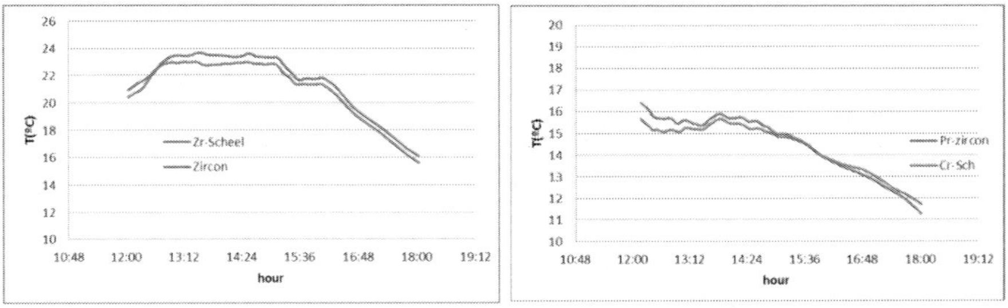

Figure 8.18. Comparison of daytime temperatures from 12 to 18 hours in white and yellow glazed coatings.

Table 8.7 compares the results obtained for the four pigments of the colorimetric parameters analyzed with the mean decrease in temperature observed in the diurnal period of one with respect to the other (ΔT_{kunak}). The results are consistent with the values of the total reflectance of the R surfaces, very high in whites (exceeding 77% and slightly higher for zircon) and around 60% in yellows (slightly higher for Pr-zircon).

Table 8.7. Results of the mean of the differences observed in Figure 8.18 (ΔT_{kunak}) with the characteristics of the alkyl paint coatings

Pigment	L*/a*/b*	band gap (eV)	R_{NIR}(%)	R(%)	ΔT_{kunak} (ºC)
White Zr-sch	90.9/0.7/0.3	3.45	75.6	77.0	-
Zircon	88.8/0/0.8	3.44	78.0	79.5	-0.5
Cr-sch	84.2/-2.2/27	2.27	74.0	61.0	-
Pr-zircon	81.9/0.4/34.8	2.26	74.0	62.0	-0.2

8.2.1. Evaluation of Photocatalytic Activity of Coatings over NO, NO_2 and CO Concentrations

Concern about air pollution is growing given the consequences on health and the environment. In the winter and summer periods, the appearance of pollution episodes by NO_x and CO is common in urban areas.

8.2.1.1. Alkyd Paints Coatings

Figure 8.13 show the two stations loaded with the white (anatase white vs. Zr-scheelite white) and yellow (diarylided yellow 12 vs. Cr-scheelite) paints. Previously, as for temperature analysis, studies of each one of the colors were carried out against the rutile base paint used (CONTROL).

8.2.1.1.1. Tracking NO

In the daily evolution of NO in the whole day (from 12 to 12 hours of the next day approximately) of pigments versus CONTROL it is found that there are only differences during daylight hours. The evolution of the NO concentration on sunny days follows the usual guidelines; at dawn NO begins to grow, reaching a maximum with peaks of 10-70 ppb to go decreasing throughout the day consumed by ozone. Figure 8.19 shows the comparisons between 12 and 18 hours, which confirms the above description.

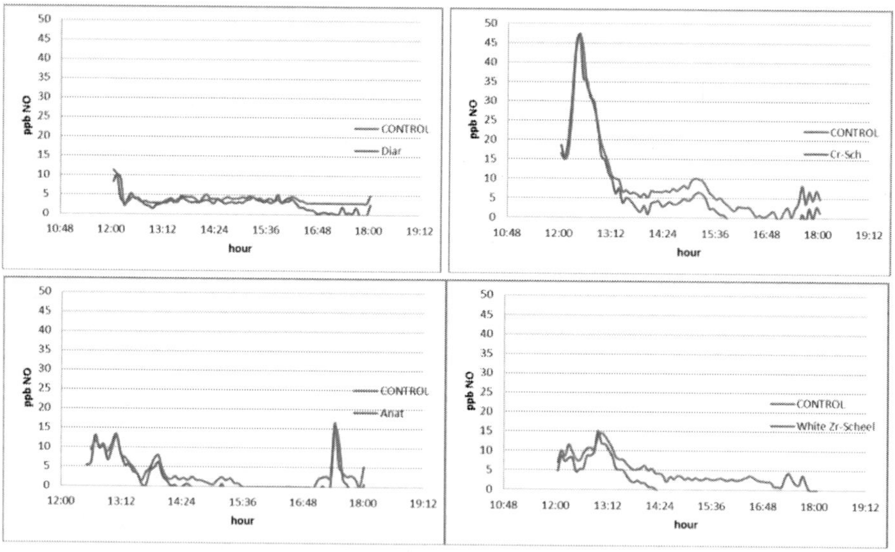

Figure 8.19. Evolution of NO in daytime from 12 to 18 hours.

These studies definitely indicate a decrease in NO in the insolated periods in the four pigments studied with respect to the control rutile paint (Figure 8.19). Likewise, scheelite pigments (Zr-scheelite and Cr-scheelite) decrease 1-1.5 ppb (6%) of NO with respect to their commercial counterparts (anatase and diarylide 12).

8.2.1.1.2. Tracking NO_2

In the daily evolution of NO_2 in full day of pigments versus CONTROL it is found that there are only differences during daylight hours. The evolution of the concentration of NO_2 in sunny days follows the usual guidelines in urban areas: in the first hours of the day there is a maximum of NO_2 that reaches a maximum of 30 ppb to decrease as it is eliminated by oxidants and reacts with the humidity, at dusk picks up and decreases slightly until it rises again at daybreak. Figure 8.20 shows the comparison of NO_2 between 12 to 18 hours of pigments paints with the control that shows the above behavior. The studies indicate, in short, a decrease in NO_2 in the insolated periods in the four pigments studied with respect to the control rutile paint; likewise scheelite pigments decrease 0 and 5 ppb (0 and 20% respectively) of NO_2 with respect to of their anatase white and diarylide yellow counterparts.

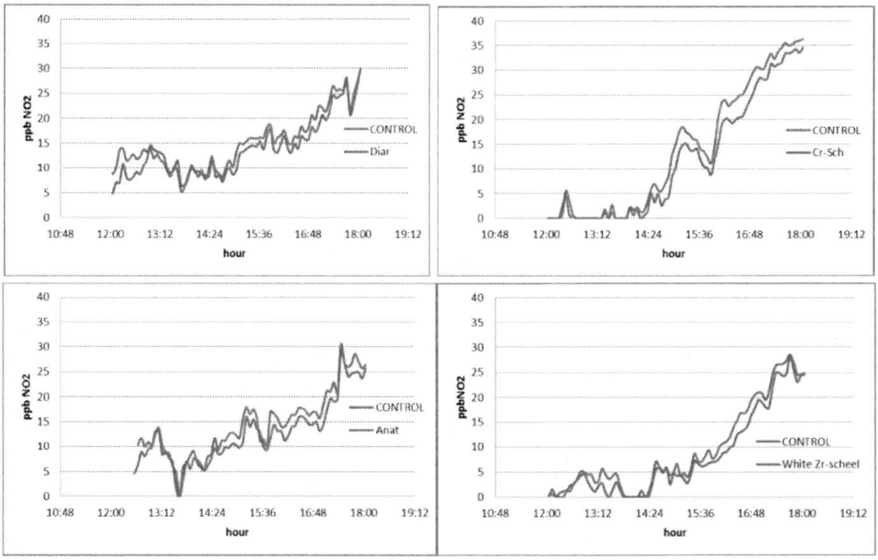

Figure 8.20. Evolution of NO_2 in the daytime from 12 to 18 hours.

8.2.1.1.3. Tracking CO

In the daily evolution of CO in full day (Figure 8.21) it is found that there are only small differences during daylight hours. The evolution of the CO concentration on sunny days begins to grow progressively at dawn until 3:00 p.m., reaching maximum of between 200-400 ppb on the days studied (December 2018), then decreases reaching minimum values throughout the night period (between 30 and 100 ppb). This pattern corresponds to a primary pollutant emitted by traffic.

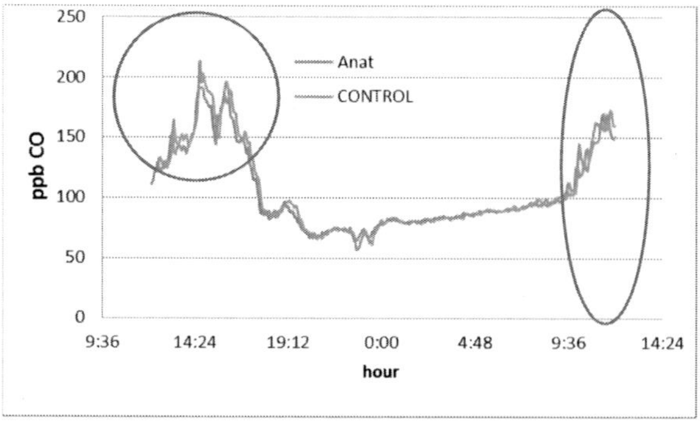

Figure 8.21. Daily evolution of CO.

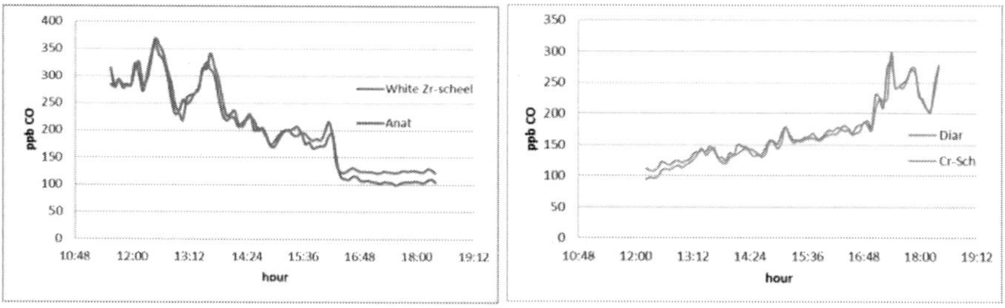

Figure 8.22. Comparison of CO in daytime hours from 12 to 18 hours in whites and yellows.

Figure 8.22 shows the comparison of CO in daytime hours from 12 to 18 hours in whites (anatase versus Zr-sche) and yellows (Diarylide versus Cr-scheelite). The study indicates, in short, a decrease in CO in sunny periods in the four pigments studied compared to the control rutile paint (not shown); likewise scheelite pigments decrease 0 and 5 ppb (0 and 2.5% respectively) of CO relative to its commercial white and yellow counterparts

8.2.1.2. Glazed Tiles in Double Firing Frit (1080°C)

Figure 8.17 shows the two stations loaded with the glazed tiles (zircon white vs. Zr-scheelite white) and yellow (Pr-zircon vs. Cr-scheelite). Previously, as for temperature analysis, studies of each one of the colors were carried out against the rutile base paint used (CONTROL).

8.2.1.2.1. Tracking NO

Figure 8.23 shows the comparison of NO in daytime from 12 to 18 hours in whites (zircon white vs. Zr-scheelite white) and yellows (Pr-zircon vs. Cr-scheelite). Preliminary studies indicate that Cr-Scheelite yellow exceeds 2 ppb (20%) from NO to Pr-zircon, but the white Zr-scheelite surprises, which in the central hours has worse behavior than zircon but in the afternoon hours, insolated with radiation that is richer in long wavelengths due to the effect of atmospheric scattering, presents better results.

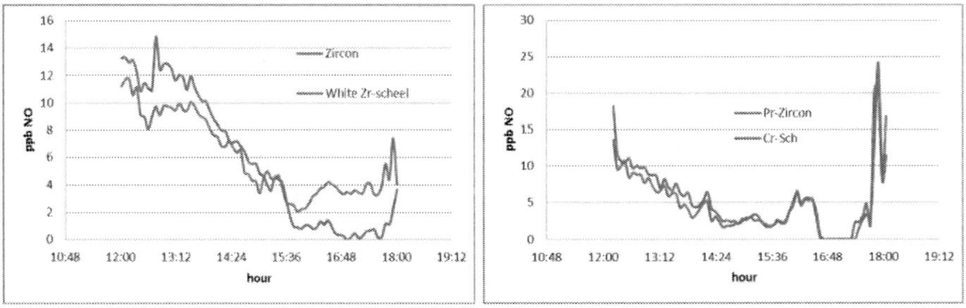

Figure 8.23. Comparison of NO in daytime from 12 to 18 hours in enameled whites and yellows.

8.2.1.2.2. Tracking NO_2

Figure 8.24 shows the comparison of NO_2 in daytime from 12 to 18 hours in whites (zircon white vs. Zr-scheelite white) and yellows (Pr-zircon vs. Cr-scheelite. The studies indicate that Cr-Scheelite yellow exceeds 1 ppb (10%) of NO_2 to Pr-zircon and the white Zr-scheelite presents, as in the case of NO, worse behavior in the central hours than zircon but in the sunny afternoon hours with richer radiation in long wavelengths due to the effect of atmospheric dispersion, it presents better results, however the zircon values are anomalously constant in this experiment.

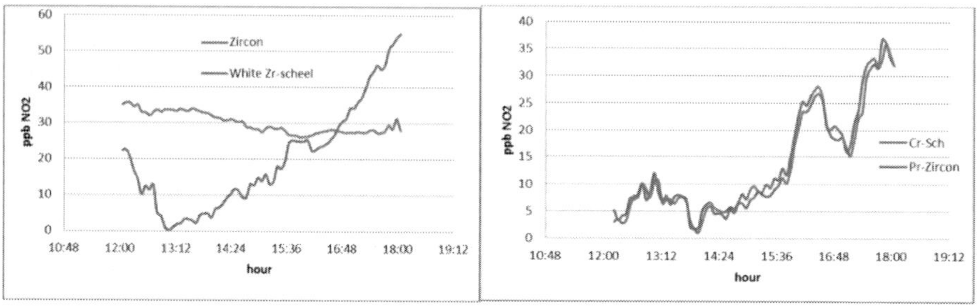

Figure 8.24. Comparison of NO_2 in daytime from 12 to 18 hours in enameled white and yellow pigments.

8.2.1.2.3. Tracking CO

Figure 8.25 shows the comparison of CO in daytime from 12 to 18 hours in whites (zircon white vs. Zr-scheelite white) and yellows (Pr-zircon vs. Cr-scheelite). Preliminary studies indicate that Cr-Scheelite has a very similar behavior to Pr-Zircon and Scheelite White decreases 30 ppb (20%) with respect to Zircon.

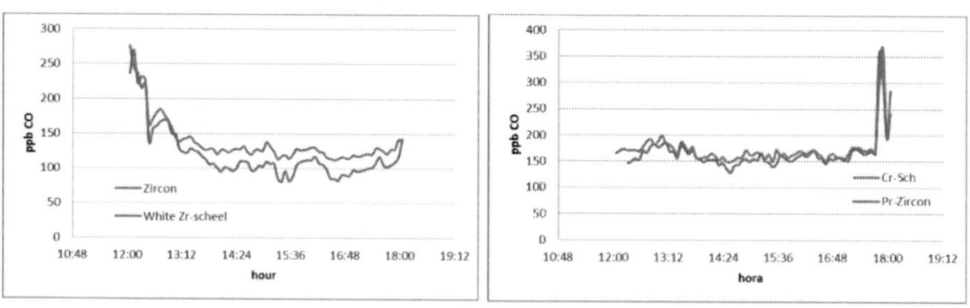

Figure 8.25. Comparison of CO in daytime from 12 to 18 hours in enameled white and yellow pigments.

REFERENCES

[1] Schabbacha, L. M., Marinoski, D. L., Güthsb, S., Bernardinc, A. M., Fredel, M. C., "Pigmented glazed ceramic roof tiles in Brazil: Thermal and optical properties related to solar reflectance index," *Solar Energy*; 159(2018):113–124. doi.org/10.1016/j.solener.2017.10.076.

[2] Molinari, C., Contea, S., Zanellia, C., Arditb, M., Crucianib, G., Dondi, M., "Ceramic pigments and dyes beyond the inkjet revolution: From technological requirements to constraints in colorant design," *Ceramics International* 46 (2020):21839–21872.

[3] Eppler, R. A., Eppler, D. R., "Which colors can and cannot be produced in ceramic glazes?," *Ceram. Eng. Sci. Proc.* 15 (1994) 281–288.

[4] Munsell, A. H., *Atlas of the Munsell color system*, (Malden, Mass., Wadsworth, Howland & Co., inc., Printers, 1915).

[5] Pfeffer, R., Dave, R. N., Dongguang, W., Ramlakhan, M., "Synthesis of engineered particulates with tailored properties using dry particle coating," *Powder Technol* 117(2001):40-67.

[6] Ouabbas, Y., Chamayou, A., Galet, L., Baron, M., Thomas, G., Grosseau, P., Guilhot, B., "Surface modification of silica particles by dry coating: Characterization and powder ageing," *Powder Technol* 190(2009):200-209.

[7] Sheethu, J., Mundlapudi, L. R., "Lanthanum strontium copper silicates as intense blue inorganic pigments with high near-infrared reflectance," *Dyes Pigm* 98(2013):540-546.

[8] Kendrick, E., Kirk C. J, Dann, S. E., "Structure and color properties in the Egyptian Blue Family, $M_{1-x}M'_xCuSi_4O_{10}$, as a function of M, M' where M, M'=Ca, Sr and Ba," *Dyes Pigm* 73(2007):13-18.

[9] García, A., Llusar, M., Calbo, J., Tena, M. A., Monrós, G., "Low-toxicity red ceramic pigments for porcelainised stoneware from lanthanide-cerianite solid solutions," *Green Chem* 3(2001):238.

[10] Vishnu, V. S., Reddy, M. L., "Near-infrared reflecting inorganic pigments based on molybdenum and praseodymium doped yttrium

cerate: Synthesis, characterization and optical properties," *Solar Energy Materials & Solar Cells* 95(2011):2685–2692.

[11] Bae, B., Takeuchi, N., Tamura, S., Imanaka, N., "Environmentally friendly orange pigments based on hexagonal perovskite-type compounds and their high NIR reflectivity," *Dyes Pigm* 147(2017):523-528.

[12] Ozel, E., Yurdakula, H., Turana, S., Ardit, M., Cruciani, G., Dondi, M. "Co-doped willemite ceramic pigments: Technological behaviour, crystal structure and optical properties," *J Eur Ceram Soc* 30, 16 (2010):3319-3329, doi.org/10.1016/j.jeurceramsoc.2010.08.013.

[13] Benda, P., Kalendová, A., "Anticorrosion properties of pigments based on ferrite coated zinc particles," *Physics* Procedia 44(2013):185-194.

[14] Sorlí, S., Tena, M. A., Badenes, J. A., Llusar, M., Monrós, G., "Study of nickel precursors in $(Ni,M,Ti)O_2$ (M = Sb, Nb) yellow ceramic pigments," *British Ceramic Transactions* 103(2004):10-14.

[15] Eliziárioa, S. A., De Andrade, J. M., Limac, S. J. G., Paskocimas, C. A., Soledade, L. E. B., Hammer, P., Longo, E., Souza, A. G., Santos, I. M. G., "Black and green pigments based on chromium–cobalt spinels," *Materials Chemistry and Physics* 129(2011):619–624.

[16] Pu, Y., Wang, G., Chang, K., Ling, Y., Lin, Y, Fitzmorris, B.C., Liu, C., Lu, X., Tong, Y., Zhang, J. Z., Hsu, Y., Li, Y., "Au Nanostructure-Decorated TiO_2 Nanowires Exhibiting Photoactivity Across Entire UV-visible Region for Photoelectrochemical Water Splitting," *Nano Lett*, 13, 8(2013):3817–3823 doi: 10.1021/nl4018385.

[17] Pozza, G., Ajò, D., Chiari, G., F. De Zuane, M. Favaro, "Photoluminescence of the inorganic pigments Egyptian blue, Han blue and Han purple," *J Cult Heritage* 1(2000):393-398.

[18] Forés, A., Llusar, M., Badenes, J. A., Calbo, J., Tena, M. A., Monrós, G., "Cobalt minimisation in willemite $(Co_xZn_{2-x}SiO_4)$ ceramic pigments," *Green Chem*, 2(2000):93–100.

[19] Dondi, M., Matteucci, F., Cruciani, G., Gasparoto, G., Tobaldi, D. M., "Pseudobrookite ceramic pigments: Crystal structural, optical and technological properties," *Solid State Sci*, 9(2007):362-369.

[20] Prim, S. R., García, A., Galindo, R., Cerro, S., Llusar, M., Folgueras, M. V., Monrós. G., "Pink ceramic pigment based on chromium doped $M(Al_{2-x}Cr_x)O_4$, M=Mg, Zn, normal spinel," *Ceram Int* 39(2013):6981-6989.

[21] Sorlí, S. Tena, M. A., Badenes, J. A., Llusar, M., Monrós, G., "Study of nickel precursors in $(Ni,M,Ti)O_2$ (M = Sb, Nb) yellow ceramic pigments," *British Ceramic Transactions* 103(2004):10-14.

[22] Bae, B., Takeuchi, N., Tamura, S., Imanaka, N., "Environmentally friendly orange pigments based on hexagonal perovskite-type compounds and their high NIR reflectivity," *Dyes Pigm* 147(2017):523-528.

[23] Katiyar, R., Kousik, S., Structural and optical properties of $Zn_{1-x}Cu_xO$ thin films, in *Handbook of Zinc Oxide and Related Materials: Volume One*, Materials. Z. C. Feng ed. (CRC Press, 2013).

[24] Monrós, G., in Sol-Gel Materials for Pigments and Ceramics, in *Sol-Gel Handbook - Synthesis, Characterization and Applications Vol. 3*, D. Levy and M. Zayat, ed., (Wiley, 2015). ISBN 9783527334865, 1145-1173;

[25] Tong, Y., Zhang, H., Wang, S., Chen, Z., Bian, B., "Highly Dispersed Re-Doped $CoAl_2O_4$ Nanopigments: Synthesis and Chromatic Properties," *Journal of Nanomaterials*, 2016; Article ID 4169673:1-7, https://doi.org/10.1155/2016/4169673.

[26] Pujadas, M., Palacios, M., Nuñez, L., Germán, M., *Nuevas metodologías para la caracterización del efecto descontaminante de los materiales fotocatalíticos en ambiente real en el marco del proyecto LIFE MINOx-STREET* [*New methodologies for characterizing the decontaminating effect of photocatalytic materials in a real environment within the framework of the LIFE MINOx-STREET project*], Centro de Investigaciones Energéticas, Medioambientales y Tecnológicas (CIEMAT), 2016, http://www.conama11.vsf.es/conama10/download/files/conama2014/CT%202014/1896711462.pdf.

[27] Ventura, C., Torres, A., Escribano, M., Estaciones Smart de calidad del aire y contaminación atmosférica [Smart Stations for air quality

and air pollution], *IV Congreso Ciudades Inteligentes*, 2016. file:///C:/Users/usuario/Downloads/Kunak.EDUSI.RivasVaciamadrid.pdf.

About the Author

Guillermo Monrós
Professor
University Jaume I of Castelló

Prof. Guillermo Monrós is Professor of Inorganic Chemistry at Jaume I University of Castellón, Coordinates the research group "Environmental Inorganic Chemistry and Ceramic Materials" made up of teachers and researchers from the Jaume I University of Castellón. Was co-director of the Master of Environment at the Jaume I University (10 editions). He has supervised 10 PhD and is author of more than 120 international publications SCI in the ceramics field and related environmental management and application. Is the author among other books of "El Color de la Cerámica" (2003), co-editor of "Waste Engineering: towards a sustainable management" (2008) or “Pigmentos termosolares ecoeficientes” (2017). His research topics focus on the synthesis and characterization of low toxicity ceramic pigments with environmental functionalities such as "cool pigments" associated with their high NIR reflectance or with photocatalytic activity on substances with difficult biodegradability or atmospheric pollutants.

INDEX

D

E

F

G

H

I

L

M

N

O

P

Q

R

S

T

U